Methods in Molecular Biology

Series Editor
John M. Walker
School of Life and Medical Sciences
University of Hertfordshire
Hatfield, Hertfordshire, AL10 9AB, UK

For further volumes:
http://www.springer.com/series/7651

Plant Epigenetics

Methods and Protocols

Second Edition

Edited by

Igor Kovalchuk

Department of Biological Sciences, University of Lethbridge, Lethbridge, AB, Canada

Humana Press

Editor
Igor Kovalchuk
Department of Biological Sciences
University of Lethbridge
Lethbridge, AB, Canada

ISSN 1064-3745 ISSN 1940-6029 (electronic)
Methods in Molecular Biology
ISBN 978-1-4899-7706-9 ISBN 978-1-4899-7708-3 (eBook)
DOI 10.1007/978-1-4899-7708-3

Library of Congress Control Number: 2016952622

Printed on acid-free paper

This Humana Press imprint is published by Springer Nature
The registered company is Springer Science+Business Media LLC New York

Preface

For many decades, scientists tried to uncover the mechanisms of inheritance and mechanisms of evolution. Work of Mendel who described laws of trait segregation, Morgan who demonstrated that genes reside on chromosomes, Beadle and Tatum who showed association of mutations with appearance of new traits and Horowitz who further synthesized this idea into "one gene-one enzyme hypothesis," Franklin whose work hinted to helical structure of DNA, Watson and Crick who proposed double-helical structure of DNA and semiconservative nature of replication, and many others made a significant contribution to the development of genetics and molecular biology as sciences.

Advancements in sequencing technologies allowed sequencing simple and more complex genomes, revealing many interesting features. Peculiar characteristics in genome composition such as significant redundancy consisting of many repetitive elements and noncoding sequences, active transcriptional units with no protein product, presence of important vital sequences in introns, and unusual sequences in promoter regions all added to the mysteries of genetic make-up and gene regulation. Even so the advent of genetics and genomics made it possible to understand the causes of many simple monogenetic traits, it still fell short of understanding the inheritance of complex genetic traits in plants and animals, especially those influenced by environment.

A new science that took into consideration many peculiarities of genome organization, gene regulation, and the effect of the environment emerged in the last two decades. Epigenetics, literally translated as "above genetics," is the science that describes the mechanisms of heritable changes in gene regulation that does not involve modifications of DNA sequence. These changes are faithfully inherited by somatic cells and can also be passed across multiple generations. Among others, epigenetic modifications include DNA cytosine methylation, histone modifications, non-histone chromatin-binding proteins, and a variety of noncoding RNAs. Despite the fact that the ideas behind epigenetics had already been developing in the late nineteenth and early twentieth centuries, and the term "epigenetics" was coined in 1940 by Conrad von Waddington, major breakthroughs have only occurred within the last 15–20 years as various epigenetic mechanisms have been uncovered and in many cases described in detail. Epigenetics has made a significant contribution towards the understanding of basic mechanisms governing the processes of fertilization, embryo development, cell differentiation, and tissue specification as well as basic mechanisms of homeostasis maintenance in normal conditions and in response to stress in many organisms. Epigenetics was also important for the better understanding of population genetics, ecology, and evolution.

Plants among other organisms benefited greatly from the progress in epigenetics and from the development of various techniques allowing to analyze various epigenetic modifications. It was in plants where the mechanisms of silencing and RNAi have been discovered, resulting in the development of powerful tools for the regulation of gene expression, used in basic science and applied to the needs of agriculture and medicine. Many curious phenomena such as plant acclimation and adaptation to stress, hybrid and heterozygote vigor (heterosis), plant tolerance to viral infection, transgenerational inheritance, and paramutations, among others, are now considered to be controlled via epigenetic mechanisms.

Epigenetics is also making great contribution to the development of better and hardier crops. The discovery of bacterial immune system CRISPRs/Cas9, an epigenetic mechanism of controlling bacteriophage infection, allowed to develop new precise genome editing system that currently revolutionizing plant agriculture. Finally, the discovery of transgenerational response to stress in plants allowed to show that many trait characteristics appearing in response to contrasting environments are due to epigenetic changes, and in many cases, it was demonstrated that such epigenetic changes (such as changes in DNA methylation) are stable across many generations. Therefore, plant breeding obtained a substantial boost in the form of epigenetic QTLs that can potentially be used for the identification of plants with new traits that arise without modification of DNA sequence.

Future studies involving various protocols for the analysis of methylation patterns, histone modifications, chromatin structure, and small RNA expression, the hallmarks of epigenetic regulation, will undoubtedly aid in making even more progress in basic and applied plant research. In this book we have collected a variety of protocols allowing to analyze various epigenetic changes, including differential expression of noncoding RNAs, changes in DNA methylation, and histone modifications in plants. Where possible and appropriate, we presented several protocols with different degrees of complexity. Since many protocols for the analysis of epigenetic modifications generate massive amount of data, we also added several protocols describing bioinformatics approaches for data processing and analysis.

Lethbridge, AB, Canada *Igor Kovalchuk*

Contents

Contributors

DAVID BAULCOMBE • *The Sainsbury Laboratory, John Innes Centre, Norwich, UK*
ANDRIY BILICHAK • *Agriculture and Agri-Food Canada, Lethbridge Research Centre, Lethbridge, AB, Canada*
BERNARD BILLOUD • *Atelier de bioinformatique, Sorbonne Universités, UMR 7205 (MNHN, UPMC, CNRS, EPHE), Museum national d'histoire naturelle, rue Buffon, Paris, France*
TODD BLEVINS • *Institut de Biologie Moléculaire des Plantes, Centre National de la Recherche Scientifique (CNRS) UPR2357, Strasbourg Cedex, USA*
MARTINE BOCCARA • *PSL Research University, Institut de Biologie de l'Ecole Normale Supérieure (IBENS), CNRS UMR 8197, INSERM U1024, Ecole Normale Supérieure, Paris, France*
AGNES BULSKI • *PSL Research University, Institut de Biologie de l'Ecole Normale Supérieure (IBENS), CNRS UMR 8197, INSERM U1024, Ecole Normale Supérieure, Paris, France*
JOSE ANTONIO CABEZAS • *Department of Forest Ecology and Genetic, Instituto Nacional de Investigación y Tecnología Agraria y Alimentaria - Centro de Investigación Forestal (INIA-CIFOR), Madrid, Spain*
MARÍA-TERESA CERVERA • *Department of Forest Ecology and Genetic, Instituto Nacional de Investigación y Tecnología Agraria y Alimentaria - Centro de Investigación Forestal (INIA-CIFOR), Madrid, Spain*
LOUISE CHAPELL • *The Sainsbury Laboratory, John Innes Centre, Norwich, UK*
VINCENT COLOT • *PSL Research University, Institut de Biologie de l'Ecole Normale Supérieure (IBENS), CNRS UMR 8197, INSERM U1024, Ecole Normale Supérieure, Paris, France*
CLELIA DE-LA-PEÑA • *Unidad de Biotecnología, Centro de Investigación Científica de Yucatá, Col. Chuburná de Hidalgo, Mérida, Yucatán, Mexico*
JULIA ENGELHORN • *Max Planck Institute for Plant Breeding Research, Köln, Germany*
ANDREY GOLUBOV • *Department of Biological Sciences, University of Lethbridge, Lethbridge, AB, Canada*
STEFAN GROB • *Department of Plant and Microbial Biology & Zürich-Basel Plant Science Center, University of Zürich, Zürich, Switzerland*
UELI GROSSNIKLAUS • *Department of Plant and Microbial Biology & Zürich-Basel Plant Science Center, University of Zürich, Zürich, Switzerland*
MARÍA ÁNGELES GUEVARA • *Department of Forest Ecology and Genetic, Instituto Nacional de Investigación y Tecnología Agraria y Alimentaria - Centro de Investigación Forestal (INIA-CIFOR), Madrid, Spain*
SARA HERNÁNDEZ-CASTELLANO • *Unidad Biotecnología, Centro de Investigación Científica de Yucatán, Chuburná de Hidalgo, Mérida, Yucatán, Mexico*
SLAVA ILNYTSKYY • *Department of Biological Sciences, University of Lethbridge, Lethbridge, AB, Canada*
PALAK KATHIRIA • *Agriculture and Agri-Food Canada, Lethbridge Research Centre, Lethbridge, AB, Canada*

IGOR KOVALCHUK • *Department of Biological Sciences, University of Lethbridge, Lethbridge, AB, Canada*

NURIA DE MARÍA • *Department of Forest Ecology and Genetic, Instituto Nacional de Investigación y Tecnología Agraria y Alimentaria - Centro de Investigación Forestal (INIA-CIFOR), Madrid, Spain*

GEOVANNY I. NIC-CAN • *Campus de Ciencias Exactas e Ingeniería, Universidad Autónoma de Yucatán, Merida, Yucatán, Mexico*

CHRISTIAN PARISOD • *Laboratory of Evolutionary Botany, Biology Institute, University of Neuchâtel, Neuchâtel, Switzerland; Bayer CropScience AG, Alfred-Nobel-Straße, Monheim am Rhein, Germany*

STÉPHANE PIEN • *Department of Plant and Microbial Biology & Zürich-Basel Plant Science Center, University of Zürich, Zürich, Switzerland; Bayer CropScience AG, Alfred-Nobel-Straße, Monheim am Rhein, Germany*

ENRIQUE SÁEZ-LAGUNA • *Department of Forest Ecology and Genetic, Instituto Nacional de Investigación y Tecnología Agraria y Alimentaria - Centro de Investigación Forestal (INIA-CIFOR), Madrid, Spain*

ALEXIS SARAZIN • *PSL Research University, Institut de Biologie de l'Ecole Normale Supérieure (IBENS), CNRS UMR 8197, INSERM U1024, Ecole Normale Supérieure, Paris, France*

PAVLINA SASHEVA • *Department of Plant and Microbial Biology & Zurich-Basel Plant Science Center, University of Zürich, Zürich, Switzerland*

MASAYUKI TSUZUKI • *Department of Life Sciences, Graduate School of Arts and Sciences, The University of Tokyo, Tokyo, Japan*

FRANZISKA TURCK • *Max Planck Institute for Plant Breeding Research, Köln, Germany*

ERIKA VARKONYI-GASIC • *The New Zealand Institute for Plant & Food Research Limited (Plant & Food Research) Mt Albert, Auckland, New Zealand*

MARÍA DOLORES VÉLEZ • *Department of Forest Ecology and Genetic, Instituto Nacional de Investigación y Tecnología Agraria y Alimentaria - Centro de Investigación Forestal (INIA-CIFOR), Madrid, Spain*

YUICHIRO WATANABE • *Department of Life Sciences, Graduate School of Arts and Sciences, The University of Tokyo, Tokyo, Japan*

WANHUI YOU • *Department of Plant and Microbial Biology & Zürich-Basel Plant Science Center, University of Zürich, Zürich, Switzerland*

Chapter 1

Chromatin Immunoprecipitation Protocol for Histone Modifications and Protein-DNA Binding Analyses in *Arabidopsis*

Wanhui You, Stéphane Pien, and Ueli Grossniklaus

Abstract

Epigenetic control of plant development via histone modifications is involved in different processes ranging from embryonic development, vegetative development, flowering time control, floral organ development, to pollen tube growth. The identification of an increasing number of epigenetically regulated processes was greatly advanced by methods allowing the survey of genome-wide histone modifications and chromatin-protein interactions. However, genome-wide approaches are too broad to access in detail a large number of histone modifications taking place at a single locus. Here, we provide a robust chromatin immunoprecipitation (ChIP) protocol, allowing in vivo analyses of multiple chromatin modifications and binding of histone modifiers in different plant organs and tissues. This method is quantitative and provides a way to study the dynamic state of chromatin during plant development and also in response to different environmental stimuli.

Key words *Arabidopsis*, Chromatin, Histone modification, Epigenetic, ChIP, Protein-DNA interaction

1 Introduction

Understanding the control of the dynamic state of chromatin is essential for unraveling the epigenetic processes involved in the regulation of the development of multicellular organisms. In eukaryotes, DNA is organized within chromatin, which consists of double-stranded DNA wrapped around nucleosomes. Nucleosomes are composed of histone proteins and contain two copies of each of the core histones H2A, H2B, H3, and H4. Histones are basic, globular proteins, whose N-terminal tails protrude from the core nucleosome. Histone tails are often modified by chromatin-modifying enzymes, which can mediate methylation, acetylation, phosphorylation, citrullination, sumoylation, ubiquitination, and ADP-ribosylation. Specific combinations of these histone modifications result in the maintenance of an active or repressed state of transcription.

Igor Kovalchuck (ed.), *Plant Epigenetics: Methods and Protocols*, Methods in Molecular Biology, vol. 1456,
DOI 10.1007/978-1-4899-7703-3_1,

Several plant *Polycomb* group (PcG) and *trithorax* group (trxG) proteins have been shown to encode chromatin-modifying enzymes or are part of modifier complexes [1]. Among the PcG and trxG proteins are SET domain proteins that control diverse plant developmental processes via their histone lysine methyltransferase activity. PcG complexes were shown to repress PcG target genes via the modification of the N-terminal tail of histones, by directly methylating H3K27. Conversely, trxG complexes maintain gene expression through the deposition of H3K4 and H3K36 methyl marks on their targets. Chromatin immunoprecipitation (ChIP) is the method of choice to monitor in vivo histone modifications and DNA-protein interactions [2]. This method allows the detection of histone modifications and chromatin-binding protein present at a specific locus. It is quantitative and provides a way to study the dynamic state of chromatin during plant development and also in response to different environmental stimuli. This last point is of importance since biotic and abiotic stresses can result in epigenetic modifications of the genome (reviewed in refs. 3, 4).

In this protocol, plant tissue is harvested and immediately subjected to a cross-linking step (Fig. 1). After in vivo cross-linkage of the chromatin and associated proteins, e.g., chromatin-modifying enzymes, chromatin is extracted and sheared into fragments ranging from 300 bp to 1000 bp. This sheared chromatin is incubated with antibodies that recognize and bind either specific histone modifications, chromatin modifiers, or other associated proteins. To reduce nonspecific binding of antibody to chromatin, antibody is firstly conjugated to magnetic protein beads, which give higher recovery, produce less nonspecific binding, and lower background compared to protein agarose. Afterwards, the covalently coupled antibody is added to the sonicated chromatin fraction to immunoprecipitate specific chromatin fragments. Subsequently, the chromatin fragment is released from the antibody and associated proteins through a reverse cross-linking step. The resulting DNA is precipitated and used in PCR reactions with specific primers designed against the gene of interest. The amplified immunoprecipitated gene fragments are detected on an agarose gel and quantified or analyzed by quantitative PCR. This last step shows whether histone modifications or binding of specific modifier proteins is associated with the investigated DNA fragments. This protocol facilitates the probing of multiple histone modifications or DNA-protein interactions in a single experiment, which allows for the simultaneous monitoring of many epigenetic marks.

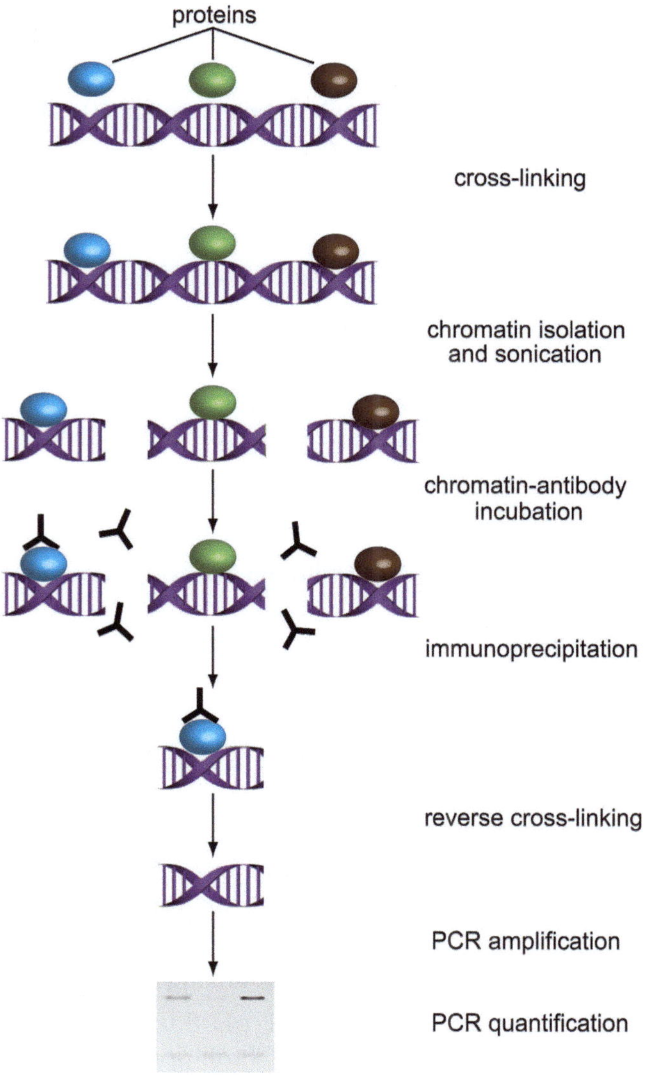

Fig. 1 Summary of the experimental steps in the ChIP protocol. A schematic representation of cross-linking, shearing, immunoprecipitation, reverse cross-linking, and PCR analysis is shown

2 Materials

2.1 Plant Growth Medium

1. Murashige and Skoog (MS) salt base (4.3 g/L) (Carolina Biological Supply Company) supplemented with sucrose (10 g/L) and adjusted to pH 5.6. Add Phytagar (9 g/L) (Sigma Aldrich) and autoclave. Pour the MS medium in Petri dishes and store plates at 4 °C.

2.2 Equipment for ChIP

1. Vacuum pump and desiccator.
2. Mortar and pestle.
3. Cooled centrifuge for 50-mL Falcon tube.
4. Cooled centrifuge for low-binding Eppendorf tubes (1.5 and 2 mL).
5. Sonicator (Bioruptor UCD-200TM, Diagenode).
6. Vortex mixer.
7. Cold room (4 °C).
8. Magnetic rack.
9. Rotating wheel.
10. Thermomixer.
11. Real-time PCR machine.
12. Agarose electrophoresis equipment.
13. UV illuminator coupled to an image acquiring system.

2.3 ChIP Solutions

All solutions with the mention "prepare fresh and keep on ice" have to be prepared just prior to processing tissue samples or chromatin material.

1. Phosphate-buffered saline (PBS): prepare 10× stock with 1.37 M NaCl, 27 mM KCl, 100 mM Na_2HPO_4.
2. Formaldehyde (37%) (Fluka). **Caution: Formaldehyde is toxic through skin contact and inhalation of vapours. Manipulations involving formaldehyde should be done in a chemical fume hood. Gloves are required to handle it.**
3. Stop cross-linking solution: 1× PBS, 0.125 M glycine.
4. Tris–HCl/pH 8.0: 1 M Tris, adjust to pH 8.0 with concentrated HCl solution.
5. Tris–HCl/pH 6.5: 1 M Tris, adjust to pH 6.5 with concentrated HCl solution.
6. Ethylene diamine tetraacetic acid (EDTA): 0.5 M $Na_2EDTA.2H_2O$, adjust to pH 8.0 with NaOH solution.
7. Nuclei extraction buffer 1 (NEB 1): 0.4 M sucrose, 10 mM Tris–HCl/pH 8.0, 10 mM $MgCl_2$, 1 mM phenyl methane sulfonyl fluoride (PMSF). **Caution: PMSF is extremely toxic to mucous membranes of the lung, eyes, and skin. Any contact by inhalation, swallowing, or contact with skin must be avoided. Face shield and gloves are required to handle it**. Prepare fresh and keep on ice. Before use, dissolve one mini-tablet protease inhibitor (Roche) in 30 mL buffer.
8. Nuclei extraction buffer 2 (NEB 2): 0.25 M sucrose, 10 mM Tris–HCl/pH 8.0, 10 mM $MgCl_2$, 1% Triton X-100, 1 mM

PMSF. Prepare fresh and keep on ice. Before use dissolve half mini-tablet protease inhibitor in 8 mL buffer.

9. Nuclei extraction buffer 3 (NEB 3): 1.7 M sucrose, 10 mM Tris–HCl/pH 8.0, 2 mM $MgCl_2$, 0.15 % Triton X-100, 1 mM PMSF. Prepare fresh and keep on ice. Before use dissolve half mini-tablet protease inhibitor (Roche) in 10 mL buffer.
10. Nuclei lysis buffer: 50 mM Tris–HCl/pH 8.0, 10 mM EDTA, 1 % SDS, 1 mM PMSF. Prepare fresh and keep on ice. Before use dissolve half mini-tablet protease inhibitor (Roche) in 5 mL buffer. **Caution: SDS is extremely toxic to mucous membranes of the lung, eyes, and skin. Any contact by inhalation and swallowing or contact with the skin must be avoided. Face shield and gloves are required to handle it.**
11. ChIP dilution buffer: 16.7 mM Tris–HCl/pH 8.0, 1.2 mM EDTA, 1.1 % Triton X-100, 167 mM NaCl. Prepare fresh and keep on ice. Before use dissolve half mini-tablet protease inhibitor (Roche) in 20 mL buffer.
12. Dynal magnetic protein A or G beads (Life Technologies). Choosing of the beads is dependent on species and subclasses of antibody. The beads should be washed three times in 1 mL pre-cold ChIP dilution buffer before use (*see* **Note 1**).
13. Loading buffer: 10× stock: 50 % glycerol; 100 mM EDTA; 0.1 % bromophenol blue; 0.1 % xylene cyanol FF. **Caution: Xylene cyanol FF causes respiratory tract, eye, and skin irritation and may be harmful if swallowed. Use under chemical hood with gloves and face shield.**
14. 1 kb DNA ladder (Life Technologies).
15. Elution buffer: 1 % SDS, 0.1 M $NaHCO_3$. Prepare freshly and keep on ice.
16. Low-salt wash buffer: 20 mM Tris–HCl/pH 8.0, 2 mM EDTA, 150 mM NaCl, 0.2 % SDS, 1 % Triton X-100. Prepare freshly and keep on ice.
17. High-salt buffer: 20 mM Tris–HCl/pH 8.0, 2 mM EDTA, 500 mM NaCl, 0.2 % SDS, 1 % Triton X-100. Prepare freshly and keep on ice.
18. LiCl wash buffer: 20 mM Tris–HCl/pH 8.0, 2 mM EDTA, 150 mM NaCl, 0.2 % SDS, 1 % Triton X-100. Prepare freshly and keep on ice.
19. Tris–EDTA buffer (TE): 10 mM Tris–HCl/pH 8.0, 2 mM EDTA.
20. 5 M NaCl: prepare 5× stock with 5 M NaCl.
21. 3 M NaOAc: 3 M NaOAc in 80 mL of ddH_2O, adjust to pH 5.2 with glacial acetic acid and add ddH_2O to 100 mL.

3 Methods

This protocol has been used for quantification of H3K4me2, H3K4me3, H3K27me2, and H3K27me3 histone modifications, as well as quantification of binding by the *Arabidopsis* TRITHORAX1 (ATX1) protein to its target DNA [5] but is expected to be suitable for the quantification of other modifications as well. *ACTIN2/7* is used as control to normalize PCR products in the case of seedling tissue, while *ACTIN11*, which is specifically expressed in reproductive tissues, is used for flower tissue before fertilization and up to 4 days after pollination.

3.1 Preparation of Plant Material

1. Wild-type seeds are stratified on MS plates in the dark at 4 °C for 2 days.
2. For seedling tissue: plates are transferred to a growth cabinet with daily cycles of 16 h light at 21 °C and 8 h darkness at 18 °C. After germination plants are grown for 10 days and subsequently harvested without the roots.
3. For collection of flower tissue: seedlings are transferred to soil and grown for 3 weeks with daily cycles of 16 h light at 22 °C and 8 h darkness at 18 °C.

3.2 ChIP

1. Seedlings (500 mg) or flowers (1000 flowers) are harvested and immediately transferred to a new 50 mL Falcon tube containing 20 mL pre-chilled 1× PBS buffer. Keep the Falcon tube on ice.
2. Cross-linking is done by adding formaldehyde to the Falcon tube containing harvested tissue into PBS buffer, to a 1 % final concentration (0.55 mL 37 % formaldehyde in 20 mL 1× PBS). At this step, the tissue should not be directly transferred to 1 % formaldehyde/PBS buffer to avoid a prolonged contact with formaldehyde, which will result in irreversible and unspecific cross-linking of the chromatin with proteins.
3. Transfer the Falcon tube into a vacuum desiccator and vacuum infiltrate tissue for 15 min. Seedlings can be gently submerged into formaldehyde/PBS buffer by using a piece of nylon mesh. Slowly release vacuum to remove air bubbles. Fully infiltrated tissue will look translucent and sink to the bottom of the Falcon tube.
4. Replace formaldehyde/PBS solution with ice-cold stop-cross-linking solution. Vacuum infiltrate tissue for 5 min to quench cross-linking.
5. Remove the solution and rinse the tissue twice with ice-cold Milli-Q water and briefly blot dry the plant tissue on the filter paper. Try to remove as much water as possible. At this stage, the tissue can be frozen with liquid nitrogen and stored at −80 °C.

Caution: Liquid nitrogen is extremely cold and may burn the skin; gloves and face shield are required to handle it.

6. Grind the tissue using a pre-chilled mortar and pestle to a very fine powder. Here it is very important that the tissue is thoroughly ground to isolate the nuclei. This step requires approximately 5 min of grinding. Make sure that the mortar and pestle are well pre-chilled using liquid nitrogen. The tissue should never thaw during the grinding step in order to preserve the integrity of the chromatin.
7. Add the tissue to a pre-chilled 50 mL Falcon tube containing 30 mL of NEB 1.
8. Thoroughly dissolve the tissue powder by keeping at 4 °C, gently invert the tube time by time till the solution is homogenous, and filter the solution twice through two layers of Miracloth into a new pre-chilled Falcon tube. Squeeze the Miracloth after washing with 2.5 mL NEB1 to get more liquid.
9. Centrifuge the solution at 3000 × *g* for 20 min at 4 °C.
10. Gently remove the supernatant and resuspend the pellet by pipetting up and down in 1 mL NEB 2. Try to avoid foaming during pipetting.
11. Transfer the solution to a 1.5 mL pre-chilled Eppendorf DNA low-binding tube.
12. Centrifuge at 12,000 × *g* for 10 min at 4 °C. Repeat wash with 1 ml NEB 2 if the pellet looks dark green rather than white or slightly green. However, additional washing step is not recommended if the sample is already clean enough, since it will lead to loss of nuclei.
13. Discard supernatant and resuspend the pellet in 300 μL NEB 3.
14. In a pre-chilled Eppendorf DNA low-binding tube, add 300 μL NEB 3. Carefully transfer the resuspended pellet from **step 13** on top of the 300 μL NEB 3 by using a pipette.
15. Centrifuge at 14,000 × *g* or maximal speed for 1 h at 4 °C.
16. In the interim, prepare the nuclei lysis buffer and ChIP dilution buffer.
17. Remove the supernatant and resuspend the chromatin pellet in 500 μL of cold nuclei lysis buffer by pipetting up and down and with short vortexing. Transfer 10 μL of this sample into a new Eppendorf tube, store on ice, which will be used to check chromatin shearing (at **step 21**).
18. Sonicate lysed nuclei samples five to ten times, 10 s each, using a sonicat or in a cold room. Keep the sonicated samples on ice for 30 s between sonications to cool down, avoiding

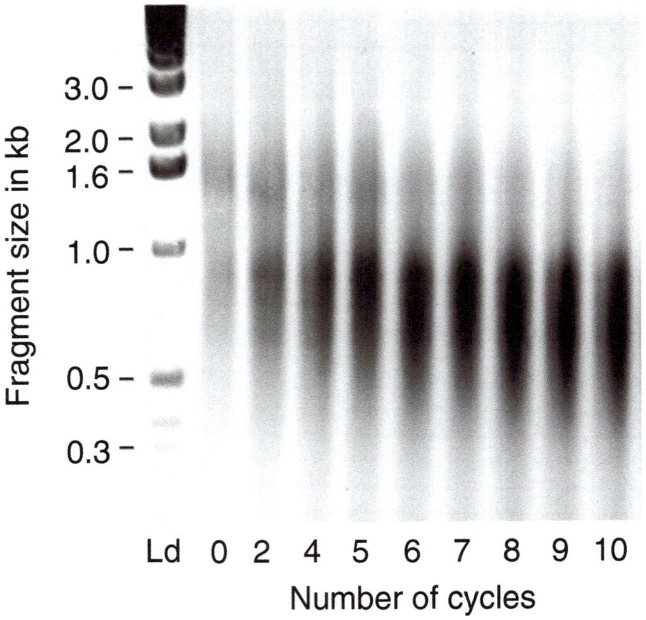

Fig. 2 Optimization of chromatin sonication using the Bioruptor UCD-200TM sonicator. Before proceeding with the ChIP experiment, the optimal number of sonication cycles required to generate DNA fragments ranging from 300 bp to 1000 bp has to be determined. Aliquots of the chromatin extract form step 17 of the protocol are each subjected to an increasing number of sonication cycles and loaded on a 1 % agarose gel. The quantity of DNA fragments with the required size increases with the number of sonication cycles, while the amount of fragments bigger than 1000 bp is reduced. After ten cycles the majority of the chromatin fragments ranges from 300 bp to 1000 bp. Ld: 1 kb DNA ladder

denaturation of proteins. The sonication steps require optimization in order to get the size range of sheared DNA fragments around 300–1000 bp. Efficient sonication will show a smear that centers around 500 bp, which is ideal for ChIP (*see* **Note 2**, Fig. 2).

19. Centrifuge the chromatin sample at 16,000 × *g* for 10 min at 4 °C and transfer the supernatant to a new ice-cold Eppendorf DNA low-binding tube. Repeat once. Pipette 10 μL of this sample into an Eppendorf tube and store one ice; this will be used to check chromatin shearing (at **step 21**).
20. Measure the remaining volume of sonicated chromatin, make two 150 μL aliquots, and add 1350 μL ChIP dilution buffer to each tube. Keep 1 % of the total volume chromatin fractions as input control, and store it in −20 °C freezer. Aliquot the remaining solution into two or three samples according to the experiment.

21. Check shearing of the chromatin using the 2 μL samples from **steps 17** (not sheared chromatin) and **19** (sheared chromatin). Prior to loading add to each sample 7 μL ddH_2O and 1 μL loading buffer. On a 1 % agarose gel supplemented with ethidium bromide (0.4 μg/mL), load side by side the two samples and 2 μL 1 kb DNA ladder to estimate the fragment sizes of the sheared chromatin. Run chromatin samples until the ladder is well resolved (*see* **Note 3**).
22. Prepare magnetic protein A or G beads by washing three times with 1 mL ChIP dilution buffer. Collect beads by using a magnetic rack. Add proper amount of antibody (*see* **Notes 4** and **5**), incubate with 40 μL equilibrated beads in 400 μL ChIP dilution buffer, and gently rotate for 1 h in the cold room. It is important to incubate a control tube containing beads without the antibody, which can be used as a mock control, at the same time.
23. Place antibody-coated magnetic beads on a magnetic rack to collect beads.
24. Add same amount of chromatin fractions onto antibody-coupled beads and also mock control. Incubate overnight with gentle rotation in the cold room.
25. Collect immune complexes by putting the samples in the magnetic rack. Pipette out and discard supernatant, be careful not to remove the magnetic protein beads.
26. Wash the beads two times with 1 mL low-salt buffer by gently rotating for 5 min in the cold room. Collect the beads with magnetic rack and remove the supernatant.
27. Wash the beads two times with 1 mL high-salt buffer by gently rotating for 5 min in the cold room. Collect the beads with magnetic rack and remove the supernatant.
28. Wash the beads two times with 1 mL LiCl wash buffer by gently rotating for 5 min in the cold room. Collect the beads with magnetic rack and remove the supernatant.
29. Wash the beads two times with 1 mL TE buffer by gently rotating for 5 min in the cold room. Collect the beads with magnetic rack and remove the supernatant.
30. Elute immune complexes by adding 250 μL of elution buffer to the pelleted beads. Resuspend the beads by briefly vortexing the tubes and incubate the three tubes at 65 °C for 15 min with gentle agitation.
31. Collect the beads with magnetic rack and transfer the supernatant into a new Eppendorf DNA low-binding tube and repeat the elution once more. Pool the two 250 μL eluted fractions.
32. Add elution buffer to the input sample from **step 20** to make a final volume around 500 μL.

33. Reverse cross-link the input control, the mock control, and immunoprecipitated sample, by adding 20 μL 5 M NaCl to each tube. Wrap the samples with parafilm and incubate at 65 °C for 6 h or overnight.
34. Treat samples with proteinase K by adding 10 μL 0.5 M EDTA, 20 μL 1 M Tris–HCl/ pH 6.5, and 2 μL proteinase K 10 mg/ mL for 1 h at 45 °C with gently shaking.
35. Recover DNA by using Qiagen PCR purification kit according to the manufacturer's manual and proceed to real-time PCR with purified DNA to quantify the samples.

3.3 PCR Quantification

1. The pair of primers used for the PCR amplification of the DNA target should be designed to amplify a genomic region corresponding to the DNA fragment length of the sonicated chromatin, the most efficient amplification of fragments is between 100 and 150 bp. Both primers should have a similar Tm and GC/AT content in order to obtain optimal primer-DNA annealing.
2. The PCR program and PCR reaction mixture have to be optimized according to each primer pair used for immunoprecipitated chromatin amplification. The primer efficiency used to amplify the chromatin region under investigation is tested with a time-point PCR reaction, where equal volumes of PCR products are loaded on an agarose gel, supplemented with ethidium bromide, and quantified after a defined number of PCR cycles. PCR product signal intensities are measured using ImageQuant software (Molecular Dynamics). Each PCR product quantification value is blotted on a semilog graph to determine the log-linear amplification stage of the PCR reaction, which determines the optimal number of PCR cycles needed for accurate PCR product quantification (Fig. 3).
3. 2 μL of eluted input control, mock control, and immunoprecipitated chromatin are used in a 12.5 μL PCR reaction mix with 1.25 μL 10× buffer (Sigma, P2192-VL), 0.25 μL dNTPs (10 mM), 0.25 μL forward target gene primer (10 μM), 0.25 μL reverse target gene primer (10 μM), 0.25 μL *ACTIN* forward primer (10 μM) 5′-CGTTTCGCTTTCCTTA GTGTTAGCT-3′, 0.25 μL *ACTIN* reverse primer (10 μM) 5′-AGCGAACGGATCTAGAGACTCACCTTG-3′, *x*μL $MgCl_2$ (25 mM) depending on primer pair, 0.125 μL DNA Taq polymerase (Sigma, D6677), and ddH_2O up to 12.5 μL. The PCR program has to be empirically established for each chromatin region investigated. The eluted input chromatin is used as a PCR control to confirm that the target DNA investigated by ChIP was present in the starting chromatin material (Fig. 4). If the target DNA is not amplified after PCR, this could be due to chromatin degradation during extraction. The mock control should not give any amplification of the target

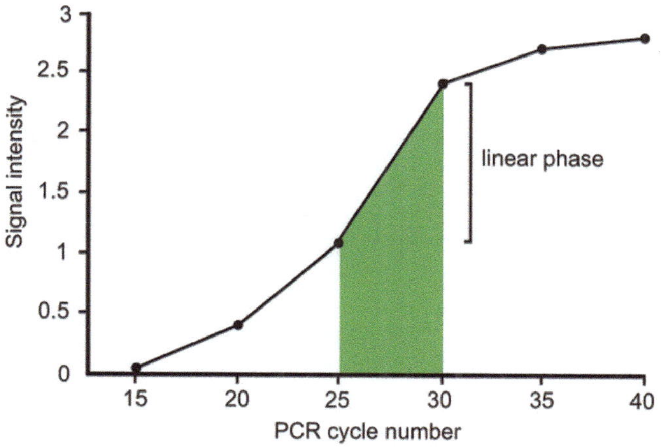

Fig. 3 Primer optimization. For each pair of primers, a time-point PCR reaction is done. Equal volumes of PCR products are loaded on an agarose gel and quantified after a defined number of PCR cycles. PCR product quantification values are blotted on a semilog graph to determine the log-linear amplification stage of the PCR reaction. The optimal number of PCR cycles needed for accurate PCR product quantification (indicated in *green*) is derived from the log-linear amplification area

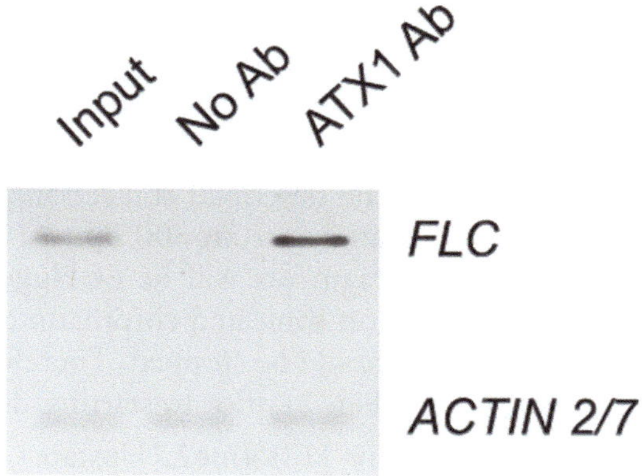

Fig. 4 Example of PCR products obtained from a ChIP assay using wild-type seedlings and an ATX1-specific antibody. ATX1 binding at the *FLC* locus is checked using *FLC*-specific primers. PCR product quantification is normalized using *ACTIN2/7*-specific primers

DNA. The amplification of the target DNA in the mock sample means that washing **steps 26–28** were not correctly done. With the immunoprecipitated chromatin used as DNA template for PCR, a band with the appropriate size should be detected on the agarose gel if the target DNA was recognized by the antibody used.

4. Quantification of PCR products is done by loading the 12.5 μL PCR product on a 1.5% agarose gel supplemented with ethidium bromide (0.4 μg/mL) (Fig. 4). PCR product signal

intensities are measured using ImageQuant software (Molecular Dynamics). Signal intensities are normalized relative to *ACTIN* PCR products. Final quantification is the result of at least three independent ChIP experiments with the corresponding standard error.

4 Notes

1. Magnetic protein A or G beads should be washed three times before use, using 1 mL pre-cold ChIP dilution buffer. Beads are resuspended by inversion in the buffer, kept on ice for 5 min, and placed on a magnetic rack for collecting.
2. In our hands, Bioruptor allows better reproducibility of chromatin shearing between independent experiments than a regular sonicator. Regular sonicators use a probe that is directly in contact with the biological sample. This has major drawbacks in terms of shearing reproducibility as the sonication energy depends on the depth of the sonication probe in the liquid. As a consequence, DNA fragment size varies from sonication to sonication, which has an impact on the quantification of histone modifications. The described sonication program produces DNA fragments ranging from 300 bp to 1000 bp, which can be visualized by loading 2 μL sonicated chromatin on an agarose gel.
3. The described sonication program generates DNA fragments ranging from 300 bp to 1000 bp. If sonication did not work, fragments will be of bigger size and will look similar to the non-sonicated chromatin control. In that case the experiment should be stopped. Therefore, sonication should be optimized (Fig. 2) prior to starting the ChIP experiments.
4. For H3K4me2 (Upstate), H3K4me3 (Upstate), H3K27me2 (Upstate), and H3K27me3 (Upstate), modifications use 4 μL antibody. The amount of the antibody to be used for ChIP depends on the supplier of the antibody; check supplier instructions. For ATX1-DNA binding, we used 5 μL ATX1 antibody (not available commercially [5]).
5. When investigating the dynamics of the deposition of repressive and permissive marks in a gene of interest, our protocol provides enough chromatin to quantify antagonistic marks in parallel using the same tissue sample.

References

1. Pien S, Grossniklaus U (2007) *Polycomb* group and *trithorax* group proteins in *Arabidopsis*. Biochim Biophys Acta 1769:375–382
2. Orlando V, Paro R (1993) Mapping *Polycomb*-repressed domains in the BX-C using *in vivo* formaldehyde crosslinked chromatin. Cell 75:1187–1198
3. Boyko A, Kovalchuk I (2008) Epigenetic control of plant stress response. Environ Mol Mutagen 49:61–72
4. Madlung A, Comai L (2004) The effect of stress on genome regulation and structure. Ann Bot 94:481–495
5. Pien S, Fleury D, Mylne JS, Crevillen P, Inzé D, Avramova Z, Dean C, Grossniklaus U (2008) *ARABIDOPSIS TRITHORAX1* dynamically regulates *FLOWERING LOCUS C* activation via histone 3 lysine 4 trimethylation. Plant Cell 20:580–588

Chapter 2

Chromatin Conformation Capture-Based Analysis of Nuclear Architecture

Stefan Grob and Ueli Grossniklaus

Abstract

Nuclear organization and higher-order chromosome structure in interphase nuclei are thought to have important effects on fundamental biological processes, including chromosome condensation, replication, and transcription. Until recently, however, nuclear organization could only be analyzed microscopically. The development of chromatin conformation capture (3C)-based techniques now allows a detailed look at chromosomal architecture from the level of individual loci to the entire genome. Here we provide a robust Hi-C protocol, allowing the analysis of nuclear organization in nuclei from different wild-type and mutant plant tissues. This method is quantitative and provides a highly efficient and comprehensive way to study chromatin organization during plant development, in response to different environmental stimuli, and in mutants disrupting a variety of processes, including epigenetic pathways regulating gene expression.

Key words 3C, 4C, 5C, Hi-C, *Arabidopsis*, Chromatin folding, Chromatin organization, Chromosomal interactions, Chromatin loops, Higher-order organization, Nuclear architecture

1 Introduction

Investigating chromosomal architecture without using microscopic techniques was first made possible with the invention of chromosome conformation capture (3C) by Job Dekker and colleagues [1]. Since then, our understanding of chromosomal architecture has made impressive progress, which was facilitated by the addition of several 3C-derivate technologies, such as circular chromosome conformation capture (4C) [2, 3] and chromosome conformation capture carbon copy (5C) [4]. By treating native chromatin with formaldehyde, chromosomal regions in close spatial proximity are cross-linked and, therefore, covalently bound to the same molecular entity. Subsequently, chromatin is digested and religated, yielding a pool of circular DNA molecules, termed 3C templates. Each 3C template is characteristic for a specific interaction between two chromosomal regions, as it is comprised of the two interacting DNA sequences, which could originally be found in close spatial

Igor Kovalchuck (ed.), *Plant Epigenetics: Methods and Protocols*, Methods in Molecular Biology, vol. 1456, DOI 10.1007/978-1-4899-7708-3_2, © Springer Science+Business Media New York 2017

proximity within the nucleus. The abundance of a given 3C template can directly be used to determine the interaction frequency between the two interacting fragments and thus provides a measure for the spatial proximity between these two chromosomal regions. The various 3C methods, although based on a similar protocol, mainly differ by the readout of the pool of 3C templates:

(a) 3C measures the abundance of a single 3C template by selectively amplifying the 3C template using specific primers, which bind to the two interacting DNA fragments. Thus, 3C represents the *one-to-one* approach of the 3C family.

(b) 5C allows a *many-to-many* analysis of chromosomal architecture, by multiplexed PCR amplification of several 3C templates and, hence, is the method of choice to investigate a confined chromosomal region.

(c) 4C is employed to investigate chromosomal architecture in a *one-to-all* manner. Thereby, all interacting partners of a given chromosomal region can be detected by inverse PCR, using primers specific to the chromosomal region of choice.

(d) Hi-C is the *all-to-all* version of the 3C family [5]. Hi-C allows simultaneous analysis of all genome-wide interactions and, hence, the determination of the three-dimensional relationship between all genomic regions.

(e) Unlike in other 3C technologies, in the Hi-C protocol (*see* Fig. 1), 3C templates are marked by a biotinylated nucleotide. Thus, the entire pool of 3C templates can later be linearized and equipped with adapter sequences used for next-generation sequencing. As the biotin marks the border between two interacting sequences, only informative DNA fragments are sequenced.

Although the Hi-C protocol involves more steps than all its sibling technologies, it does not necessarily represent the most difficult technology of the 3C family to perform. Whereas reliable data acquisition by PCR, such as in 3C, can be tedious, Hi-C data can quite easily be gathered by next-generation sequencing. However, the subsequent Hi-C data interpretation should not be underestimated, as profound bioinformatics analyses are required. Although Hi-C data analysis is not covered in this chapter, we recommend Hi-Cdat, which only requires limited bioinformatics expertise (knowledge in R), can be run on all platforms, unifies Hi-C data preprocessing and data analysis, and is, in principle, applicable to all organisms [6, 7].

Hi-C was first employed to study chromosomal architecture in human cells [5] but was then quickly applied to other metazoans, such as *Drosophila* [8] and mouse [9], and was finally adapted to plants [6, 10–12]. Here we describe a robust Hi-C protocol,

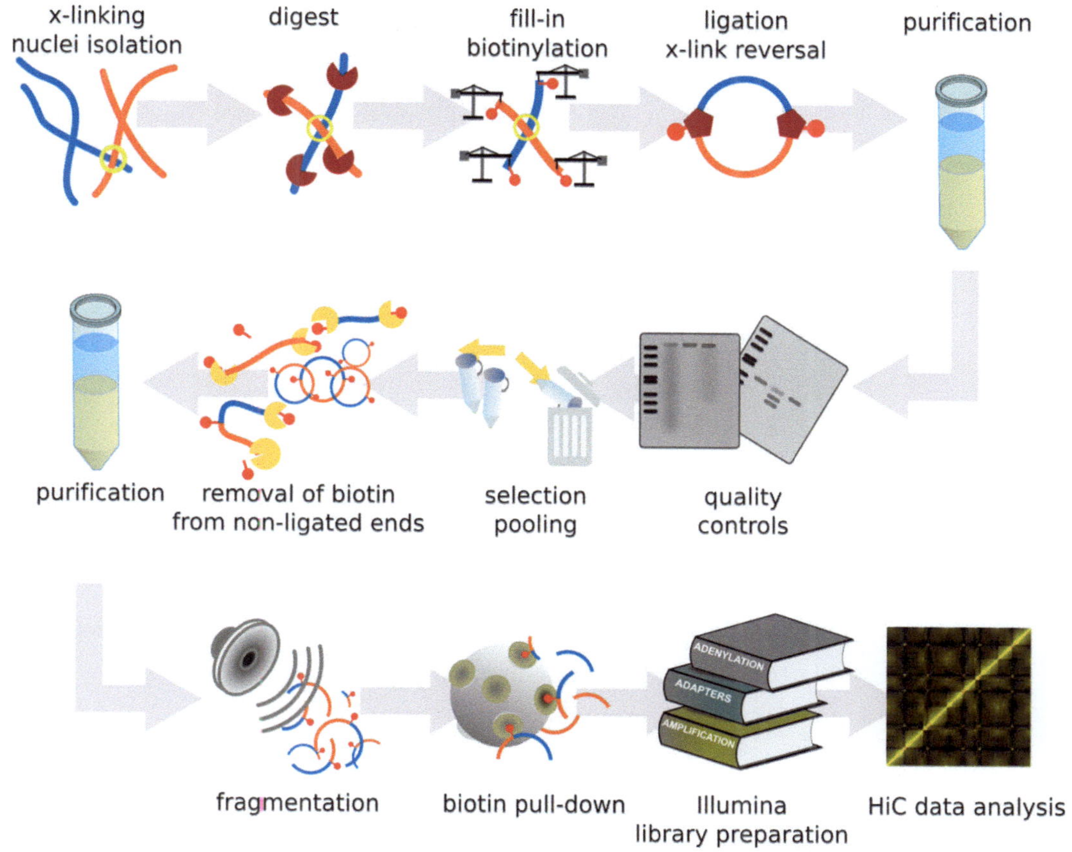

Fig. 1 HiC workflow. *Top row*: Experimental steps to prepare preliminary HiC templates. *Middle row*: Quality assessment of preliminary HiC templates. *Bottom row*: Sequencing library preparation

optimized for *Arabidopsis thaliana* seedling tissue but adaptable to other sources of chromatin, which can be performed by any experienced molecular biologist. The protocol has also been successfully performed in other plant species.

2 Materials

2.1 Reagents

Prepare reagents using ultrapure water (ddH_2O, e.g., Milli-Q) and store at room temperature (RT) if not otherwise stated.

1. Nuclei isolation buffer (NIB): 20 mM HEPES (pH 8), 250 mM sucrose, 1 mM $MgCl_2$, 0.5 mM KCl, 40% glycerol, 0.25% Triton X-100, 0.1 mM phenylmethanesulfonyl fluoride (PMSF), 0.1% 2-mercaptoethanol. NIB should be prepared freshly prior to the start of the experiment. To prepare 150 ml NIB, mix 60 ml of glycerol with 63.375 ml ddH_2O and pour into a 250 ml glass bottle. Subsequently, working under a fume hood, add 18.75 ml of 2 M sucrose (sterile filtered), 3.75 ml

of 10% Triton X-100 (store in the dark), 3 ml of 1 M HEPES buffer (pH 8), 0.75 ml of 1 M KCl, 0.15 ml of 1 M $MgCl_2$, 0.075 ml of 0.2 M PMSF (dissolved in isopropanol, store at −20 °C), and 0.15 ml of 2-mercaptoethanol (be aware of the toxicity of 2-mercaptoethanol). Mix the NIB by gentle swirling, avoiding the formation of bubbles and keep at RT under a fume hood.

2. 4% formaldehyde solution (dilute from 36% formaldehyde stock solution at the day of the experiment).
3. 2 M glycine.
4. 20% SDS.
5. 20% Triton X-100 (store at RT, keep in the dark).
6. Complete protease inhibitor tablets (Roche).
7. Restriction enzyme of choice.
8. Restriction enzyme buffer for the restriction enzyme of choice (e.g., CutSmart, NEB).
9. T4 ligase: purchase in high concentration in Weiss units (WU).
10. Klenow large fragment polymerase.
11. T4 polymerase.
12. T4 polynucleotide kinase.
13. 0.4 mM biotin-14-dCTP.
14. Streptavidin C1 magnetic beads.
15. 10 mM dATP, 10 mM dGTP, 10 mM dTTP.
16. 10 mM dNTP mix.
17. Ligation buffer 10×: 0.5 M Tris–HCl, 0.1 M $MgCl_2$, 0.1 M DTT, pH 7.5. Although commercial ligation buffer can be used, it is advisable to prepare your own ligation buffer due to large quantities needed.
18. 100 mM ATP.
19. Molecular biology grade bovine serum albumin (BSA): 10 mg/ml stock, store at −20 °C.
20. Proteinase K: 10 mg/ml stock, dissolved in ddH_2O, store aliquots at −20 °C.
21. DNase-free RNase A: To prepare a 10 mg/ml stock, weigh 30 mg of lyophilized RNase A powder and add 30 μl of 1 M Tris–HCl (pH 8), 45 μl of 5 M NaCl, and 2925 μl of ddH_2O. Boil for 10 min, cool down to RT, and aliquot. Store at −20 °C.
22. Phenol (liquefied); **Caution: Phenol is toxic; use under a fume hood and wear gloves and goggles.**

23. Chloroform, be aware of the toxicity and only use under a fume hood.
24. Isoamyl alcohol.
25. 3 M sodium acetate.
26. Glycogen: 20 mg/ml.
27. 5 M sodium chloride.
28. 0.5 M EDTA.
29. 1 M Tris–HCl.
30. Tween-20.
31. Ethanol: 100% and 70%.
32. Agarose.
33. Ethidium bromide, be aware of the toxicity and handle correspondingly.
34. Illumina library preparation kit.
35. AMPure XP PCR purification beads (Agencourt).
36. Tween wash buffer (TWB) (1×): 5 mM Tris–HCl, 0.5 mM EDTA, 1 M NaCl, 0.05% Tween-20.
37. Binding buffer (BB) (2×): 10 mM Tris–HCl, 1 mM EDTA, 2 M NaCl.

2.2 Equipment

1. Vacuum pump.
2. Mortar and pestle.
3. Whatman filter paper.
4. Miracloth.
5. 15 ml and 50 ml conical tubes (Falcon tubes).
6. 1.5 ml reaction tubes.
7. Netting material (e.g., nylon mesh, mesh size approx. 2–3 mm).
8. Thermomixer, ideally equipped for 15 ml reaction tubes.
9. Cooled water bath.
10. Refrigerated microcentrifuge (for 1.5 ml tubes).
11. Refrigerated centrifuge equipped for 15 ml and 50 ml conical tubes.
12. Thermal cycler.
13. Fume hood.
14. Sonicator (Covaris S2).
15. Q-bit spectrometer.
16. Magnetic rack.

3 Methods

3.1 Cross-Linking and Isolation of Plant Nuclei

1. Collection of plant material and vacuum infiltration (1.5–2 h). Collect approx. 5–10 g of 2-week old seedlings (*see* **Note 1**). Distribute the seedlings equally to four 50 ml conical tubes. Working under a fume hood at RT, add 15 ml of NIB and 15 ml of 4% formaldehyde solution to each tube (*see* **Note 2**). Submerge the seedlings by pressing them down, using a nylon mesh (*see* Fig. 2a). Place the conical tubes in a desiccator connected to a vacuum pump and apply vacuum for 1 h. To ensure efficient vacuum infiltration, turn the vacuum on and off (by releasing the vacuum) a few times within the first 5 min of vacuum infiltration. After 1 h of vacuum infiltration, quench the formaldehyde by adding 1.9 ml of 2 M glycine and applying the vacuum for 5 min. Subsequently, remove the NIB/formaldehyde mixture and discard it into the appropriate chemical waste container. To remove residual formaldehyde, the seedlings are washed three times in ddH_2O. During the vacuum infiltration, place the residual NIB on ice and add two tablets of protease inhibitor (complete protease inhibitor, Roche).
2. Grinding and filtration of plant material (1–1.5 h). Dry the seedlings by using absorbent paper towels (*see* **Note 3**). Pool the plant material of all four conical tubes, snap freeze it in liquid nitrogen, and grind it to a fine powder using mortar and pestle (*see* **Note 4**). Distribute the powder evenly into two 50 ml conical tubes and immediately place them on ice. Note that from this point on, all steps should be performed on ice

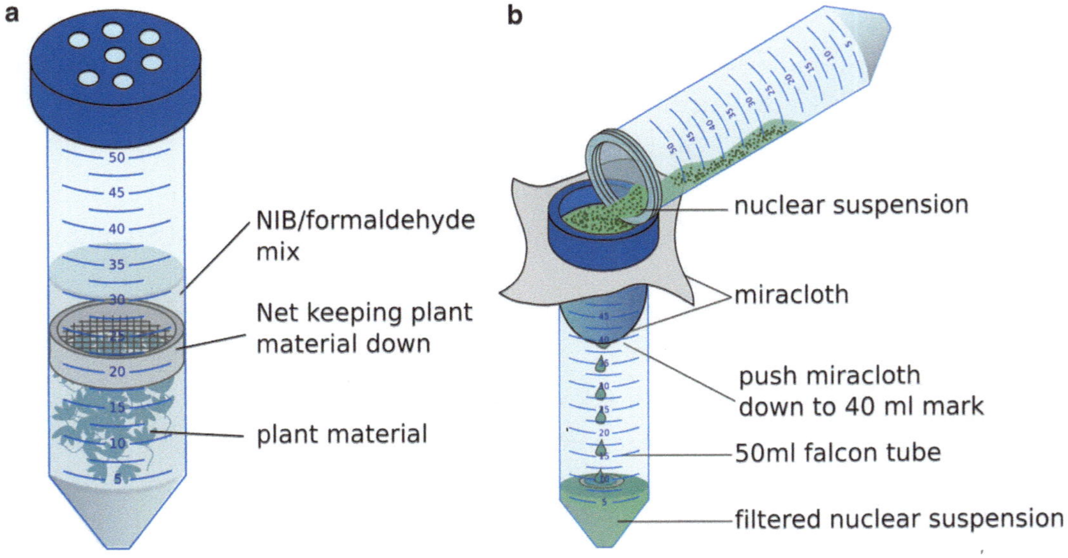

Fig. 2 Illustrations on how to set up (**a**) vacuum infiltration and (**b**) filtration of nuclei

and under a fume hood if not otherwise stated. Resuspend the ground plant material in 20 ml of cold NIB each, by gently swirling the tubes. Install a piece of Miracloth (approx. 7 cm × 7 cm) onto two fresh 50 ml conical tubes each (*see* Fig. 2b) and filter the suspension through the Miracloth (*see* **Note 5**). Repeat this step using a fresh piece of Miracloth installed onto a fresh 50 ml conical tube.

3. Purification of nuclei (1 h). Spin down the filtrate at 3000 × *g* at 4 °C for 15 min. Remove the supernatant and discard in an appropriate chemical waste container. Resuspend the two pellets in 2 ml of NIB each, by gently swirling the conical tubes. If necessary, the pellet may be resuspended by slowly pipetting up and down using a cut-off pipette tip. Be aware that vigorous pipetting may cause the nuclei to disrupt. Pool the resuspended nuclei and evenly distribute into four 1.5 ml reaction tubes (*see* **Note 6**). Spin down the nuclei at 1900 × *g* at 4 °C for 5 min. Discard the supernatant and resuspend in 1 ml of NIB. Repeat the centrifugation and resuspension step twice or more times (total of two to three times, *see* **Note 7**). After the final purification step, remove the supernatant and resuspend the isolated nuclei in 400 μl of 1.2× restriction enzyme buffer (e.g., CutSmart® (NEB); *see* **Note 8**). Centrifuge the nuclei suspension at 1900 × *g* at 4 °C for 5 min. Repeat this step once (total two times). Finally, resuspend the nuclei in 500 μl of 1.2× restriction enzyme buffer.

3.2 Digestion, Fill-In Reaction, and Ligation

1. Permeabilization of the nuclear membrane (2 h). Add 5 μl of 20% SDS to the nuclei suspension and incubate for 40 min at 65 °C under constant shaking (ideally 900 rpm on an Eppendorf thermal shaker). Subsequently transfer the subsamples to 37 °C and incubate for another 20 min under constant shaking. To quench the SDS, add 50 μl of 20% Triton X-100 and incubate 1 h at 37 °C under constant shaking (*see* **Note 9**).
2. Digestion (2 h + overnight + 2 h). Set aside 60 μl of each subsample as a pre-digest control. Dilute the sample by adding 15 μl of 10× restriction enzyme buffer and 115 μl of ddH_2O. Add 100 U of the restriction enzyme of choice (e.g., this corresponds to 5 μl of 20,000 U/ml high-fidelity (HF) *Hind*III-HF) and incubate for 2 h at 37 °C under constant shaking (900 rpm). After that, add 200 U of restriction enzyme (10 μl of *Hind*III) and incubate overnight at 37 °C under constant shaking. Next morning, add another 100 U of restriction enzyme (5 μl of *Hind*III) and incubate for 1–2 h. Set aside 60 μl from each subsample as a post-digest control (*see* **Notes 10** and **11**).

3. Fill-in reaction (1.5 h). Set aside one of the subsamples, which will later serve as a no-fill-in control. Continue with the other three subsamples by adding 40 μl of 0.4 mM biotin-14-dCTP and 1.6 μl of each, 10 mM dATP, 10 mM dGTP, and 10 mM dTTP. Start the reaction by adding 60 U Klenow polymerase (corresponds to 12 μl of NEB Large Klenow Fragment) and incubate at 37 °C for 45 min. To stop the reaction, add 43 μl of 20% SDS and incubate for 30 min at 65 °C (*see* **Note 12**).

4. Ligation (6 h). Continue with all four subsamples including the no-fill-in control subsample. To sequester the SDS, place all four reaction tubes on ice and transfer the subsamples to four 15 ml conical tubes. Add to each tube 750 μl 10% Triton X-100, 750 ml 10× ligation buffer, 75 μl of 10 mg/ml molecular biology grade BSA, and 5.3 ml of ddH_2O, and then incubate for 30 min at 37 °C. Add 75 μl of 100 mM ATP and 50 WU (*see* **Note 13**) of T4 ligase to each biotinylated subsample and 75 μl of 100 mM ATP and 10 WU of T4 ligase to the non-biotinylated no-fill-in control subsample (*see* **Note 14**). Incubate 5 h at 16 °C.

5. Reversal of the cross-linking and RNase treatment (overnight + 3 h). To reverse the formaldehyde cross-linking, add 50 μl of 10 mg/ml proteinase K to all four subsamples. Simultaneously, add 5 μl of proteinase K and 130 μl of ddH_2O to all eight, pre- and post-digestion control samples. Incubate all samples (Hi-C and controls) at 65 °C overnight under constant shaking. Next morning add another 50 μl (5 μl for the controls, respectively) of 10 mg/ml proteinase K and incubate for another 2 h at 65 °C under constant shaking (*see* **Note 15**). To remove residual RNA from the Hi-C samples, add 30 μl (5 μl for the controls, respectively) of 10 mg/ml RNase A and incubate for 45 min at 37 °C.

6. DNA purification (1–2 h). All the following steps should be carried out under a fume hood at RT. To all four Hi-C samples, add 7.5 ml of phenol/chloroform/isoamyl alcohol (25:24:1 v/v) mixture and shake vigorously for 40 s. Spin for 15 min at 4500 × *g* at RT. Carefully remove the upper phase and transfer it to a fresh 15 ml conical tube. Discard the remaining phenol/chloroform mixture into the appropriate chemical waste container. Add 7.5 ml of chloroform/isoamyl alcohol (24:1 v/v) and shake vigorously for 40 s. Separate the phases by spinning 15 min at 4500 × *g* at RT. Gently transfer the upper phase to a fresh 50 ml conical tube and add 7 ml of ddH_2O, 1.4 ml of 3 M NaOAc, and 30 μl of 20 mg/ml glycogen (*see* **Note 16**). The DNA is precipitated by filling the conical tube with ice-cold ethanol up to the 50 ml mark, mixing by inverting the tubes, and storing overnight at −80 °C. The following morning, spin the tubes at maximal speed for 1 h at 4 °C, discard the supernatant, and wash the pellet once with

70% ethanol. Air-dry the pellets and resuspend in 150 µl of ddH_2O. The control samples can be purified in parallel. To reach a working volume of final 200 µl, add 35 µl of ddH_2O to each control sample. Add 200 µl of phenol/chloroform/isoamyl alcohol (25:24:1 v/v), shake vigorously for 40 s, and spin at full spin for 5 min. Collect the upper phase and repeat this step by adding 200 µl of chloroform/isoamyl alcohol (24:1 v/v). Finally, transfer the upper phase to a fresh 1.5 ml reaction tube; add 2 µl of 20 mg/ml glycogen, 20 µl of 3 M NaOAc, and 550 µl of ice-cold ethanol. Store the controls alongside with the Hi-C samples at –80 °C overnight and then pellet the DNA by spinning the control samples at 4 °C at maximal speed for 45 min. Wash once with 70% ethanol, air-dry, and finally resuspend the control samples in 30 µl of ddH_2O.

3.3 Quality Controls

1. Assessing the digestion and ligation efficiency (2 h). To faithfully compare pre-digest controls, post-digest controls, and Hi-C samples, the DNA concentration has to be accurately measured. This can be best achieved with a Q-bit measuring device (*see* **Note 17**). Following the measurement of the DNA concentration, 180 ng of DNA from each Hi-C and control samples are loaded onto a 1% agarose gel, containing 4 µl of ethidium bromide (medium-sized gel, run for 60–90 min at 100 V). Assess the quality of chromatin (pre-digest control), the efficiency of the restriction digest (post-digest control), and the ligation efficiency (Hi-C sample) (*see* **Notes 18** and **19** and Fig. 3).
2. Quality control for the fill-in reaction (4 h). Set up a PCR using a set of parallel primers, which amplify a specific 3C template of your choice (*see* **Note 20** and Fig. 4). For each PCR reaction, use approx. 50 ng of the remaining Hi-C samples, which passed the prior quality control. Purify the PCR product using a standard PCR product purification kit and elute in 60 µl. Transfer 25 µl each to two separate 1.5 ml reaction tubes and set aside the remaining 10 µl of PCR product. If *Hind*III was used as a restriction enzyme in **step 5** of this protocol, set up two restriction digest reactions, using *Hind*III HF and *Nhe*I HF restriction enzymes. For both reactions, add 25 µl purified PCR product, 4 µl of 10× CutSmart buffer, 0.5 µl of the respective restriction enzyme, and 10.5 ml of ddH_2O to a 1.5 ml reaction tube. Incubate for 2 h at 37 °C. Inactivate the restriction enzymes by incubating at 65 °C (*Hind*III) or 80 °C (*Nhe*I) for 15 min. Subsequently, load the remaining purified PCR product, the *Hind*III, and the *Nhe*I digestion products next to each other on a 1% agarose gel. Successful fill-in reactions can be recognized if the PCR product is by the majority digested in the *Nhe*I digest but not in the *Hind*III digest (*see* **Note 21** and Fig. 4b).

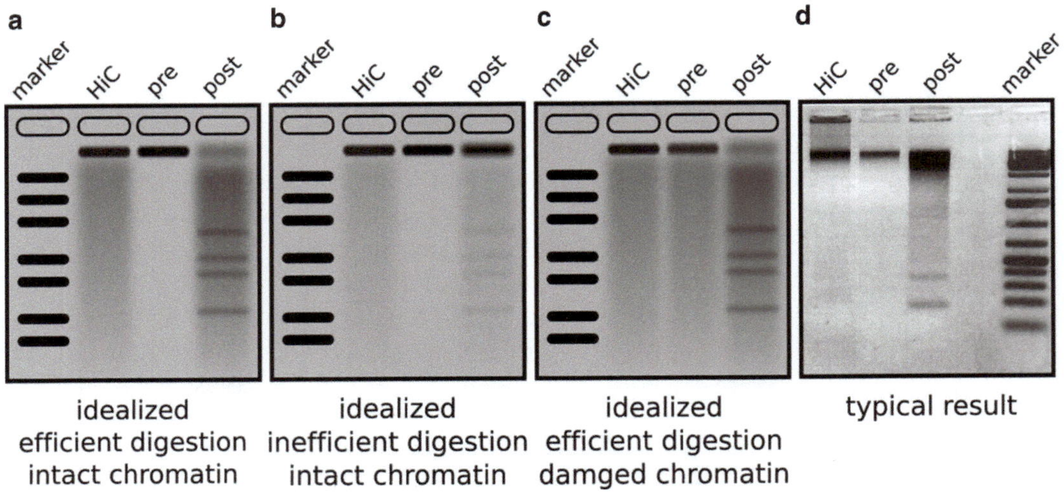

Fig. 3 HiC quality controls. (**a**) Successful Hi-C template generation. The cross-linked chromatin is intact prior to digestion (*see* lane "pre") and is subsequently efficiently digested (*see* lane "post"). The majority of digested chromatin is religated (compare lanes "post" to "HiC"). (**b**) Unsatisfactory Hi-C template generation. Although the cross-linked chromatin is intact, the sample should be discarded, as the digestion efficiency is not satisfactory (*see* lane "post"). (**c**) Unsatisfactory Hi-C template generation. Chromatin integrity is compromised prior to digestion (*see* lane "pre"); hence, the sample should be discarded. Note that (**a**–**c**) show idealized gel pictures and were produced using a drawing software. (**d**) Typical Hi-C result with satisfactory chromatin integrity (*see* lane "pre"), efficient digestion (*see* lane "post"), and efficient ligation (*see* lane "HiC"), as the "smear" in the lane "post" nearly disappeared in lane "HiC". Note that it is not always possible to acquire a gel image, which clearly shows the smaller digested fragments ("smear") and at the same time is not overexposed for longer fragments. To estimate both it is recommended to acquire several pictures with different exposure times

3.4 Hi-C Sample Finalization

1. Pooling and second purification of Hi-C samples (3 h). Pool all Hi-C subsamples that passed both quality control steps and perform another DNA purification, working under a fume hood. For this, add an equal volume of phenol/chloroform/isoamyl alcohol (25:24:1 v/v) mixture to the pooled Hi-C sample, shake vigorously for 40 s, and spin for 5 min in a benchtop centrifuge at full speed at RT. Transfer the upper phase to a fresh 1.5 ml reaction tube and add an equal volume of chloroform/isoamyl alcohol (24:1 v/v). After shaking vigorously for 40 s, spin again full speed at RT for 5 min. To precipitate the Hi-C DNA, transfer the upper phase to a fresh 1.5 ml reaction tube, add 0.1 volumes of 3 M NaOAc, 2 μl of 20 mg/ml glycogen, and 2.5 volumes of ice-cold 100% ethanol. Mix by inverting the tube and store it for 1.5 h at −80 °C. Subsequently pellet the Hi-C DNA by spinning at full speed at 4 °C for 30 min. Wash once with 70% ethanol and air-dry the pellet. Finally, resuspend the DNA in 80 μl ddH_2O.
2. Removal of biotin from unligated ends (5 h). We make use of the exonuclease activity of the T4 DNA polymerase to remove biotinylated cytosines from unligated ends (*see* **Note 22**). Add 2 μl of 10 mg/ml molecular biology grade BSA, 20 μl of

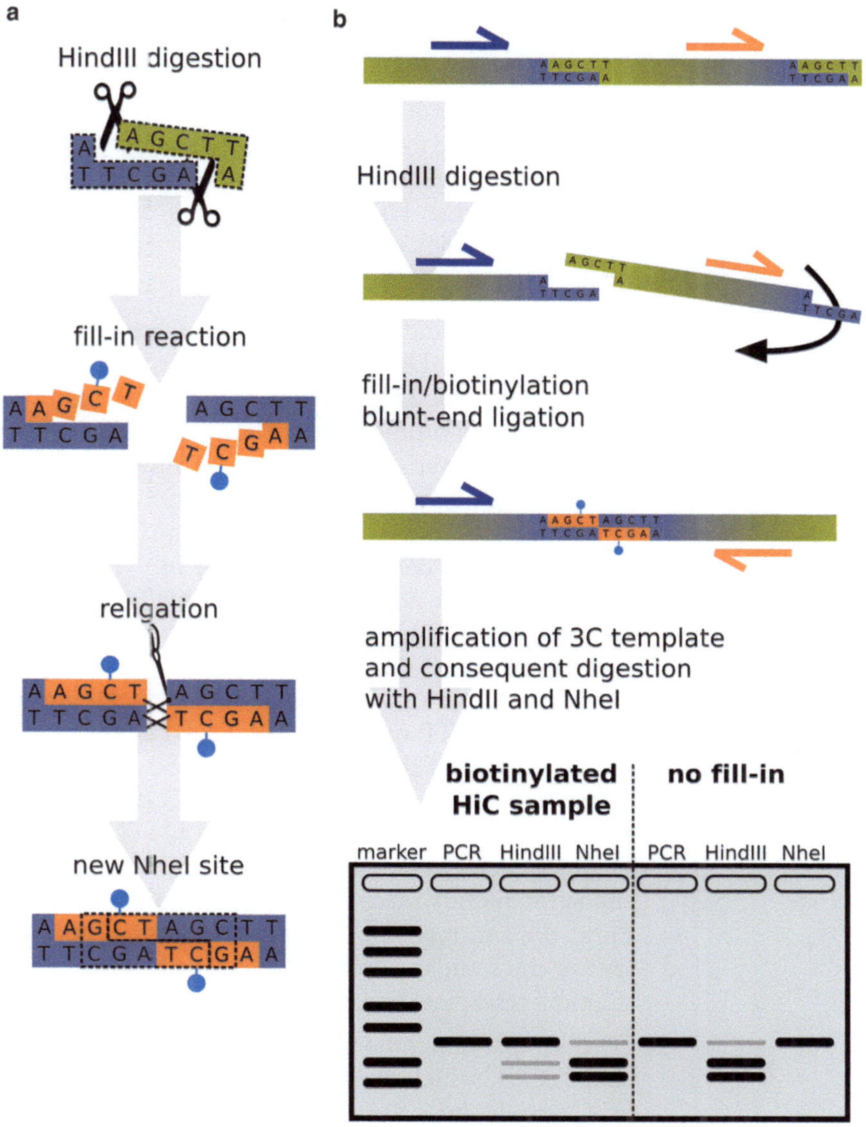

Fig. 4 Rationale of fill-in reaction and its subsequent quality control. (**a**) The fill-in reaction incorporates a biotinylated nucleotide and mutates the original *Hind*III restriction site into an *Nhe*I site. (**b**) A 3C PCR, followed by *Hind*III or *Nhe*I digestion, is used to assess the efficiency of the fill-in reaction. In a successful fill-in reaction, the majority of PCR product can be digested by *Nhe*I but not by *Hind*III

10× NEB 2 buffer, 2 μl of each, 10 mM dATP and 10 mM dGTP, and finally 10 units of T4 DNA polymerase (corresponds to 3.3 μl of the NEB product). Add ddH_2O to reach a final volume of 200 μl. Then incubate the reaction for 2 h at 12 °C and subsequently stop the reaction by the addition of 4 μl of 0.5 M EDTA (pH 8). Repeat the phenol/chloroform/isoamyl alcohol (25:24:1 v/v) purification described in **step 1** in Subheading 3.4 of this protocol and resuspend the precipitated DNA in 100 μl of ddH_2O.

3.5 Sequencing Library Preparation

Normally, for most projects involving next-generation sequencing, sequencing library preparation is carried out by specialized facilities. However, the enrichment of successfully biotinylated Hi-C fragments requires major adaptations to the standard library preparation protocols. The following protocol worked well in our hands. However, since reagents and sequencing technology evolve rapidly, we suggest that Hi-C experimenters discuss the following steps with the experts of their sequencing facility and adapt the protocol if necessary.

1. Fragmentation and end-repair (1 h). For the final Illumina sequencing, the DNA should be sheared to a mean size of 300 bp (*see* **Note 23**). Split the purified Hi-C sample into two 1.5 ml reaction tubes (50 μl each) (*see* **Note 24**). For shearing, we utilize a Covaris S2 sonicator with the following settings: duty cycle 5; intensity 5; cycle burst ratio 200; 5 cycles of 55 s. Following the sonication, the DNA ends have to be repaired to obtain blunt-ended DNA fragments. This can either be achieved with the commercially available End Repair Mix from Illumina or by using standard reagents found in most molecular biology labs: Pool the two 50 μl Hi-C samples and add 1 μl of ddH_2O, 14 μl of 10× ligation buffer containing ATP, 14 μl of 2.5 mM dNTP mix, 15 U of T4 DNA polymerase (corresponds to 5 μl of the NEB product), 50 U of polynucleotide kinase (corresponds to 5 μl of the NEB product), and 5 U of Klenow polymerase (corresponds to 1 μl of NEB Large Klenow Fragment). Incubate the reaction for 30 min at RT (20 °C) and subsequently inactivate the enzymes at 75 °C for 20 min. Purify the end-repaired Hi-C samples using AMPure beads following the standard protocol and, finally, resuspend in 300 μl Illumina resuspension buffer (RSB).
2. Hi-C sample enrichment (1.5 h). In the first step, the streptavidin beads, which are later used to enrich the biotinylated Hi-C sample, have to be washed. Add 60 μl of streptavidin C1 magnetic beads to a low-binding 1.5 ml reaction tube and mix with 400 μl Tween wash buffer (TWB). Recover the streptavidin beads by placing the tube on a magnetic rack for 1 min and pipette away all the liquid from the tube. Repeat this washing step once (total two times). Subsequently, add 300 μl of 2× binding buffer (BB) (*see* **Note 25**) and the 300 μl of Hi-C sample. Incubate the mixture for 15 min with gentle rotation at RT. Reclaim the beads, which should now bind the biotinylated Hi-C fragments, by placing the tube on a magnetic rack for 1 min. Remove the supernatant and resuspend the beads in 400 μl of 1× BB and transfer the suspension to a fresh low-bind 1.5 ml reaction tube. Again, reclaim the beads using a magnetic stand, remove the supernatant, and resuspend the beads in 60 μl of RSB. Repeat this step once (total twice), finally resuspend in 35 μl of RSB and transfer to 0.2 ml PCR tube.

3. Adenylation and adapter ligation (1 h). Add 25 μl of A-Tailing Mix (ATL) to the enriched Hi-C sample and incubate at 37 °C for 30 min in a thermal cycler. To ligate Illumina paired-end sequencing adapters to the Hi-C fragments, add 2.5 μl of DNA adapters and 5 μl of Illumina ligation mix (LIG) and mix gently by slowly pipetting up and down ten times. Incubate the mixture for 10 min at 30 °C. To stop the ligation reaction, remove the Hi-C sample from the thermal cycler and add 10 μl Stop Ligation Mix (STL) and mix thoroughly by pipetting up and down.
4. Cleanup and amplification (2 h). Place the tube on a magnetic rack and remove the supernatant, resuspend in 200 μl of TWB, and transfer to a fresh low-bind 1.5 ml reaction tube. Add an additional 200 μl of TWB and place the tube on a magnetic rack for 5 min. Remove the supernatant and repeat this washing step with 400 μl of TWB. After removing the supernatant, recover the beads in 200 μl of 1× BB and transfer to a fresh tube. Finally, reclaim the beads on a magnetic stand and resuspend them in 200 μl of RSB. Repeat this step once (total two times) and finally resuspend the beads in 50 μl RSB. Your Hi-C sample is now ready for PCR amplification. First, a trial PCR should be set up to determine the linear phase of your PCR amplification process. To this aim, perform standard Illumina amplification PCR reactions with 9, 12, 15, and 18 cycles. Measure the resulting DNA concentration by Q-bit and check whether the Hi-C DNA is amplified in a linear fashion. For final PCR amplification, use the highest cycle number that still confers linear amplification of the Hi-C DNA. Purify your PCR product using AMPure beads, following the standard protocol. Your sample is now ready for sequencing!

4 Notes

1. Seedlings can be grown in plastic petri dishes containing standard Murashige and Skoog (MS) culture media for 2 weeks. To facilitate harvesting, remove parts of the rim of the petri dish. This can be best accomplished by making several ca. 2 cm-spaced vertical incisions on the rim of the petri dish using scissors. Subsequently, brake off the single rim fragments. Now an even surface is available to cut off the aerial tissue using a small scissor. Be careful not to contaminate the plant material with pieces of culture media, as this will interfere with the later grinding and may also affect other steps of the experimental procedure.
2. In the first publication describing the 3C methodology in plants (maize) [13], formaldehyde crystals were directly dissolved in NIB. Thus, in this protocol, NIB is diluted by a factor

of two compared to the latter study. However, we did not observe any negative effects of the dilution.

3. Wet seedlings will make the grinding step considerably harder. Thus, place the seedlings onto Whatman filter paper, cover the plant material with another filter paper, and subsequently place the two filter papers holding the cross-linked plant material in-between several layers of absorbent paper towels. By applying slight pressure (e.g., putting a plastic rack on top of the stack), the water will quickly be absorbed by the paper towels. Eventually, exchange the paper towels, leaving the filter paper stack intact.
4. Take your time to grind the plant material properly as this will facilitate the later filtration step. It is advisable to include several grinding steps, with new liquid nitrogen added in-between.
5. The filtration step can be quite tedious, as the suspension is likely to clog the Miracloth. It is advisable to let the plant powder resuspend completely in NIB; thus, leave the tubes on ice for a while. If the Miracloth clogs anyway, an additional 5 ml of NIB can be added to help unclogging. If not successful, carefully remove the suspension by pipetting slowly using a 1 ml pipette tip, from which the tip was cut to widen the opening. Subsequently, filter the removed suspension through a fresh piece of Miracloth and add, if necessary, an additional 5 ml of NIB. In total, up to 35 ml of NIB can be used per individual conical tube. By using 35 ml of NIB for each conical tube, expect to obtain approx. 25–30 ml of filtrate after the second filtration step.
6. Prior to this point, separating the plant material into different tubes had only practical reasons. However, from this point on, we aim at producing four individual Hi-C subsamples; thus, the four 1.5 ml reaction tubes should be properly labeled, and their content should not be mixed anymore. Several steps (such as digestion, fill-in reaction, and ligation) are difficult to perform; thus, usually not all subsamples reach a satisfactory quality and will later be discarded. All successful subsamples will be pooled prior to the library preparation.
7. Sufficient purification of nuclei is key for successful digestion of the chromatin as cellular debris can inhibit the restriction enzyme. The exact number of repetitions can be best estimated with some experience. The nuclei suspension should finally reach a light green to (ideally) whitish color to ensure efficient digestion. Of course, adding more repetitions will decrease the final yield; thus, the researcher performing the Hi-C experiment has to subjectively judge the trade-off between purity and yield.

8. The choice of the restriction enzyme is crucial for a successful Hi-C experiment. Try to use enzymes that are not blocked by methylation, can be purchased in high concentrations, and are known to digest efficiently. In various Hi-C studies *Hind*III or *Bam*HI were used, which fulfill all these criteria.
9. SDS allows the permeabilization of the nuclear membrane; however, it also efficiently denatures enzymes, such as the restriction enzyme used later. Thus, the SDS has to be efficiently quenched to ensure complete digestion of the chromatin. In various 3C, 4C, and Hi-C studies, higher concentrations of SDS were employed; however, we observed that using a smaller amount of SDS significantly enhances digestion efficiency while confering sufficient permeabilization of the nuclear membrane. Similarly, it is not advisable to significantly enhance the time during which the nuclei are incubated at 65 °C, as this might result in a reversal of the cross-linking.
10. The digestion is both one of the most crucial and most challenging steps in the Hi-C protocol. If the digestion efficiency is not satisfactory, one can play with following factors:
    (a) Type and amount of restriction enzyme
    (b) Number of washing steps with both NIB and restriction enzyme buffer
    (c) Concentration of SDS used to permeabilize the nuclear membrane
11. Pre- and post-digestion controls will be analyzed later (after initial DNA purification) to assess the efficiency of the restriction digest.
12. Digestion, fill-in, and ligation reactions are the most challenging part of the Hi-C protocol and might not work successfully in a first attempt. It is advisable to perform a trial Hi-C sample preparation without using biotinylated cytosines (replace with 10 mM dCTP), as 14-Biotin-dCTP is an expensive reagent.
13. Be careful by choosing the appropriate T4 ligase. The unit definitions of ligases vary among different manufacturers. For example, Promega sells ligase in "Weiss units" (WU), whereas New England Biolabs (NEB) offers their ligase in "cohesive end units" (CEU). The two unit definitions differ significantly from each other. We advise to choose ligases that are sold in WU and at high concentrations.
14. Biotinylated samples are blunt-ended, and thus, considerably more T4 ligase is needed for efficient ligation, compared to the no-fill-in control, which exhibits sticky-end DNA. Generally, condensing agents, such as PEG and Ficoll, can facilitate blunt-end ligation [14]. However, such agents that lead to molecular crowding are not suitable for Hi-C, as *intra*molecular ligations should be favored over *inter*molecular ligation.

15. In our hands, the reversal of cross-linking and proteinase K treatment has a significant impact on the final yield of the Hi-C sample. DNA, which is still cross-linked to proteins, cannot be optimally extracted because such DNA might not dissolve in the aqueous phase but rather in the phenol/chloroform phase.
16. The DTT in the ligation buffer can interfere with the DNA precipitation [13]; therefore, we add ddH_2O to lower the DTT concentration.
17. Do not use direct spectrometric devices such as the NanoDrop. Traces of phenol and other impurities within your sample will lead to wrong results using a NanoDrop. The Q-bit uses a double-stranded DNA specific dye; thus, impurities distort the measurement less.
18. The pre-digest control should be clearly visible as a single high molecular band; a smear below indicates that the chromatin was not intact before digestion and, thus, the corresponding Hi-C sample should be discarded (*see* Fig. 3). The post-digestion sample should appear as a smear on the gel, and the high molecular band should not be visible. Typically, after efficient digestions, one can observe a ladderlike pattern within the smear that stems from highly repetitive sequences. If the post-digest control does not show satisfactory digestion efficiency, the corresponding Hi-C subsamples should be discarded. Finally, the digested and religated Hi-C samples should resemble the pre-digest controls, exhibiting a high molecular band somewhat smaller in size than in the pre-digest control.
19. In Hi-C, digestion and ligation efficiency are somewhat less critical than in 3C and 4C experiments, as the final enrichment will select for successful religation products. Still, the quality control described above should be taken seriously.
20. Primers should be designed in parallel, in order to specifically amplify a 3C template [1]. Thus, primers should occur on the same strand ("pointing" in the same direction) and be located close (approx. 100 bp) to the restriction site of two neighboring or closely located restriction fragments. Therefore, the primer pair should not be able to amplify genomic DNA. The advantage of analyzing the efficiency of the fill-in reaction using a 3C template lies in the potential need for troubleshooting. If primers would simply amplify across a restriction site ("pointing" toward each other), one could not tell whether the fill-in reaction did not work itself or whether the chromatin was not digested efficiently, which would not offer any free DNA ends for the fill-in reaction.
21. A successful fill-in reaction and following blunt-end ligation will mutate the original *Hind*III restriction site. At its place, we can now find an *Nhe*I restriction site (*see* Fig. 4a).

Thus, biotinylated Hi-C fragments can only be digested by *Nhe*I but not by *Hind*III. The no-fill-in control, however, will still be digested by *Hind*III but not by *Nhe*I. Do not expect that the band representing the undigested PCR product will fully disappear. The quality of the Hi-C is sufficient if the bands representing the digested PCR product appear significantly brighter than the undigested PCR product.

22. Biotin should only mark the borders that arose by the ligation of two interacting sequences. These borders contain the information we are seeking for. Biotinylated free DNA ends are not informative; thus, the biotin has to be removed.

23. Large (larger than 700 bp) and small (smaller than 100 bp) DNA fragments were not efficiently sequenced in Illumina machines at the time of writing this protocol. Therefore, try to achieve sonication that results in fragment lengths that do not deviate too much from the mean 300 bp, as this will result in a lower number of useful sequencing reads. If larger insert sizes are required, we propose to perform size selection at the very end of the library preparation protocol. The biotin tag might hamper size selection using AMPure beads, which would normally be performed directly after sonication. However, after library enrichment, the biotin tag will not be present.

24. The Hi-C sample was mainly split into two for one reason: The glass vials for sonication can break in rare cases and, thus, there is a certain risk to lose a Hi-C sample during sonication. Therefore, having two samples can rescue your experiment.

25. Be aware that the binding buffer (BB) is used in two concentrations (2× BB and 1× BB) throughout the library preparation. The initial stock is 2×.

References

1. Dekker J, Rippe K, Dekker M et al (2002) Capturing chromosome conformation. Science 295:1306–1311
2. Simonis M, Klous P, Splinter E et al (2006) Nuclear organization of active and inactive chromatin domains uncovered by chromosome conformation capture-on-chip (4C). Nat Genet 38:1348–1354
3. Zhao Z, Tavoosidana G, Sjölinder M et al (2006) Circular chromosome conformation capture (4C) uncovers extensive networks of epigenetically regulated intra- and interchromosomal interactions. Nat Genet 38: 1341–1347
4. Dostie J, Richmond TA, Arnaout RA et al (2006) Chromosome Conformation Capture Carbon Copy (5C): a massively parallel solution for mapping interactions between genomic elements. Genome Res 16:1299–1309
5. Lieberman-Aiden E, Van Berkum NL, Williams L et al (2009) Comprehensive mapping of long-range interactions reveals folding principles of the human genome. Science 326: 289–293
6. Grob S, Schmid MW, Grossniklaus U (2014) Hi-C analysis in *Arabidopsis* identifies the *KNOT*, a structure with similarities to the *flamenco* locus of *Drosophila*. Mol Cell 55: 678–693
7. Schmid MW, Grob S, Grossniklaus U (2015) HiCdat: a fast and easy-to-use Hi-C data analysis tool. BMC Bioinformatics 16:277
8. Sexton T, Yaffe E, Kenigsberg E et al (2012) Three-dimensional folding and functional

organization principles of the *Drosophila* genome. Cell 148:458–472
9. Dixon JR, Selvaraj S, Yue F et al (2012) Topological domains in mammalian genomes identified by analysis of chromatin interactions. Nature 485:376–380
10. Moissiard G, Cokus SJ, Cary J et al (2012) MORC family ATPases required for heterochromatin condensation and gene silencing. Science 336:1448–1451
11. Feng S, Cokus SJ, Schubert V et al (2014) Genome-wide Hi-C analyses in wild-type and mutants reveal high-resolution chromatin interactions in *Arabidopsis*. Mol Cell 55:694–707
12. Wang C, Liu C, Roqueiro D et al (2014) Genome-wide analysis of local chromatin packing in *Arabidopsis thaliana*. Genome Res 25:246–256
13. Louwers M, Splinter E, Van Driel R et al (2009) Studying physical chromatin interactions in plants using Chromosome Conformation Capture (3C). Nat Protoc 4:1216–1229
14. Green MR, Sambrook J (2012) Molecular cloning. Cold Spring Harbor Laboratory Press, Cold Spring Harbor

Chapter 3

Meta-analysis of Genome-Wide Chromatin Data

Julia Engelhorn and Franziska Turck

Abstract

Genome-wide analyses of chromatin factor-binding sites or histone modification localization generate lists of up to several thousand potential target genes. For many model organisms, large annotation databases are available to help with the characterization and classification of genomic datasets. The term meta-analysis has been coined for this type of multi-database comparison. In this chapter, we describe a workflow to perform a transcriptional and functional analysis of genome-wide target genes. Sources of transcription data and clustering tools to subdivide genes according to their expression pattern are described. For a functional analysis, we focus on the Gene Ontology (GO) vocabulary and methods to uncover over- or underrepresented functions among target genes. Genomic targets of the histone modification H3K27me3 are presented as a case study to demonstrate that meta-analysis can uncover functions that were hidden in genome-wide datasets.

Key words Meta-analysis, AtGenExpress, Hierarchical clustering, K-means clustering, Gene Ontology, Functional enrichment analysis

1 Introduction

The output of genome-wide methods such as chromatin immunoprecipitation followed by whole-genome tiling array hybridization (ChIP-chip) or high-throughput sequencing (ChIP-seq) often consists of long lists of genes that require further characterization. In the case of general regulatory mechanisms such as histone modifications, the set of target genes may encompass several thousands of genes that cannot be analyzed at a single-gene level. For model organisms like *Arabidopsis thaliana*, well-organized databases for gene functions and transcriptional profiles are available and can facilitate the functional characterization of target genes identified by genome-wide ChIP. The term meta-analysis is applied to these multi-database/experiment comparisons aiming to uncover correlations between high-throughput datasets.

In the following, the analysis of gene targets of the histone mark H3K27me3 identified by ChIP-chip is used as an illustrated example for transcriptional and functional meta-analysis [1]. Although the

Igor Kovalchuck (ed.), *Plant Epigenetics: Methods and Protocols*, Methods in Molecular Biology, vol. 1456,
DOI 10.1007/978-1-4899-7708-3_3, © Springer Science+Business Media New York 2017

example described in this chapter refers to publicly available data from *A. thaliana*, the general procedures also apply to individually generated transcriptome and genome-wide ChIP data.

As a first step, differential expression patterns direct the subdivision of target genes. For this subdivision, we apply clustering algorithms to publicly available expression microarray datasets. The advantage of using microarray data generated by the same platform is their high level of consistency between experiments and laboratories. However, the methods can also be applied to the analysis of public or user-provided RNA-seq data.

Hierarchical clustering of expression data provides a preliminary overview of the main groupings within the set [2]. The algorithm starts by calculating the distance between expression values of all genes for all experiments. Then, an iterative process is started, in which the two genes with the closest expression values are joined to one branch of a tree. Mean expression values of joint genes are compared, and the most similar pairs grouped together until all genes are assigned. The result is displayed as a tree with genes assigned to branches and branch length reflecting the similarity between the connected genes. The interpretation of the tree becomes too complex because several hundreds or thousands of genes are to be clustered. Therefore, it makes sense to perform hierarchical clustering on randomly selected subsets of genes to estimate the number of major branches/patterns. This number can indicate the number of patterns (k) to be selected in a subsequent k-means clustering approach.

The advantage of k-means clustering lies in the generation of a delimited list of genes co-expressed in certain conditions or tissues. The k-means algorithm searches for k average expression profiles (centers) that represent the expression patterns of genes in a group by minimizing the sum of distances to the centers. Therefore, this approach searches for both the best position for and the best assignment of genes to centers. Initially, cluster centers are randomly set, and genes are assigned to the center with the lowest distance. The average for each cluster is calculated and becomes a new center to which genes are assigned. The process is repeated for a number of iterations specified in advance. Due to the arbitrary initiation, there is no unique solution in k-means clustering, and genes might be assigned to different clusters in two consecutive calculations run on the same expression dataset. In practical terms, it is reasonable to repeat the calculation several times (e.g., ten times) and to choose only genes that are stably assigned to comparable clusters of interest for a subsequent analysis such as enrichment of Gene Ontology (GO) terms.

GO is a functional vocabulary that was built to standardize gene descriptions within three categories which are "biological processes," "molecular functions," and "cellular components" [3]. The advantage of GO terms is that one unique GO term defines a function, thus omitting different synonyms (i.e., programmed cell death and apoptosis). In addition, each GO term is represented by a number which makes the statistical analysis easier to handle computationally. GO annotations and terms are continuously curated and adapted by teams coordinated through the GO consortium (http://www.geneontology.org). A plant-specific ontology (PO) is under development to better describe plant-specific developmental processes (http://www.plantontology.org/).

From the three general GO categories, up to six more specific (lower) sublevels are present in the GO vocabulary. For example, a "biological process" can be a "developmental process" that is then further divided into 53 processes among which are "reproductive developmental process," "anatomical structure development," and "multicellular organismal development." A gene product annotated to a low-level "child" term is also automatically assigned to all higher level "parent" terms. Since a child can belong to more than one parent term, the complete GO vocabulary has the structure of a direct acyclic graph. With an increasing complexity, visualizing and presenting the relationships and terms in a meaningful way become rather challenging. GO slim terms were therefore established to obtain a more concise overview on large datasets. In a GO slim, GO terms of lower levels are all joined to one higher, more general level.

The method of investigation leading to the annotation of a gene with a GO term is indicated by an evidence code. For example, codes are called "IDA" for "inferred from direct assay" or "ISS" for "Inferred from sequence or structural similarity." Based on these codes, the user can decide whether or not an annotation is trustworthy. Some tools for GO enrichment analysis allow the user to only consider terms with predefined evidence codes.

A complete description of the GO project can be viewed on the website of the GO consortium [4]. The GO annotation for *Arabidopsis thaliana* is curated by "The Arabidopsis Information Resources" (TAIR). A review which considers plant-specific GO aspects was written by Clark et al. [5].

2 Materials

A spreadsheet program such as Microsoft (MS) Excel or OpenOffice/LibreOffice (OO) Calc and database tools such as MS Access or OO Base. Assure that your software uses a dot as decimal limiter, e.g., 100.00 for one hundred instead of 100,00. A recent web browser and a fast Internet connection are recommended. The installation of the software Genesis is described below.

A preprocessed table of expression data from the developmental AtGenExpress set for H3K27me3-positive genes can be downloaded here (http://www.mpipz.mpg.de/16431/Teaching).

3 Methods

3.1 A Preliminary Analysis Using Web Tools and Expression Data Download

For *Arabidopsis thaliana*, numerous expression arrays have been analyzed and made publicly available. We describe here the analysis using the developmental AtGenExpress dataset created by Schmid et al. [6]. AtGenExpress data were generated using the Affymetrix ATH1 array as a common platform and are available for different tissues, developmental stages, responses to biotic and abiotic stress, ecotypes, and other variables. Each experiment was performed in biological replicates with standardized quality control criteria. The preprocessed and raw AtGenExpress data can be downloaded from TAIR, NCBI "Gene Expression Omnibus" (GEO), and the European Bioinformatics Institute (EBI) (*see* **Note 1** and Table 1).

3.1.1 Single-Gene Analysis with BAR

The Arabidopsis eFP browser from BAR (http://bbc.botany.utoronto.ca/) offers a visualization tool for expression patterns of single genes [7]. The eFP browser displays expression data from the AtGenExpress dataset for *A. thaliana*, and it also offers display features for other dicot and monocot plants. AGI locus codes can be entered at "Primary AGI ID," and "Data Source" will select the different expression datasets which will be displayed with a color code. The maximum value can be adjusted by checking the "Signal Threshold" button. Depending on the biological question, one can either choose the "relative" or "absolute" values as a display mode. Relative values are generated by dividing expression values per gene by their median. At the very bottom of the page, a button is offered to view expression values in a table.

3.1.2 Virtual Northern Analysis with BAR

The Arabidopsis e-Northern option from BAR retrieves the preformatted expression data from a large number of publicly available microarray datasets including AtGenExpress. Expression

Table 1
Summary of expression set download

Source	URL	Format
TAIR	https://www.arabidopsis.org/portals/expression/microarray/ATGenExpress.jsp	.zip archive per arraym containing raw .CEL and RMA-normalized .txt files
EBI ArrayExpress	https://www.ebi.ac.uk/arrayexpress/search.html?query=AtGenExpress	.zip archive per series containing either raw. CEL or RMA-normalized .txt file (matrix for entire series)
NCBI GEO	http://www.ncbi.nlm.nih.gov/gds?term=AtGenExpress	As above

profiles can be clustered by pattern either as absolute values or in comparison to a control. There is no restriction as to the number of genes that can be submitted, but longer lists tend to time out the connection to the website.

To obtain an overview on the expression patterns of H3K27me3 target genes, a subset of genes can be analyzed in BAR.

1. Download a list of H3K27me3 target genes: http://genomebiology.com/content/supplementary/gb-2012-13-12-r117-s2.xlsx.
2. Open the file in MS Excel and select the first column.
3. Follow the option "Random ID List Generator" on the BAR homepage and paste the AGI list into the box. The tool generates random lists that can be copied into a new worksheet.
4. Select the first 300 AGI codes from the random list.
5. Select the e-Northern option from the BAR homepage and paste the gene list in the provided location. Run the analysis after selecting the options "AtGenExpress-Tissue Series, RMA normalization" and checking the boxes "Average of replicate treatments relative to average of appropriate control."
6. Visualize the tree as log-transformed clustered data.
7. Repeat the analysis with the next random set and compare output.

3.1.3 Downloading Data Files from Archives

For a more complete and flexible analysis of data, it is preferable to perform the analysis locally.

1. Download the processed expression data from EBI ArrayExpress: https://www.ebi.ac.uk/arrayexpress/experiments/E-TABM-17/ by following the links to "Processed Data," "Array Design," and "Sample and data relationship" (*see* **Note 2**).
2. The extraction of the zip archive containing processed data generates a tab-separated text file containing Affymetrix probe identifiers (affy-IDs) in the first and mean values of three biological replicates and their standard deviation in the following columns. The array data are normalized and scaled across all samples and indicated as log-transformed relative expression values of each gene per array. The first and second rows contain sample codes in the form of "ATGE_*number*" as headers. The experiment key corresponding to "ATGE_*number*" sample codes can be found in the table provided through the link "Sample and data relationship" (*see* **Note 3**).
3. Open the Expression data file in MS Excel to create a version containing only mean values and only one informative header row. Save the file under a new name.

4. Open the array design file in MS Excel and delete all rows not containing Affymetrix probe ID to AGI information. Save the file under a new name.
5. Download the current gene annotation file from TAIR:
   (a) ftp://ftp.arabidopsis.org/home/tair/Genes/TAIR10_genome_release/TAIR10_gene_lists/TAIR10_gene_type and save as tab limited .txt file.

3.1.4 Extracting Expression Data for Genes of Interest

The downloaded expression sets contain data for all genes present on the ATH1 array, while the list of H3K27me3 target genes may include genes not represented on the array. Depending on the application, we would like to compare only data present on both lists. An MS Access database can be used to extract data only for genes of interest.

1. Open MS access.
2. Create a blank database following default suggestions after opening the program.
3. Select the "External data" tab and import:
   (a) The expression data table
   (b) The list of H3K27me3-positive genes
   (c) The reformatted array design table
   (d) The text file containing all AGI annotations of the current genome version

During the import, check boxes "first row contains header" whenever this applies and "no primary key" among the options.

4. Select the "Database Tools" tab and the "Relationships" option.
5. Drag the symbols for all tables indicated on the left to the Relationship space in the middle.
6. Select the "Design" tab and the "Edit Relationships" option.
7. Within the pop-up window, select "Create New" and indicate the table containing all current AGI codes and every other table containing AGI codes one at a time. Select which column contains AGI codes in the second pop-up window. The new relationship is indicated by a line connecting the two columns between the related tables.
8. Use the "Edit Relationship" option to create a relationship between the first column of the expression data table and the first column of the array design table, both containing Affymetrix probe identifiers.
9. Select the "Create" tab and the "Query wizard" option. In the pop-up window, select "Simple Query." Select the table containing the H3K27me3 target genes and the column containing the

AGI codes. Select the expression data table and all columns except the first containing affy-IDs. The query result will contain the expression data for all H3K27me3 genes that were represented on the array. Instead of affy-IDs, the first column will indicate AGI codes. Select "External Data" and export Query result as a tab-limited .txt or Excel file for further editing.

10. Create a random subset of expression values for 300 genes by inserting a column to the data table in MS Excel. In the second row, insert the function RANDBETWEEN and indicate select to choose a number between 1 and the number of target genes. Use the MS Excel fill down column to complete the row. Use "Copy" and "Paste Special" options "Values" to copy the numbers into a new column. Use the "Sort" → "Custom sort" to sort the values according to a new column. Use copy and paste into a new worksheet to create a subset containing the first 300 sorted genes. Delete the random index column from the subset and save the contents of the worksheet as a tab-limited text file.

3.1.5 Clustering and Visualization with WebGimm

The software Cluster and its visualization tool "Mapletree" were originally developed by the Eisen lab [8]. Several extensions of both tools are available as a downloadable program and web tools (Table 2). The WebGimm site offers a JavaWebstart version that can be run without the need to locally install software [9].

1. Go to the WebGimm start site http://eh3.uc.edu/gimm/webgimm/ and launch the WebGimm start.
2. Upload expression data as a tab-limited .txt file.
3. Tick the "Center Rows" box.
4. Tick the "Hierarchical Clustering" box.
5. Select the default options "Average Linkage" and "Euclidean Distance."
6. Launch by selecting "Cluster" tab.
7. Once the analysis is finished, visualize the results by selecting "View Results."
8. A FunctionalTreeView window opens.
9. Adjust pixel settings by selecting tabs "Settings" → "Pixel Settings" and sliding the "Contrast" bar.
10. Groups of genes can be selected by clicking on the nodes of the left dendrogram or by opening the yellow window over a heatmap in the left graphical panel. Selected genes appear in the middle panel; AGI codes are indicated in the right panel.
11. Export subsets of data by selecting the tabs "Export" → "Save lists" or → "Save Data."
12. Go back to the cluster window.

Table 2
Summary of clustering platforms

Tool	URL	Type	Advantages	Disadvantages	References
WebGimm	http://eh3.uc.edu/gimm/webgimm/	Java WebStart	Nothing to install. Offers also clustering according to Gaussian Infinite Mixture Model	Limited options for data preprocessing	[9]
Cluster 3.0	http://bonsai.hgc.jp/~mdehoon/software/cluster/software.htm	Installer for Windows, Linux, Mac-OS	Easy to use, fast	Needs separate visualization software	[8]
JavaTreeeView	http://jtreeview.sourceforge.net/	Java	Easy to use, nice graphics	No expression profile summaries	[16]
Genesis	http://genome.tugraz.at/genesisclient/genesisclient_download.shtml	Java	Clear layout for k-means clustering	Tree not suited to view many genes	[10]

13. Select tab "K-Means Clustering" and indicate 9 as a number of clusters.
14. Export results as indicated above.

3.1.6 Clustering with Genesis Software

The software Genesis offers additional display and normalization options while implementing the same algorithms for hierarchical and k-means clustering as Cluster3.0 or WebGimm [10].

1. Download Genesis software at http://genome.tugraz.at/genesisclient/genesisclient_download.shtml.
2. For system requirements, installation, and a detailed description of all functions in Genesis, *see* Genesis Operation Manual (can be downloaded from the web page) and Sturn et al. [10]. Please be aware that a license for this software has to be requested prior to the installation. The license is free for non-commercial users. For some applications (e.g., hierarchical clustering of the whole genome), the memory used by Genesis has to be enlarged (default 512 MB). The Genesis Operation Manual describes how this is performed for different operating systems (*see* **Note 4**.)
3. Load expression file (must be in a tab-delimited format).
4. Expression is visualized as a heatmap in red color (if only positive values are present in the dataset) or red and green color (if positive and negative values are present).
5. Per default, the maximum color intensity is set to 3.0/−3.0. All higher/lower values are also displayed with a maximum intensity. Adjust the maximum value to the real maximum of expression values or to a customized value (main menu "view" → "adjust to maximum"/"set maximum").
6. Genes can be searched for in clusters and in a complete set. The AGI locus code of search gene will be highlighted in pink.
7. Depending on the original data and the biological implication, the data might need to be adjusted: "adjust" → "different options" (for example, a transformation from a linear scale to a log scale). Please note that an adjustment will not be reverted. To change the analysis, the dataset needs to be reloaded. If the goal of clustering is an analysis of patterns rather than absolute expression values, divide the values per gene by root mean square (rms). The division by rms leads to the adjustment of absolute values between the genes, but it retains the expression pattern.
8. Hierarchical clustering: Click on the "HCL" button in the toolbar to open a dialog window. Choose "average linkage clustering" (*see* **Note 5**) and "cluster genes" and "cluster experiments" (*see* **Note 2**). The tree can then be displayed by

clicking on "Tree" ("average linkage") in the program tree (the left part of the window, data tree to navigate through the results).

9. K-means clustering: Click on the "KMC" button on the left in the analysis tool bar to open the dialog. Choose the required number of clusters and the maximum number of iterations (the default 50 is recommended by the author of Genesis [10]). The result is displayed in the program tree in different ways:
   (a) "Expression Images": shows the color-coded expression for all genes in the chosen cluster.
   (b) "Cluster Information": provides the number of genes per cluster and their percentage of the whole set.
   (c) "Centroid Views": returns a graphic showing samples of the expression set on the *x*-axis and the average expression on the *y*-axis. The variation inside the samples is indicated with bars (*see* Fig. 1).
   (d) "Expression Views": the same axes as for "Centroid Views" but a line for every gene.

   Centroid and expression views can also be displayed for all clusters at once (*see* Fig. 1). This is useful to compare clusters and to find clusters with special attributes (e.g., the expression only in one tissue type).
10. Saving the data: a complete project and all images can be saved by using the menu "file" → "save project"/"save expression image." Gene lists from single clusters can be saved by clicking on the cluster with the right mouse button and choosing the data type (for example, "save cluster gene list").

3.2 Gene Ontology Analysis

Several web tools offer the analysis of GO. These tools can be divided into simple annotation displays and into more sophisticated tools that calculate whether certain GO terms are statistically over- or underrepresented.

3.2.1 Basic Annotation Using TAIR

The easiest way to access Arabidopsis GO data is to use the TAIR webpage (http://www.arabidopsis.org). Complete annotations for gene lists can be displayed via "Home > Tools > Bulk Data Retrieval > GO Annotations." The result displays a complete listing of GO terms associated with the submitted gene list, including evidence codes, the corresponding GO slim categories, and links to references. To provide users with an overview about the functions present in a gene set, the "Functional Categorization" function can be used instead of "Get all GO Annotations." A list of all GO slim categories with the number of genes in the submitted list belonging to this category is then displayed. For a graphical representation, follow "Gene Bar Chart" → "Draw." Bar charts are

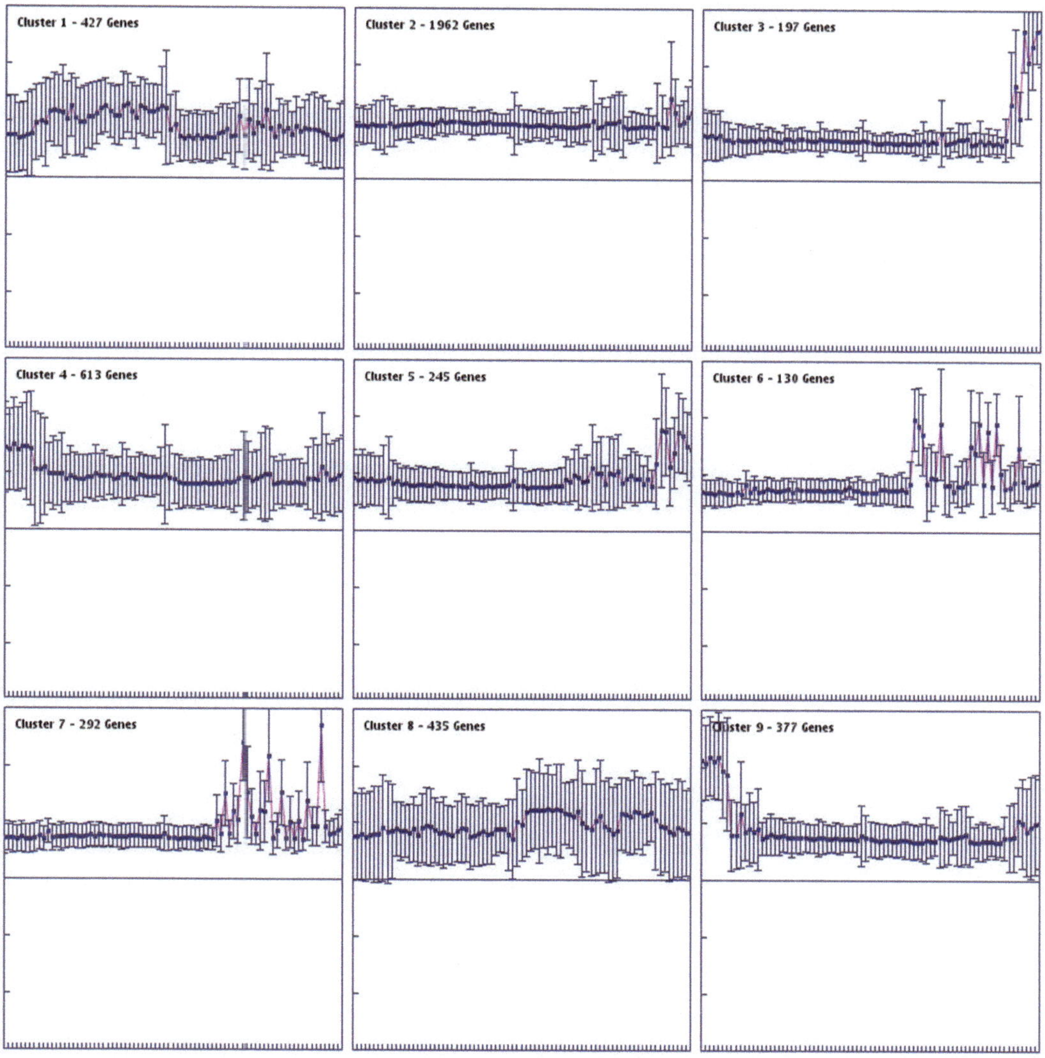

Fig. 1 K-means clustering of expression data from genome-wide H3K27me3 targets. *Each box* represents one cluster. Different tissues are presented on the *x*-axis (from *left* to *right*: root, stem, and whole plants, leaves, the shoot apex, flowers, and seed samples), expression values on the *y*-axis. Squares represent the average expression of one tissue, bars indicate the variance within one tissue (output "Expression Centroid View" from Genesis, note that blank spaces appear in the output because no negative values are present in the dataset)

drawn that show the percentage of GO slim categories in the submitted dataset. An impression of the general distribution of GO categories in the entire genome can be provided by performing a "Whole Genome Categorization."

3.2.2 Functional Enrichment Analysis

The functional enrichment analysis tools employ statistical methods to test whether a certain GO term appears in a gene list at higher or lower than expected ratios. To perform such analysis, a reference set of genes is needed to calculate the expected distribution of GO terms. The tools supply users with a statement whether there are overrepresented (terms appearing more often than expected) or underrepresented (terms appearing less often in the query list than expected) GO terms in the query list compared to the reference list. Tools testing for the over/underrepresentation of GO terms are available as web interfaces and as freely available stand-alone programs. They differ in the statistical method employed, their input (GO slim versus GO), and their graphical output. Since all tools are updated and well documented, it does not make sense to describe how to submit data to them. Rather, we will present an overview and indicate their particularities (*see* Table 3).

Choosing your reference set: Depending on the gene set of interest, different reference sets need to be chosen. Some tools offer gene sets for the commonly used arrays as reference sets, while others require them to be supplied by the user. Otherwise, the whole genome is used for comparison. However, if a microarray was involved (like the ATH1 Affymetrix in the case of AtGenExpress data), it makes sense to compare the genes of interest (e.g., all seed-expressed genes) to the genes present on that array rather than comparing them to the whole genome. Similarly, if RNA-seq data are analyzed, only the genes detected as present in the dataset should be used for statistical computation. The steps to create a custom list using MS Access have been described above (*see* Subheading 3.1, **step 4**).

Tests and statistics: All tools start with the null hypothesis that a certain GO term appears as often in the query list as in a randomly picked list of genes from the reference set. As a result, *p*-values are generated that reflect the probability of the null hypothesis to be true. If these *p*-values fall below a certain threshold (usually 0.05), the null hypothesis can be rejected, and the GO term is then called over- or underrepresented in the query gene set. One difference between the available tools is the distribution assumed for a given hypothetically randomly selected set. The ones offered are a hypergeometric distribution, a binomial distribution, and a χ^2 distribution. It should be mentioned that the hypergeometric distribution gives a correct description of the situation (in this distribution, genes can only be sampled once for the random distribution), whereas the binomial and the χ^2 distributions are only approximations for larger reference sets (binomial-distribution) and large reference sets with large sample sets (χ^2 distribution) [11]. The *p*-values obtained by all three tests are usually corrected by false discovery rate (FDR) estimation. FDR correction estimates the probability of a wrong rejection of the null hypothesis and adjusts the *p*-values according to this probability. There are several methods to estimate the FDR correction term, and several tools offer more than one. Among those methods, the Bonferroni family-wise error rate is the most conservative. In contrast, FDR methods (e.g., Benjamini & Hochberg, Benjamini & Yekutieli) are less conservative but well suited, especially if dependencies exist as in the case of GO terms [11, 12]. Simulation methods should not be used, if only a few categories are involved [11].

Drawbacks of GO: One should always keep in mind that in a large annotation file like the GO annotation, there will always be errors [3]. Usually, these wrong annotations are corrected by TAIR in the next release of the annotation file. Therefore, one should only use the latest version of the GO annotation. Functional annotations for single genes should be verified by considering evidence codes, publications, and BLAST. In the case of gene families, a wrong annotation of one member can be transferred to the entire family because of sequence similarity. Misannotation errors are avoided if only the experimentally verified evidence codes are included in the analysis.

Example analysis: We aim to find genes involved in the embryonic development among the H3K27me3 targets. For all H3K27me3 targets obtained by ChIP-chip, a GO slim analysis using the "Classification super viewer" from the BAR web interface [13] shows only a slight enrichment of the term "development." Since the embryonic development mainly takes place in developing seeds, GO analysis is performed on a cluster of genes highly expressed in seeds (*see* Fig. 1, cluster 3). The gene list is submitted to the web tool "FatiGO" [14] using all genes on the ATH array as a reference set (since this array was used in the AtGenExpress project). The analysis reveals an overrepresentation of GO terms involved in the embryonic development (e.g., GO term: "embryonic development ending in seed dormancy") (*see* Fig. 2 for an overview of the procedure). To ensure that this is not uniquely an attribute of seed-specific expression, we perform a clustering analysis for the whole-expression dataset and submit a cluster with a similar seed-specific expression to FatiGO. In this case, no "developmental" GO terms are observed. As a second control, we submit all H3K27me3 targets for which expression data is available in the developmental series (these are the ones that had a chance to be in the cluster) to a GO analysis. Among these genes, the developmental GO terms are enriched but not embryo-specific ones. Thus, the clustering analysis followed by the functional enrichment analysis enables us to identify a group of genes (the seed-expressed H3K27me3 targets) that are probably involved in the embryonic development.

Table 3
Summary of GO tools

Name	URL	Type	Statistic test method	Test correction	Select evidence codes?	Input IDs	Reference sets	Visualization	GO-level indicated?	Advantages	Disadvantages	Comment	References
Classification super viewer	http://bar.utoronto.ca/	Web-tool	Ratio between frequencies	Simulation	No	AGI	Whole genome	Bar chart	No	Fast overview	Uses GO slim (only general functions)		[13]
AmiGO	http://amigo.geneontology.org/amigo	Web-tool	Binominal	Bonferroni (optional)	No	AGI	Genome	Table	No	Very fast	Limited visualization options		[17]
BiNGO	http://www.cytoscape.org/download.php	Java	Hypergeometric, binomial	Benjamini & Hochberg, Bonferroni	Yes	AGI	Whole genome	Table, network	No	Very fast, easy to handle	No automatic update, under-representation has to be tested separately	Plugin of Cytoscape	[18]
FatiGO	http://babelomics.bioinfo.cipf.es/	Web tool	Fisher's exact test	Westfall & Young, Benjamini & Hochberg, Benjamini & Yekutieli	No	AGI	User list, whole genome	Table	Yes	Jobs are stored under login name, easy to handle	Calculation times can be very long at certain times of day	Cave: "remove all duplicates" removes duplicates from both lists	[14]
GOEAST	http://omicslab.genetics.ac.cn/GOEAST/	Web-tool	Hypergeometric, Fisher's exact test, χ^2 test	Benjamini & Hochberg, Benjamini & Yekutieli	No	Affy-IDs, AGI	ATH1 array, whole genome	Table, tree	No	Clicking on "details" shows information about each gene in query		Use "Batch-genes" for whole genome	[19]

GOTerm Finder	http://go.princeton.edu/cgi-bin/GOTermFinder	Web tool	Hypergeometric	Bonferroni, simulation	Yes	AGI	User list, whole genome	Table, tree	No	Clear tree showing which genes are assigned to each term	Every category of GO has to be tested separately		[20]
AgriGO	http://bioinfo.cau.edu.cn/agriGO/	Web tool	Hypergeometric, binomial, χ^2 test	Benjamini & Yekutieli	Yes	Affy-ID, AGI	ATH1 array	Text-tree, graphic	Yes	Overview shown first, deeper levels can be visualized stepwise	Every branch of the GO has to be tested separately		[21, 22]
Gostat	http://gostat.wehi.edu.au/cgi-bin/goStat.pl	Web tool	Fisher' exact test	Benjamini & Hochberg, Benjamini & Yekutieli	No	AGI	User list, whole genome	Table	No	Gene names displayed	GO terms only displayed as numbers		[23]
Genecodis	http://genecodis.cnb.csic.es/	Web tool	Hypergeometric, χ^2	Benjamini & Hochberg, simulation	No	AGI	User list, whole genome	Table, pie chart	Levels can be selected	Additional returns genes that share a group of GO terms		User GO-annotation file can be used	[24]

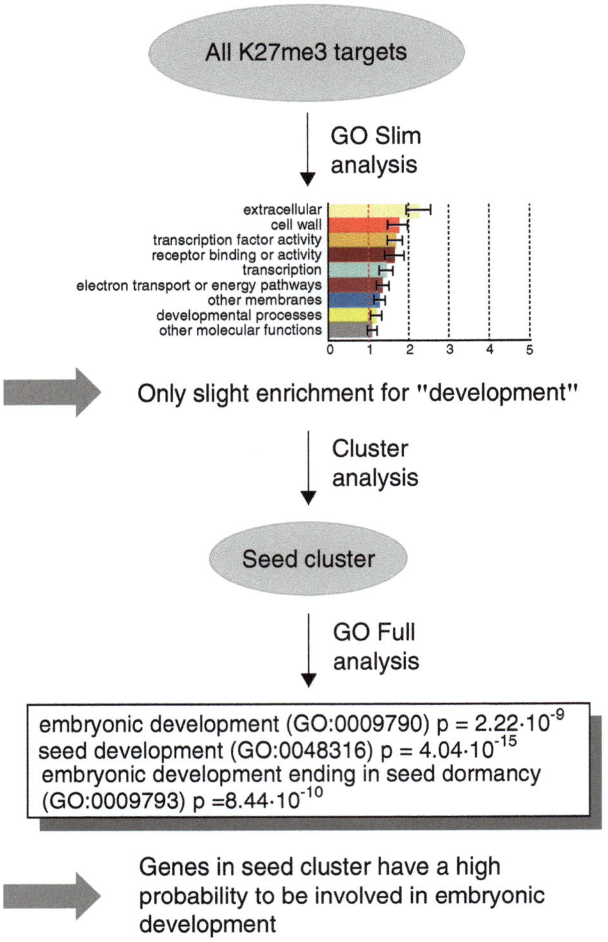

Fig. 2 Overview of example procedure. GO slim analysis was performed with the "Classification super viewer" from the BAR's web interface [13]; displayed is a selected part of the output. GO full analysis was performed by the tool FatiGO [14]; only a selection of enriched GO terms is shown. The *p*-values given here are the FDR-adjusted *p*-values

4 Notes

1. Affymetrix microarray data can be normalized by different methods. The Affymetrix® Microarray Analysis Suite 5.0 (Mas 5.0) calculates the mean intensity of all except the upper and lower 2% of values on the array and scales the data so that the mean value is equal to 100. The cgRMA (robust multi-array average) algorithm [15] uses a global background and takes the GC content of probes into account for the correction. Expression values are calculated using a log_2 scale.
2. To shortcut the following steps, directly download "ready-to-cluster" data here (https://www.mpipz.mpg.de/35320/At_H3K27me3_targets_Development_gcRMA.rtf).

3. Order samples according to tissue and developmental stage. One possible order (from root to flowers) can be obtained by using the "AtGenExpress Visualization Tool (AVT)" (http://jsp.weigelworld.org/expviz/expviz.jsp). By submitting any AGI locus code, the ordered tab-delimited file containing the slide numbers, tissue descriptions, and expression values can be downloaded. Another option is to cluster the experiments. Clustering the experiments will group tissues with similar expression patterns.
4. A higher number k results in clusters that show a similar expression pattern.
5. Administrator rights are needed to install Genesis. The program needs unrestricted write access to its own program folder because it needs to be able to save temporary files.

References

1. Dong X, Reimer J, Göbel U et al (2012) Natural variation of H3K27me3 distribution between two Arabidopsis accessions and its association with flanking transposable elements. Genome Biol 13:117
2. Juan HF, Huang HC (2007) Bioinformatics: microarray data clustering and functional classification. Methods Mol Biol 382:405–416
3. Rhee SY, Wood V, Dolinski K et al (2008) Use and misuse of the gene ontology annotations. Nat Rev Genet 9:509–515
4. Blake JA, Dolan M, Drabkin H et al (2012) The Gene Ontology enhancements for 2011. Nucleic Acids Res 40:D559–D564
5. Clark JI, Brooksbank C, Lomax J (2005) It's all GO for plant scientists. Plant Physiol 138:1268–1279
6. Schmid M, Davison TS, Henz SR et al (2005) A gene expression map of Arabidopsis thaliana development. Nat Genet 37:501–506
7. Winter D, Vinegar B, Nahal H et al (2007) An "electronic fluorescent pictograph" browser for exploring and analyzing large-scale biological data sets. PLoS One 2, e718
8. de Hoon MJL, Imoto S, Nolan J et al (2004) Open source clustering software. Bioinformatics 20:1453–1454
9. Joshi VK, Freudenberg JM, Hu Z et al (2011) WebGimm: an integrated web-based platform for cluster analysis, functional analysis, and interactive visualization of results. Source Code Biol Med 6:3
10. Sturn A, Quackenbush J, Trajanoski Z (2002) Genesis: cluster analysis of microarray data. Bioinformatics 18:207–208
11. Khatri P, Draghici S (2005) Ontological analysis of gene expression data: current tools, limitations, and open problems. Bioinformatics 21:3587–3595
12. Benjamini Y, Yekutieli D (2001) The control of the false discovery rate in multiple testing under dependency. Ann Stat 29:1168–1188
13. Provart NJ, Zhu T (2003) A browser-based functional classification SuperViewer for Arabidopsis genomics. Curr Comput Mol Biol 271–272
14. Al-Shahrour F, Minguez P, Tarraga J et al (2007) FatiGO +: a functional profiling tool for genomic data. Integration of functional annotation, regulatory motifs and interaction data with microarray experiments. Nucleic Acids Res 35:W91–W96
15. Irizarry RA, Hobbs B, Collin F, Beazer-Barclay YD, Antonellis KJ, Scherf U, Speed TP (2003) Exploration, normalization, and summaries of high density oligonucleotide array probe level data. Biostatistics 4(2):249–264
16. Saldanha AJ (2004) Java Treeview—extensible visualization of microarray data. Bioinformatics 20:3246–3248
17. Carbon S, Ireland A, Mungall CJ et al (2009) AmiGO: online access to ontology and annotation data. Bioinformatics 25:288–289
18. Maere S, Heymans K, Kuiper M (2005) BiNGO: a Cytoscape plugin to assess overrepresentation of gene ontology categories in biological networks. Bioinformatics 21:3448–3449
19. Zheng Q, Wang XJ (2008) GOEAST: a web-based software toolkit for Gene Ontology enrichment analysis. Nucleic Acids Res 36:W358–W363

20. Boyle EI, Weng S, Gollub J et al (2004) GO::TermFinder—open source software for accessing Gene Ontology information and finding significantly enriched Gene Ontology terms associated with a list of genes. Bioinformatics 20:3710–3715
21. Du Z, Zhou X, Ling Y et al (2010) agriGO: a GO analysis toolkit for the agricultural community. Nucleic Acids Res 38:W64–W70
22. Zhou X, Su Z (2007) EasyGO: Gene Ontology-based annotation and functional enrichment analysis tool for agronomical species. BMC Genomics 8:246
23. Beissbarth T, Speed TP (2004) GOstat: find statistically overrepresented Gene Ontologies within a group of genes. Bioinformatics 20:1464–1465
24. Carmona-Saez P, Chagoyen M, Tirado F et al (2007) GENECODIS: a web-based tool for finding significant concurrent annotations in gene lists. Genome Biol 8:R3

Chapter 4

Localization of miRNAs by In Situ Hybridization in Plants Using Conventional Oligonucleotide Probes

Sara Hernández-Castellano, Geovanny I. Nic-Can, and Clelia De-la-Peña

Abstract

Among the epigenetic mechanisms studied with a greater interest in the last decade are the microRNAs (miRNAs). These small noncoding RNA sequences that are approximately 17–22 nucleotides in length play an essential role in many biological processes of various organisms, including plants. The analysis of spatiotemporal expression of miRNAs provides a better understanding of the role of these small molecules in plant development, cell differentiation, and other processes; but such analysis is also an important method for the validation of biological functions. In this work, we describe the optimization of an efficient protocol for the spatiotemporal analysis of miRNA by in situ hybridization using different plant tissues embedded in paraffin. Instead of LNA-modified probes that are typically used for this work, we use conventional oligonucleotide probes that yield a high specificity and clean distribution of miRNAs.

Key words Epigenetics, miRNAs, *Coffea canephora*, *Arabidopsis thaliana*, In situ hybridization

1 Introduction

MicroRNAs (miRNAs) are small noncoding RNAs (17–21 nucleotides (nt) in length) that play an important role in plant growth and development. In plants, miRNAs can control gene expression throughout posttranscriptional regulation, usually by degradation of target mRNA transcripts or translational inhibition [1, 2]. Unlike animal miRNAs, plant miRNAs display a high complementarity between their sequence and target mRNAs [3]. The degree of complementarity between miRNAs and mRNAs (full or partial) has allowed a better prediction and identification of miRNA targets in plants. In addition, according to the similarity of their sequences, miRNAs have been classified into families which can be conserved or unique among different plant species [4, 5]. This feature is a key starting point for a comparison between monocotyledonous and dicotyledonous plants [3, 6]. Furthermore, the analysis of miRNAs has been focused on the study of plant development such as organ boundary formation; radial paternity; the

Igor Kovalchuck (ed.), *Plant Epigenetics: Methods and Protocols*, Methods in Molecular Biology, vol. 1456,
DOI 10.1007/978-1-4899-7708-3_4, © Springer Science+Business Media New York 2017

development of roots, stems, and floral organs; as well as somatic embryogenesis regulation [6–9].

Even though most studies have been realized in *Arabidopsis*, maize, and rice, the current technological advances on small RNA have allowed the discovery of many miRNAs in important crops [8, 10]. At the same time, the accumulation of massive sequencing data has already been deposited in published databases such as miRbase [11, 12], PMRD (Plant MicroRNA Database) [13], and PMTED (Plant MicroRNA Target Expression Database) [14]. These data can be used to predict and classify different families of miRNAs, precursor sequences, and targets of action. However, only a small percentage of miRNAs have been experimentally validated. Among the methods that are widely used for studying the temporal expression of miRNAs are Northern blot analysis, RT-qPCR, microarrays, and RNA ligase-mediated amplification of cDNA ends [15–19]. Although they provide quantitative data, expression patterns are considered to be global ones because of the different cell types analyzed through these techniques.

The visualization of miRNAs by in situ hybridization can be used to obtain information about where and to what degree these molecules are distributed throughout cells of a particular tissue, revealing the cellular localization of miRNA-mediated gene regulation [20, 21]. However, studies examining the spatial expression of miRNAs in specific cell tissues, particularly in plants, have been limited, and most of the available protocols have been used only in model plants, and thus sometimes they cannot be applied to other plant species [17, 20, 22]. For this reason, we have developed a method to visualize miRNAs in situ in the model plant *Arabidopsis thaliana* and an important crop plant, *Coffea canephora* (coffee). The method described in details here includes the fixation of samples (a crucial step to in situ hybridization) and the embedding of samples in paraffin (these steps are usually omitted). In this protocol, we explain how to achieve the correct hybridization using conventional only oligonucleotide probes, instead of using LNA-modified probes labeled with DIG-UTP, throughout the introduction of TEA treatment [23] for the better hybridization of oligonucleotide probes. In addition, we describe the recommended times to achieve a better detection and visualization of miRNAs that can be visualized by immunohistochemistry using anti-DIG and NBT/BCIP. This step is important for increasing the signal strength that can be visualized with a bright-field microscope (*see* Fig. 1). This method is easy and reproducible and can be used to validate novel miRNAs for a better understanding of plant development.

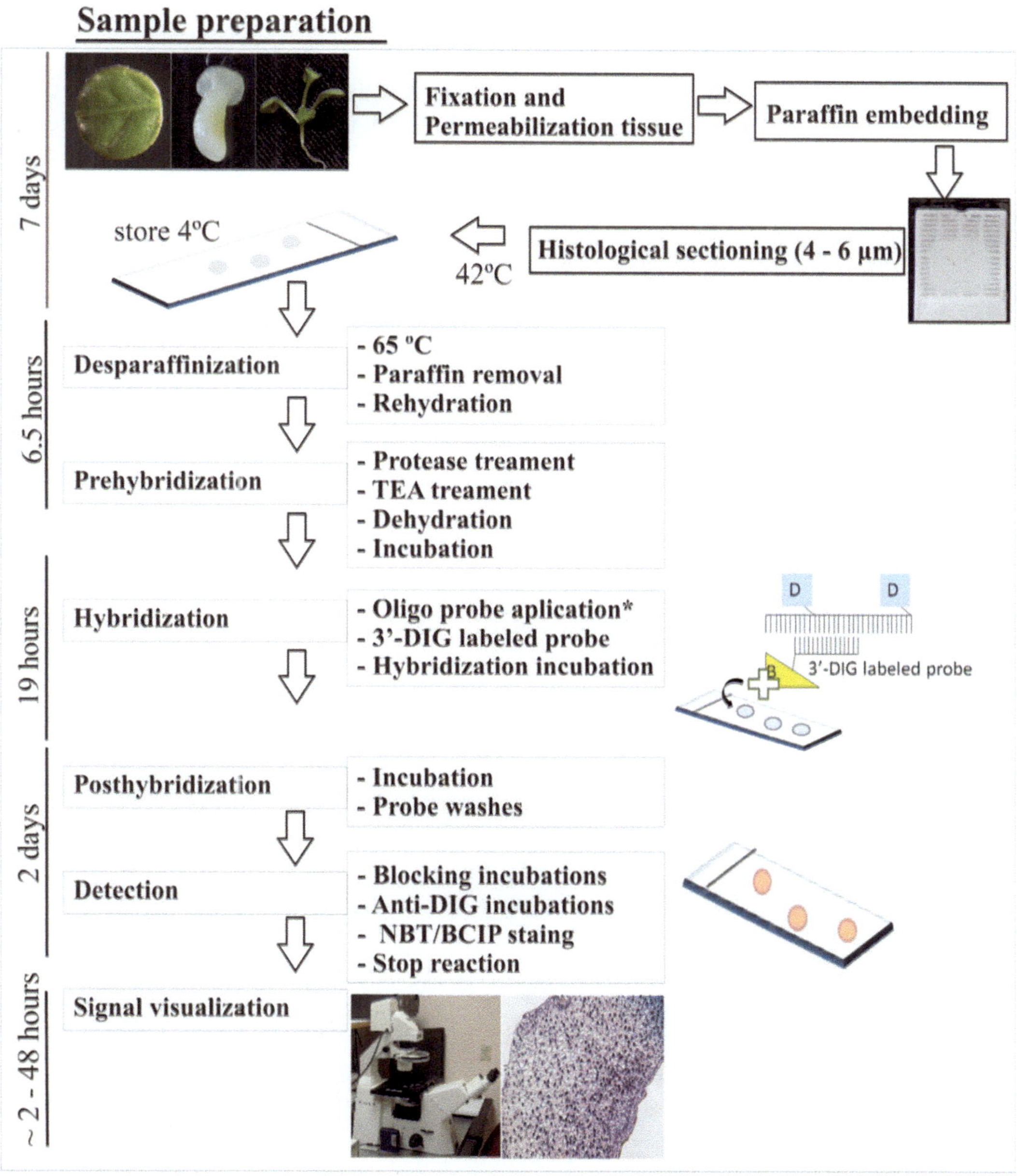

Fig. 1 The general scheme of principal steps in the protocol for in situ hybridization and localization of miRNAs. Tissue samples were collected and fixed with formalin solution and embedded in paraffin. Tissue sections were dewaxed, treated with protease and TEA, and incubated with DIG-labeled oligonucleotide conventional probes. The signal was detected using anti-DIG and NBT/BCIP staining, and the distribution of miRNAs was observed under a bright-field microscope. *Concentrations of the conventional oligonucleotide probe and the temperature of hybridization must be optimized according to the plants or tissue used for each miRNA

2 Materials

2.1 In Vitro Plantlets

The plant material used was leaf explants and somatic embryos at the cotyledonary stage obtained from *C. canephora* plantlets as described elsewhere [24] as well as 14-day-old seedlings of *A. thaliana* (wild-type, ecotype Col-0) grown at 25 ± 2 °C under photoperiod conditions (16/8-h light/darkness) (*see* Fig. 2).

2.2 Fixation and Permeabilization

Before starting, all buffers and solutions should be prepared with diethyl pyrocarbonate (DEPC)-treated water to avoid any RNase activity.

1. Formalin solution, neutral buffered, 10% (Sigma-Aldrich).
2. A 4% buffered formalin solution: this solution is hypotonic. Adjust pH to 6.8 with HCl and later add to 0.1% (vol/vol) methanol.
3. 1× PBS buffer: 7 mM Na_2HPO_4, 3 mM NaH_2PO_4, and 130 mM NaCl, adjust pH to 7.0 with HCl.
4. Ethanol series of 10, 20, 30, 50, 70, 85, and 96% in RNase-free water.

Fig. 2 Plant material used for the methodology of miRNA in situ hybridization. (**a**) Leaf explants of *Coffea canephora*. (**b**) A somatic embryo at the cotyledonary stage obtained from leaves of *C. canephora*. (**c**) 14-Day-old seedlings of *Arabidopsis thaliana* (wt, ecotype Col-0)

5. 1-Butanol.
6. Glass tubes (5 mL).
7. Pipettes for volumes 0.5–1000 μL.
8. A vacuum pump.
9. A timer.

2.3 Embedding Tissue in Paraffin

1. Paraplast plus tissue embedding medium.
2. An orbital shaker.
3. Stainless steel base molds.
4. Cassettes to hold tissue samples.
5. An oven at 37 °C and 65 °C.

2.4 Sectioning and Deparaffinization of Tissue Samples

1. A microtome.
2. Low-profile blades.
3. A heating block.
4. A thermometer.
5. RNase-free water.
6. Cover glass 22 mm × 50 mm.
7. Glass staining dish (Wheaton®).
8. An incubator at 37, 42, and 65 °C.
9. Xylene, ACS reagent.
10. Ultra-Clear.
11. Protease: Dilute the powder in RNase-free water to a final concentration of 50 mg/mL and predigest the protease by incubating it at 37 °C for 2 h; store aliquots of 650 μL at −20 °C. To prepare the protease solution, mix 650 μL of protease (50 mg/mL) and 250 mL of 1× TE solution previously warmed at 37 °C.
12. Ethyl alcohol absolute.
13. 100 % Ethanol and a decreasing series of ethanol solutions (10, 20 30, 50, 70, 85, and 95 % (v/v)) in RNase-free water.
14. TE solution: 100 mM Tris-HCl (pH 8.0). 50 mM EDTA.
15. TEA solution: Prepare 0.1 M triethanolamine–HCl (pH 8.0) buffer by mixing 393 mL of H_2O, 5.2 mL of triethanolamine, and 1.6 mL of HCl.
16. Oxide anhydride.
17. 1× PBS buffer.
18. 0.2 % Glycine (wt/vol) in 1× PBS buffer.

2.5 Pre-hybridization and Hybridization

1. Oligonucleotide conventional probes: Prepare an appropriate concentration of oligonucleotide probe (try from 10 to 100 pmol), and mix with 10 μL formamide and RNase-free

water up to a final volume of 20 μL. Heat the oligo probe mix at 85 °C for 5 min and place it on ice immediately before use.

2. In situ hybridization salts: 3 M NaCl, 100 mM Tris–HCl (pH 8.0), 100 mM sodium phosphate dibasic (pH 6.8), 50 mM EDTA; store aliquots of 1 mL at −20 °C.
3. In situ hybridization buffer: To prepare 10 mL, mix 1.25 mL of in situ hybridization salt solution, 5 mL of deionized formamide, 2.5 mL of 50% (wt/vol) dextran sulfate, 125 μL of tRNA (100 mg/mL), 250 μL of 50× Denhardt's solution, and 875 μL RNase-free water; store aliquots at −20 °C.
4. 50% (wt/vol) Dextran sulfate: To prepare this solution, the dextran sulfate should be heated at 80 °C in order to be dissolved; store aliquots at −20 °C.
5. Recombinant terminal transferase RNA (tRNA): Dilute the powder from one vial of tRNA in 1 mL of RNase-free water to obtain a concentration of 100 mg/mL; store at −20 °C.
6. 50× Denhardt's solution.
7. TE solution.
8. Oligonucleotide conventional probes (*see* Table 1).
9. Deionized formamide.
10. A DIG oligonucleotide 3′-end labeling kit, second generation.
11. An incubator at 37 °C and 42–55 °C.

2.6 Post-hybridization and Detection

1. 1× SSC: Mix 0.3 M NaCl and 0.03 M sodium citrate, adjust to pH 7.0.
2. 1× TBS buffer: Mix 100 mM Tris–HCl (pH 7.5) and 150 mM NaCl.
3. 1× TN buffer: Mix 100 mM Tris–HCl (pH 9.5) and 100 mM NaCl.
4. Blocking buffer: Mix 1% (wt/vol) blocking reagent in 1× TBS buffer. Prepare a fresh solution every time.

Table 1
Conventional oligonucleotide probe sequences used for miRNA in situ hybridization

miRNA	Probe sequence	Probe Tm (°C)	Hybridization temperature (°C)
miR156	5′-GTCCTCTCTATCTTCTGTCAA-3′	56.2	55
miR390	5′-GGCGCTATCCCTCCTGAGCTT-3′	69.7	57
miR535	5′-GCGTGCTCTCTCTCGTTGTCA-3′	63	56
U6snRNA	5′-AGGGGCCATGCTAATCTTCTC-3′	65.4	57

5. Washing buffer: Mix 0.5 % (wt/vol) albumin from bovine serum (BSA) and 0.3 % (vol/vol) Triton X-100 in 1× TBS buffer. Prepare just before use.
6. Anti-DIG-AP, Fab fragments from sheep: Dilute anti-DIG solution at 0.75 U/mL in washing buffer. This solution must be prepared fresh.
7. NBT/BCIP stock solution.
8. Staining solution: Mix to 20 μL of NBT/BCIP stock solution in 1 mL of 1× TN buffer. Prepare in fresh.
9. In situ mounting medium.
10. TE solution.
11. Microscopy.

3 Methods

3.1 Fixation and Permeabilization of Plant Tissue

1. Tissue preparation: Collect tissue (embryos or sections of plants) and pre-fix it for 15 min in 10 % neutral buffering formalin at room temperature (RT) applying vacuum for 5 min (*see* **Note 1**).
2. Discard the first formalin solution and add a new 4 % formalin solution to the samples; apply vacuum for 10 min and keep the sample at RT for 12 h (*see* **Note 2**).
3. Discard the 4 % formalin solution of tissue sample, add the 1× PBS buffer, apply vacuum for 10 min, and maintain at RT for 12 h.
4. Remove the 1× PBS buffer and dehydrate tissue samples through a series of increasing concentrations of absolute ethanol of 30, 50, 70, and 85 % (2×2 h each) and 96 % (2×30 min each), applying vacuum for 5 min at each step and maintaining the samples at RT.
5. Maintain tissue samples in 100 % 1-butanol for 24 h at RT two times (48 h in total).

3.2 Paraffin Embedding and Sectioning Plant Tissues

1. Add fresh 1-butanol and 10–15 flakes of Paraplast plus® to the samples and incubate them overnight at RT under gentle agitation (60 rpm).
2. Incubate tissue samples at 65 °C and add 20–30 flakes of paraffin every 2–3 h, three times.
3. Remove an excess of 1-butanol by changing the liquid paraffin every 12 h, four times.
4. Place tissue samples in the center of stainless steel base molds that were previously heated to 65 °C, and embed them in paraffin to fill the molds and cover with cassettes for holding tissue. Then the samples are maintained at RT for at least 4 h.

5. Remove the cassettes from steel base molds and store the tissue samples at 4 °C until use.
6. Dissect the paraffin-embedded sample into 4–6 μm slices using a retracting microtome with low-profile blades (*see* **Note 3**).
7. Place sample sections in an RNase-free water bath at 42 °C to allow the correct expansion of tissues, and then place them on micro-slides.
8. Incubate the tissue sections attached to the micro slides at 37 °C for at least 2 h. The slides with tissue sections can be stored at 4 °C until use.

3.3 Deparaffinization and Rehydration of Tissue Samples

1. Deparaffinize tissue sections as follows: Incubate the slides at 65 °C for 15 min and place tissue samples in slide-staining jars with xylene three times for 10 min per rinse, and with Ultra-Clear™ (Histo-grade; J.T. Baker®) three times for 10 min each per rinse with slow movement.
2. Rinse the slides twice with 100% ethanol for 5 min.
3. Rehydrate the tissue samples through a decreasing ethanol concentration series of 96, 80, 70, 50, 30, and 10% (vol/vol) ethanol, for 5 min per incubation.
4. Place the slides in the 1× PBS buffer for 10 min.
5. Incubate the slides with a protease solution for 1 h at 37 °C (*see* **Note 4**).
6. Stop and neutralize the activity of protease with 0.2% (wt/vol) glycine in the 1× PBS buffer, twice for 10 min each at RT.
7. Rinse the slides twice in the 1× PBS buffer for 10 min each.
8. Rinse the slides in TEA solution for 1 h at RT, slowly stirring (*see* **Note 5**).
9. Rinse the slides twice in the 1× PBS buffer for 10 min per rinse.
10. Dehydrate the tissue section slides through a gradually increasing ethanol concentration series of 10, 30, 50, 70, 80, 95, and 100% (vol/vol) ethanol, for 5 min per incubation. The slides in 100% ethanol can be stored for 2 h at 4 °C.

3.4 Pre-hybridization and Hybridization

1. After the dehydration step, allow the tissue samples to dry completely.
2. To perform the pre-hybridization, incubate the tissue samples with 50–100 μL of hybridization in situ buffer in a wet chamber for 1 h at RT.
3. Mix 20 μL of a conventional oligonucleotide probe, using DIG-labeling according to the data in Table 2, with 80 μL of hybridization in situ buffer. Mix the samples slowly and carefully to avoid bubbles.

Table 2
Preparation of the reaction mix using oligo 3′-digoxigenin labeling. Mix the components in a sterile Eppendorf tube, incubate at 37 °C for 1 h, and place them on ice. Stop the reaction with 4 μL of 0.1 M EDTA (pH 8.0)

Component	Amount	Final concentration
5× Reaction buffer	4 μL	1×
25 mM $CoCl_2$	4 μL	5 mM
Oligo probe of interest[a]	100 pmol	10 pmol/μL
1 mM DIG-ddUTP	1 μL	50 nM
Recombinant terminal transferase	1 μL	2.5 U
H_2O	Up to 20 μL	

[a] *See* Subheading 2.5, **item 1**

4. Let the tissue samples dry, add 20 μL of hybridization solution (described in **step 3**) to the tissue sections on the slides, and carefully place them in a wet chamber.
5. Incubate the tissue sections for 18 h at the appropriate hybridization temperature. Consider that it might vary according to the melting temperature of each oligo probe (typically between 42 °C and 57 °C).

3.5 Post-hybridization and Detection of Samples

1. After the hybridization time, incubate the tissue samples in a warm 0.2× SSC solution for 10 min at RT.
2. Wash the slides twice with 0.2× SSC for 1 h per wash at the same temperature as used in the hybridization step (*see* Subheading 3.4, **step 5**).
3. Remove the slides from 0.2× SSC and incubate them twice in 1× PBS solution for 10 min each time.
4. Rinse the slides with 1× TBS solution for 10 min at RT.
5. Incubate each tissue sample on the slides with 100 μL of blocking buffer (*see* Subheading 2.6, **item 4**) for 1 h in a moist chamber at RT.
6. Remove the blocking buffer and incubate the tissue sections with 200 μL of washing buffer for 1 h in a moist chamber at RT.
7. Discard the washing buffer and carefully place 50–100 μL of the mix anti-DIG solution (*see* Subheading 2.6, **item 6**) directly onto the samples, ensuring that the solution uniformly covers all the tissue sections. Incubate the samples in a moist chamber at 4 °C overnight.
8. Wash the slides three times in washing buffer for 25 min per wash.

9. Rinse the slides with 1× TBS for 10 min at RT.
10. Wash the slides twice in 1× TN solution for 10 min at 37 °C to equilibrate the samples and improve the alkaline phosphatase activity.
11. Cover the tissue samples with 20 μL of NBT/BCIP staining solution and incubate them in the dark at 4 °C in a moist chamber.
12. Replace NBT/BCIP staining solution every 12 h, at least four times.
13. Monitor the development of a signal under a microscope (*see* **Note 6**).
14. When the signal is visualized, discard NBT/BCIP staining solution and wash the slides twice in 1× TE solution for 5 min per wash.
15. Let the tissue samples dry completely, add 100 μL of in situ mounting medium, mount with a slide cover glass, and seal the perimeter with nail polish or a plastic sealant.
16. Observe the location and distribution of miRNAs of interest on the tissue under a bright-field microscope (*see* Fig. 3).

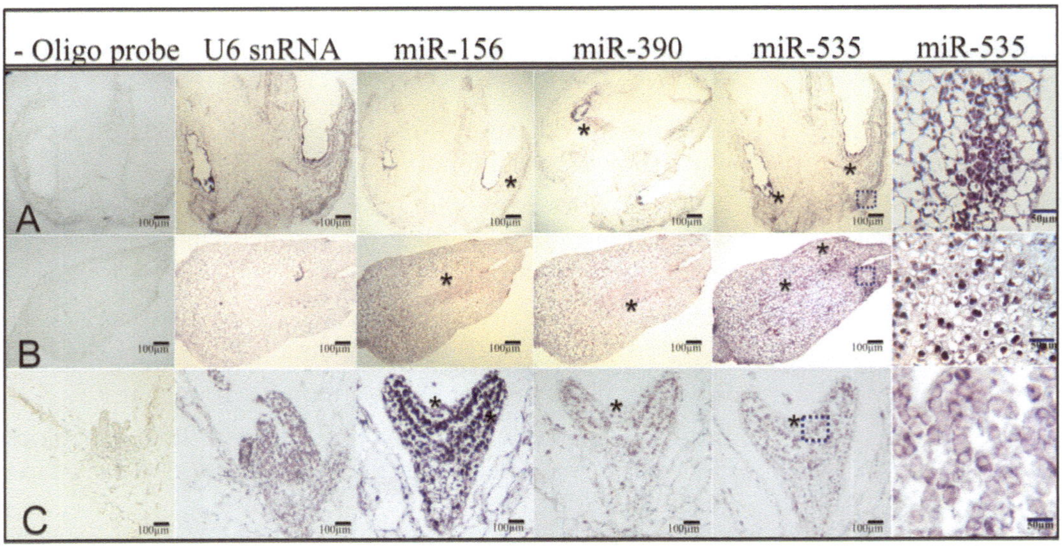

Fig. 3 Visualization of the cellular distribution of different miRNAs. miRNA in situ hybridization was performed as showed in Fig. 1 in different tissues from *C. canephora* (**a**) leaf and (**b**) a somatic embryo at the cotyledonary stage, and (**c**) 14-day-old seedlings of *Arabidopsis* with oligonucleotide probes for U6 snRNA (as a control) and miR390 hybridized at 57 °C. miR156 and miR535 were hybridized at 55 °C and 56 °C, respectively. For all miRNAs, a concentration of 10 pmol/μL was used. At this concentration, the miR156 probe showed an accumulation in the margin of the leaf and in the vascular procambium of the coffee embryo. In *Arabidopsis*, an increased accumulation in the shoot meristem was observed. A close-up of the dashed square in miR535 distribution is shown in great detail. *Asterisk* represents zones of high levels of hybridization. U6 snRNA and the absence of oligonucleotide probes were used as positive and negative controls, respectively

4 Notes

1. The fixation time in both 10 and 4 % formalin must be adjusted to the size and diameter of samples to be analyzed. Formalin has a penetration rate of 5 mm of tissue per hour.
2. To stabilize the 4 % solution of formalin, add 0.1 % (vol/vol) methanol in order to prevent the conversion of formalin to formic acid.
3. During histological sectioning of samples and their placement on slides, the use of gloves is very important. Also, all material must be sterile and RNase free to avoid contamination that might interfere with the in situ hybridization.
4. The optimum concentration of protease should be between 35 and 50 mg/mL. It should be predigested for 4 h at 37 °C and mixed in 250 mL of 1× TE fresh solution. The incubation time of protease digestion is 1 h at 37 °C in a moist chamber.
5. TEA treatment is used to acetylate the positively charged amino groups into the samples in order to reduce the nonspecific binding of probes that is required when oligo conventional probes are used, thus enabling a better hybridization.
6. Color development occurs after adding NBT/BCIP staining solution and incubation at 4 °C in a moist chamber for 2 h, reaching a maximum coloration after 48 h. The development of this colorimetric reaction depends on the pH of the TN buffer.

Acknowledgement

This work was supported by grants from CONSEJO NACIONAL DE CIENCIA Y TECNOLOGÍA (CONACYT) to C.D. (178149), CONACYT-scholarship to S.H.C. (271240), and Cátedras-CONACYT ICC1 to G.N.C.

References

1. Chen X (2005) MicroRNA biogenesis and function in plants. FEBS Lett 579:5923–5931
2. Carthew RW, Sontheimer EJ (2009) Origins and mechanisms of miRNAs and siRNAs. Cell 136:642–655
3. Axtell MJ (2013) Classification and comparison of small RNAs from plants. Ann Rev Plant Biol 64:137–159
4. Rhoades J, Bartel DP, Bartel B (2006) miRNAs and their regulatory roles in plants. Ann Rev Plant Physiol 57:19–53
5. Brodersen P, Sakvarelidze-Achard L, Bruun-Rasmussen M, Dunoyer P, Yamamoto YY, Sieburth L, Voinnet O (2008) Widespread translational inhibition by plant miRNAs and siRNAs. Science 320:1185–1190
6. Xie Z, Khanna K, Ruan S (2010) Expression of microRNAs and its regulation in plants. Sem Cell Dev Biol 21:790–797
7. Bartel DP (2004) MicroRNAs: genomics, biogenesis, mechanism, and function. Cell 116:281–297

8. Lin Y, Lai Z (2013) Comparative analysis reveals dynamic changes in miRNAs and their targets and expression during somatic embryogenesis in Longan (*Dimocarpus longan Lour*). PLoS One 8, e60337
9. Liu X, Huang J, Wang Y, Khanna K, Xie Z, Owen HA, Zhao D (2010) The role of floral organs in carpels, an Arabidopsis loss-of-function mutation in MicroRNA160a, in organogenesis and the mechanism regulating its expression. Plant J 62:416–428
10. Wu XM, Liu M, Ge X, Xu Q, Guo W (2011) Stage and tissue-specific modulation of ten conserved miRNAs and their targets during somatic embryogenesis of Valencia sweet orange. Planta 233:495–505
11. Kozomara A, Griffiths-Jones S (2011) miRBase: integrating microRNA annotation and deep-sequencing data. Nucleic Acids Res 39:D152–D157
12. Griffiths-Jones S, Saini HK, van Dongen S, Enright AJ (2008) miRBase: tools for microRNA genomics. Nucleic Acids Res 36:154–158
13. Zhang Z, Yu J, Li D, Liu F, Zhou X, Wang T, Ling Y, Su Z (2010) PMRD: plant microRNA database. Nucleic Acids Res 38:D806–D881
14. Sun X, Dong B, Yin L, Zhang R, Du W, Liu D, Shi N, Li A, Liang Y, Mao L (2013) PMTED: a plant microRNA target expression database. BMC Bioinformatics 14:174
15. Nelson PT, Baldwin DA, Scearce LM, Oberholtzer JC, Tobias JW, Mourelatos Z (2004) Microarray-based, high-throughput gene expression profiling of microRNAs. Nat Methods 1:155–161
16. Eldem V, Okay S, Ünver T (2013) Plant microRNAs: new players in functional genomics. Turk J Agric For 37:21
17. Unver T, Namuth-Covert D, Budak H (2009) Review of current methodological approaches for characterizing microRNAs in plants. Inter J Plant Gen 1:1–11
18. Tran N (2009) Fast and simple micro-RNA northern blots. Biochem Insights 2:1–3
19. Alastair W, Hye-Jin L, Wark D (2008) Multiplexed detection methods for profiling microRNA expression in biological samples. Angew Chem Int Ed 47:644–652
20. Kidner C, Timmermans M (2006) In situ hybridization as a tool to study the role of microRNAs in plant development. In: Ying SY (ed) MicroRNA protocols. Humana Press, Totowa, pp 159–179
21. Song R, Ro S, Yan W (2010) *In situ* hybridization detection of microRNAs. Methods Mol Biol 628:287–294
22. Javelle M, Timmermans MC (2012) *In situ* localization of small RNAs in plants by using LNA probes. Nat Protocols 7:533–544
23. Nuovo GJ (2008) *In situ* detection of precursor and mature microRNAs in paraffin embedded, formalin fixed tissues and cell preparations. Methods 44:39–46
24. Quiroz-Figueroa FR, Monforte-González M, Galaz-Avalos RM, Loyola-Vargas VM (2006) Direct somatic embryogenesis in *Coffea canephora*. In: Loyola-Vargas VM, Vázquez-Flota FA (eds) Plant cell culture protocols. Humana Press, Totowa, NJ, pp 111–117

Chapter 5

The Combined Bisulfite Restriction Analysis (COBRA) Assay for the Analysis of Locus-Specific Changes in Methylation Patterns

Andriy Bilichak and Igor Kovalchuk

Abstract

DNA methylation is a heritable but reversible epigenetic mechanism of control over gene expression. The level of DNA methylation of specific genomic regions correlates with chromatin condensation, the level of gene expression, and in some cases genome stability and the frequency of homologous recombination. Here, we describe the combined bisulfite restriction analysis (COBRA) assay that allows analyzing the methylation status at a specific locus. The protocol consists of the following major steps: bisulfite conversion of non-methylated cytosines to uracils, the locus-specific PCR amplification of converted DNA, restriction digestion, the analysis of restriction patterns on the gel, and the quantification of these restriction patterns using ImageJ or a similar program.

Key words Locus-specific DNA methylation, The combined bisulfite restriction analysis (COBRA), Bisulfite conversion

1 Introduction

DNA methylation is one of the major epigenetic mechanisms of regulation of chromatin condensation, transcription initiation and transcription rate, transposon activity, and inheritance of epigenetic traits, to name a few [1–3]. DNA demethylation commonly results in chromatin decondensation, the activation of transposable elements [4], and transcription activation [5]; the latter one depends on the sequence context and tissue specificity of gene expression. Although a snapshot of DNA methylation levels in certain cells of a given tissue can be obtained, DNA methylation is a dynamic process that varies from one genomic region to another, cell to cell, tissue to tissue, and organism to organism. Moreover, DNA methylation changes during its development and in response to environmental stimuli. While genes usually have several discrete methylated regions, transposons are methylated uniformly [5].

Igor Kovalchuck (ed.), *Plant Epigenetics: Methods and Protocols*, Methods in Molecular Biology, vol. 1456,
DOI 10.1007/978-1-4899-7708-3_5, © Springer Science+Business Media New York 2017

Table 1
A list of restriction endonucleases commonly used for the COBRA assay

Restriction endonuclease	Restriction site
*BsiW*I	C/GTACG
*BspD*I	AT/CGAT
*BstB*I	TT/CGAA
*BstU*I	CG/CG
*Cla*I	AT/CGAT
*Hpy*CH4IV	AC/GT
*Mlu*I	A/CGCGT
*Nru*I	TCGCGA
*Pvu*I	CGAT/CG
*Taq*I	T/CGA

Cytosine methylation most commonly analyzed is cytosine methylation in a CpG sequence context. To avoid difficulties during CpG methylation analysis, the restriction enzymes selected for the assay should have cytosine residues in their recognition sites only in the CpG sequence context

However, even within a genic region, there are variations in DNA methylation that depend on whether the gene is expressed in all tissues or it is expressed in a tissue-specific manner [6, 7]. In addition, cellular mechanisms of regulation of gene expression, such as imprinting, X-chromosome inactivation, and dosage compensation, are also methylation dependent [8, 9]. Finally, methylation plays an important role in the inheritance from cell to cell and from generation to generation of the chromatin structure and gene expression pattern under normal conditions and in response to stress [10–13].

A number of methods are available for the qualitative and quantitative detection of global genome- and locus-specific changes in DNA methylation patterns. Here, we present a method for the analysis of locus-specific changes in DNA methylation—combined bisulfite restriction analysis (or COBRA). This technique provides reliable quantitative results regardless of DNA methylation levels at the target locus [14]. The potential weaknesses of the COBRA assay include its limitation to the genomic regions which carry the restriction cites listed in Table 1.

The assay consists of *three major steps*: treatment of genomic DNA with sodium bisulfite, PCR amplification, and restriction digestion (Fig. 1). *The first step*, treatment with sodium bisulfite, converts non-methylated cytosines to uracil residues while sparing the methylated cytosines. The conversion of cytosines to uracil

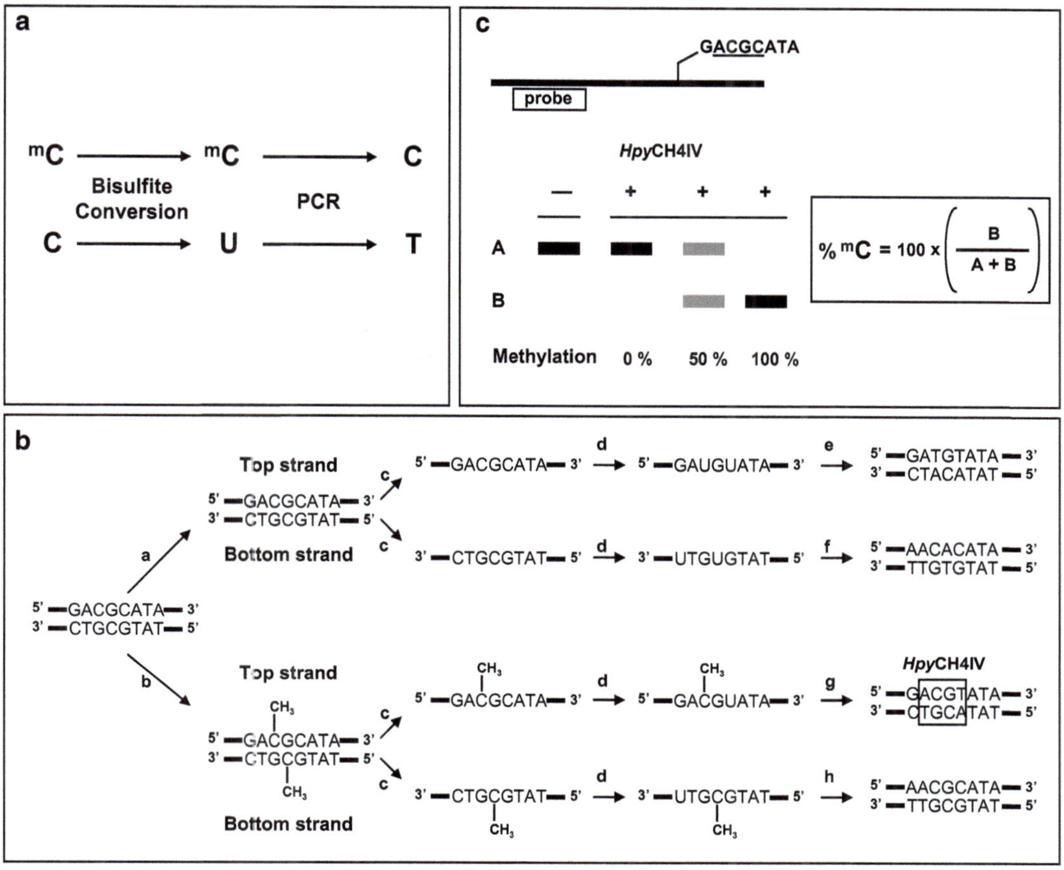

Fig. 1 A general outline of the COBRA assay. **A**—A general mechanism of methylated cytosine detection. The bisulfite treatment converts all non-methylated cytosine residues to uracil residues. Next, PCR amplification substitutes uracil for thymine. In contrast, methyl-cytosine residues remain unchanged. **B**—The generation of new restriction sites upon bisulfite conversion and PCR amplification. The original DNA sequence that was chosen for the COBRA analysis contains a precursor of a recognition site for *Hpy*CH4IV restriction endonucleases (ACGC). The original DNA sequence can exist in two forms: (**a**) the non-methylated (ACGC) cytosine nucleotide, and (**b**) the methylated (A^{m}CGC) cytosine nucleotide in a CpG sequence context. Denaturation separates the top and bottom DNA strands (**c**). The native DNA sequence is modified in a methylation-dependent manner upon bisulfite conversion of single-stranded DNA (**d**). The top (**e**, **g**) and bottom (**f**, **h**) DNA strands are amplified in different PCR reactions using distinct sets of PCR primers. In our example, only the amplification of the top DNA strand is informative. It leads to the generation of the *Hpy*CH4IV recognition site (ACGT) in a methylation-dependent manner (A^{m}CGC → A^{m}CGU → ACGT versus ACGC → AUGU → ATGT). **C**—The quantification of cytosine methylation by probe hybridization. Note that the probe sequence does not cover the restriction enzyme recognition site. The ratio of the cleaved PCR product (*B*) and the total amount of the PCR product (*A* + *B*) shows a percentage of methylated *Hpy*CH4IV recognition site precursors in the original DNA

results in the disappearance of restriction recognition sites containing cytosines and/or the appearance of new restriction sites containing thymines but lacking cytosines. Since the initial pool of DNA used for bisulfite conversion contains methylated and non-methylated cytosines at a given genomic locus, bisulfite conversion leads to the formation of a mixed population of DNA fragments containing cytosines and thymines at this specific position. Currently, there are a number of kits available on the market for fast and efficient bisulfite conversion of gDNA with the subsequent purification and desalting steps followed. *The second part*, PCR, allows the amplification of each of the sequence variants without affecting the relative ratio between them and thus preventing a bias in the comparison of different methylation patterns. Hence, the resulting PCR products usually represent a collection of DNA sequences that have the same length, but differ in sequence composition at the sites of potential DNA methylation. These differences are then revealed by *the third step*—the digestion by restriction endonucleases that recognize DNA sequences that are affected by methylation. For the visualization of the cleaved PCR products, the reaction mix is separated by gel electrophoresis. The quantification can be done in two ways, depending on the amount of DNA on the gel. If the amount is high, the image processing software such as ImageJ can be used directly, but if it is low, the cleaved PCR products can be transferred to the membrane and hybridized with a specific probe, much like in Southern blot analysis. Finally, the ratio between the cleaved and remaining PCR products corresponds to the ratio between methylated and non-methylated cytosine residues that are originally present in genomic DNA before bisulfite conversion.

The COBRA assay is characterized by high sensitivity, and in contrast to other site-specific methylation analysis techniques such as methylation-specific PCR (MSP), it has a very low possibility of false-positive results [14]. It can efficiently work with low amounts of input DNA, and it also permits analysis of cytosine methylation in two DNA strands separately [15]. Overall, due to the appearance of bisulfite conversion kits on the market, the assay is time efficient and provides a high degree of quantitative accuracy.

2 Materials

2.1 Sodium Bisulfite Conversion

1. EZ DNA Methylation-Gold™ Kit (Zymo Research, USA).

2.2 PCR Amplification

1. Nuclease-free water.
2. Primers.
3. 10× *Ex Taq*™ Buffer (Takara Bio USA).
4. dNTP Mixture (Takara Bio USA).
5. Takara Ex Taq™ DNA Polymerase (Takara Bio USA).
6. 100 and 70% ethanol.

2.3 Restriction Enzyme Digestion

1. Restriction enzyme with suitable 10× reaction buffer.
2. Nuclease-free water.

2.4 Gel Electrophoresis and Methylation Analysis

1. Agarose, electrophoresis grade.
2. 1× TBE: 90 mM Tris-HCl, pH 8.0, 90 mM boric acid 2 mM EDTA.
3. 6× DNA gel loading buffer.
4. DNA ladder (Fermentas).
5. Gel image analysis software (Image J, National Institutes of Health, USA).

3 Methods

3.1 Sodium Bisulfite Treatment

Treatment with sodium bisulfite converts non-methylated cytosines to uracils. In contrast, methylated cytosine residues remain unchanged due to a lower reactivity of 5-methylcytosine [16]. The reaction proceeds through several steps and requires that DNA remains in a single-stranded form. Incomplete DNA denaturation prevents sulfonation of cytosines at the C-6 position and results in an incomplete conversion.

1. Prepare 2 μg aliquots of genomic DNA in a final volume of 20 μl (*see* **Notes 1–3**).
2. Follow the protocol for bisulfite conversion, DNA desulfonation, and cleanup described in the manual for EZ DNA Methylation-Gold™ Kit (*see* **Note 4**).
3. Samples can be stored at −20 °C until needed.

3.2 PCR Amplification

PCR amplification of bisulfite-treated DNA is more technically challenging than PCR on native DNA. First, the remaining bisulfite may inhibit the PCR reaction. However, passing a sample through a desalting column helps to solve this problem. Other difficulties are primer design and the optimization of thermal-cycling parameters.

Bisulfite conversion significantly alters the native DNA sequence by depleting cytosine nucleotides. Following bisulfite treatment, two originally complementary DNA strands become non-complementary single-stranded molecules. Thus, two different sets of PCR primers are needed to analyze DNA methylation at both sense and antisense DNA strands. Since bisulfite conversion may generate/retain restriction sites in one strand but not in another, the right choice of a DNA strand for analysis is very important. This is especially important when methylation at non-symmetrical sites (CpHpH) is of interest. Once a DNA strand is selected, PCR primers can be designed.

There are three important rules for primer design. First, primers should be designed in such a way that they contain no CpG dinucleotides and have a low cytosine content. This insures that PCR equally amplifies the originally methylated and unmethylated sequences. Second, it is preferable to use longer primers (24 bases in length and longer); this will help compensate for the reduced sequence complexity of PCR products generated from the bisulfite-treated DNA template. Finally, design primers that would amplify a relatively small amplicon, preferably less than 500 bp, which may help improve PCR quality and reduce bias during sequence amplification.

Similarly, thermal cycling parameters may require optimization. In the first cycle, the preliminary extended denaturation time (up to 5 min) is recommended. If the amount of template DNA is sufficient, 30 cycles of PCR should be enough to produce the product for the subsequent restriction digestion. If the amount of input DNA is very low, then the number of cycles can be increased. Alternatively, a secondary or nested PCR can be performed. A nested PCR is a conventional PCR that uses the second or even third PCR with different sets of primers by using the product of the first (second) PCR as a template.

1. Use 2–3 μl of bisulfite-treated DNA per a 25 μl PCR reaction.
2. Prepare PCR reactions. Each reaction should contain 0.63 units of *Takara Ex Taq*™ *DNA polymerase*, *1*× Ex Taq™ *Buffer* (contains 2 mM $MgCl_2$), *dNTP mixture* (2.5 mM each dNTP), and 50 μM of each primer in a final volume of 25 μl. If more than five samples have to be analyzed, master mix can be used.
3. After PCR cycling is completed, keep samples on ice. If necessary, perform a secondary PCR.
4. Ethanol-precipitate the PCR product. Resuspend DNA in 20 μl of sterile distilled water. Quantify DNA (*see* **Note 5**).
5. Samples can be stored at −20 °C until needed.

3.3 Restriction Enzyme Digestion

Since bisulfite conversion generates new restriction sites and retains the original restriction sites in a methylation-dependent manner, it is possible to analyze cytosine methylation using various restriction endonucleases (Table 1). The general opinion is in favor of using newly generated restriction sites for restriction analysis. Using the newly created restriction sites allows the verification of a complete bisulfite conversion of the fully methylated cytosine. It is also possible to use methylation-sensitive restriction endonucleases because PCR products do not contain methylated cytosine residues. The application of different restriction enzymes permits the analysis of cytosine methylation in a different sequence context, including symmetrical CpG and CpHpG and non-symmetrical CpHpH methylation.

1. Digest 1 μg of purified PCR-amplified DNA with a tenfold excess of restriction enzyme in a final volume of 20 μl according to the manufacturer's protocol (*see* **Note 6**). Incubate samples overnight (*see* **Note 7**).
2. The digested DNA samples can be stored at −20 °C until needed.

3.4 Gel Electrophoresis and Methylation Analysis

1. Separate the digested PCR products on agarose gel (*see* **Note 8**).
2. Measure the intensity of DNA bands using the available software tools (e.g., Image J, National Institutes of Health, USA) (*see* **Note 9**).
3. Calculate a percentage of cytosine methylation at a given locus by relating the intensity of the cleaved and remaining undigested PCR products (*see* **Note 10**).

4 Notes

1. A precise quantification of genomic DNA is essential for the assay. Samples can be quantified using a spectrophotometer or NanoDrop. However, an equal sample loading must be confirmed using gel electrophoresis, and if necessary, it should be adjusted accordingly.
2. Sometimes it can be recommended to digest genomic DNA with restriction enzymes before bisulfite conversion. This may increase the completeness of DNA denaturation and bisulfite conversion. The restriction enzymes used for digestion should not cut within the region selected for analysis (i.e., for PCR amplification). Following restriction digestion, DNA should be ethanol-precipitated to prevent the interference with a bisulfite reaction.
3. It is possible to use much less DNA for conversion. Refer to the EZ DNA Methylation-Gold™ Kit protocol for detailed guidelines.
4. An easy and inexpensive way to check for the completeness of bisulfite conversion is to amplify the region that is known to be fully non-methylated (for instance, a part of the mitochondrial genome) and digest the converted PCR-amplified DNA with a restriction enzyme that recognizes the sequence that contains only adenine or thymine (e.g., *Dra*I: TTT/AAA) [15]. If conversion is successful, then new restriction sites will be created and no uncleaved product should be visible.

5. It is necessary to clean up a PCR reaction before restriction digestion. Residual salts from PCR buffers may inhibit the restriction enzyme activity. A commercial PCR cleanup kit can be used at this stage. If a PCR reaction produces an unspecific product, then gel extraction of the main product is recommended.
6. It may be necessary to try to use different ratios of enzyme to DNA, such as a 5- or 20-fold excess. If a tenfold excess of enzyme does not result in an increase in the amount of product of digestion, a fivefold excess may be sufficient.
7. Overnight digestion time is more practical, but it is possible that 6 h may be enough, or you may need as long as 24 h. For better results, it is recommended to set up several digestions with different digestion times.
8. If the size difference between the digested and undigested PCR product is too small, then DNA should be separated using either a higher concentration low-melting agarose gels or even polyacrylamide gel.
9. If an initial amount of PCR product is low and cannot be seen in the gel directly, then hybridization using a specific probe (Southern blotting) should be performed. The probe should be designed according to the same guidelines as for primers. The probe should not be designed for the recognition of potentially methylated sequences such as CpG dinucleotides and sequences with a high cytosine content. Also, the probe should not overlap with restriction sites of enzymes used for COBRA analysis (Fig. 1c).
10. If the intensity of PCR products after restriction digestion is visualized by probe hybridization, then a percentage of methylated cytosine residues can be calculated using the formula $\% = 100 \times B/(A+B)$ where A and B are intensities of the remaining undigested and digested PCR products, respectively (Fig. 1c). In this example, PCR product digestion can occur only if a cytosine residue in the CpG dinucleotide sequence is methylated before bisulfite conversion (Fig. 1c).

References

1. Law JA, Jacobsen SE (2010) Establishing, maintaining and modifying DNA methylation patterns in plants and animals. Nat Rev Genet 11(3):204–220
2. Saze H, Tsugane K, Kanno T, Nishimura T (2012) DNA methylation in plants: relationship to small RNAs and histone modifications, and functions in transposon inactivation. Plant Cell Physiol 53(5):766–784
3. Shibuya K, Fukushima S, Takatsuji H (2009) RNA-directed DNA methylation induces transcriptional activation in plants. Proc Natl Acad Sci U S A 106(5):1660–1665
4. Kato M, Miura A, Bender J, Jacobsen SE, Kakutani T (2003) Role of CG and non-CG methylation in immobilization of transposons in Arabidopsis. Curr Biol 13(5):421–426
5. Zilberman D, Gehring M, Tran RK, Ballinger T, Henikoff S (2007) Genome-wide analysis of Arabidopsis thaliana DNA methylation uncovers an interdependence between methylation and transcription. Nat Genet 39(1):61–69

6. Lokk K, Modhukur V, Rajashekar B, Martens K, Magi R, Kolde R, Koltsina M, Nilsson TK, Vilo J, Salumets A, Tonisson N (2014) DNA methylome profiling of human tissues identifies global and tissue-specific methylation patterns. Genome Biol 15(4):r54
7. Zhang M, Xu C, von Wettstein D, Liu B (2011) Tissue-specific differences in cytosine methylation and their association with differential gene expression in sorghum. Plant Physiol 156(4):1955–1966
8. Heard E, Disteche CM (2006) Dosage compensation in mammals: fine-tuning the expression of the X chromosome. Genes Dev 20(14):1848–1867
9. Holmes R, Soloway PD (2006) Regulation of imprinted DNA methylation. Cytogenet Genome Res 113(1-4):122–129
10. Chandler VL, Eggleston WB, Dorweiler JE (2000) Paramutation in maize. Plant Mol Biol 43(2-3):121–145
11. Choi Y, Gehring M, Johnson L, Hannon M, Harada JJ, Goldberg RB, Jacobsen SE, Fischer RL (2002) DEMETER, a DNA glycosylase domain protein, is required for endosperm gene imprinting and seed viability in Arabidopsis. Cell 110(1):33–42
12. Zilberman D, Henikoff S (2005) Epigenetic inheritance in Arabidopsis: selective silence. Curr Opin Genet Dev 15(5):557–562
13. Penterman J, Zilberman D, Huh JH, Ballinger T, Henikoff S, Fischer RL (2007) DNA demethylation in the Arabidopsis genome. Proc Natl Acad Sci U S A 104(16): 6752–6757
14. Xiong Z, Laird PW (1997) COBRA: a sensitive and quantitative DNA methylation assay. Nucleic Acids Res 25(12):2532–2534
15. Sadri R, Hornsby PJ (1996) Rapid analysis of DNA methylation using new restriction enzyme sites created by bisulfite modification. Nucleic Acids Res 24(24):5058–5059
16. Wang RY, Gehrke CW, Ehrlich M (1980) Comparison of bisulfite modification of 5-methyldeoxycytidine and deoxycytidine residues. Nucleic Acids Res 8(20):4777–4790

Chapter 6

Analysis of Global Genome Methylation Using the Cytosine-Extension Assay

Andriy Bilichak and Igor Kovalchuk

Abstract

DNA methylation is a reversible covalent chemical modification of DNA intended to regulate chromatin structure and gene expression in a cell- and tissue-specific manner and in response to the environment. Cytosine methylation is predominantly occurring in plants, and cytosine nucleotides in plants can be methylated at symmetrical (CpG and CpHpG) and nonsymmetrical sites. Although there exists a number of various methods for the detection of cytosine methylation, most of them are either laborious or expensive or both. Here, we describe a quick inexpensive method for the analysis of global genome methylation using a cytosine-extension assay. The assay can be used for the analysis of the total level of CpG, CpHpG, and CpHpH methylation in a given sample of plant DNA.

Key words Global genome methylation, Cytosine extension, Symmetrical methylation, CpG, CpHpG, CpHpH, Asymmetrical methylation, *Hpa*II, *Msp*I

1 Introduction

DNA methylation plays a crucial role in controlling a variety of cellular processes, including proper chromatin folding, the organization of chromatin loops, the regulation of transcription, silencing, imprinting, the suppression of transposon activity, a defense against foreign DNA sequences, and many others [1–3]. Recent reports also suggest the importance of DNA methylation in the inheritance of gene expression patterns [4, 5]. The maintenance of DNA methylation may even be important for a proper DNA repair or/and the maintenance of genome stability [6, 7]. DNA hypomethylation is frequently associated with an activation of transposons [8] and an increased frequency of chromosomal rearrangements [9, 10].

Cytosine methylation is catalyzed by enzymes known as DNA methyltransferases (MTases) which utilize *S*-adenosyl-methionine as a methyl donor [11]. In mammals, symmetric CpG sites are usually preferred as targets for methylation, whereas in the plant genomes, the occurrence of methylated cytosines appears to arise

Igor Kovalchuck (ed.), *Plant Epigenetics: Methods and Protocols*, Methods in Molecular Biology, vol. 1456, DOI 10.1007/978-1-4899-7708-3_6, © Springer Science+Business Media New York 2017

virtually in any sequence, including symmetric methylation at both CpG and CpHpG sites (where H = A, C, or T) and asymmetric methylation at CpHpH sites. Consequently, only 2–8 % of mammalian DNA is methylated, compared to up to 50 % DNA methylation observed in higher plants [12].

A genome-wide analysis of the occurrence of methylated cytosines in plants has revealed the enrichment of DNA methylation predominantly in repeat and transposon regions (on average, 90 % of all sequences are methylated) where a transcriptionally quiescent chromatin state is maintained [13–15]. A comparison of transcriptome and methylome data in *Arabidopsis* has indicated that both the low- and high-expressed genes have a low level of methylation throughout a gene body. At the same time, the moderately expressed genes possess a significantly higher level of DNA methylation distributed throughout the coding sequences [16–18]. The occurrence of the methylated cytosines in the terminal regions of the gene (the 5′ region of the promoter and a part of the transcribed sequence and the 3′ part of the coding sequence and the 3′ UTR) is negatively correlated with gene expression and may be involved in tissue-specific gene expression and pathogen response [16, 19–21].

To date, a number of methods allowing the detection of DNA methylation changes have been developed. Here, we present an inexpensive, sensitive, and rapid method for the detection of changes in global DNA methylation [22].

The assay needs methylation-sensitive restriction enzymes that produce 5′ guanine overhangs upon cleavage. Digestion with such enzymes is only possible when cytosines that are a part of recognition sequence are not methylated. Restriction digestion is followed by single-nucleotide primer extension with $[^3H]$dCTP (Fig. 1). As the number of $[^3H]$dCTP incorporations should be proportional to the number of 5′ guanine overhangs produced upon cleavage, the quantitative detection of non-methylated restriction sites is possible. Importantly, the cleavage efficiency of the enzyme is not affected by methylation density [22]. To prorate for the possible background readings from the broken genomic DNA, a single-nucleotide primer extension assay with DNA that was not digested with a restriction endonuclease is performed. Next, the undigested DNA readings can be subtracted from the digested DNA readings to obtain the relative number of true non-methylated restriction sites available. Some of the advantages of the assay are its sensitivity, the fact that it requires low amounts of input DNA, cost efficiency, simplicity, and high-throughput capability (several hundred of samples can be done in 2 days). In addition, it can be used for methylation analysis of the significantly damaged DNA templates that contain various DNA adducts, strand breaks, and abasic sites [22].

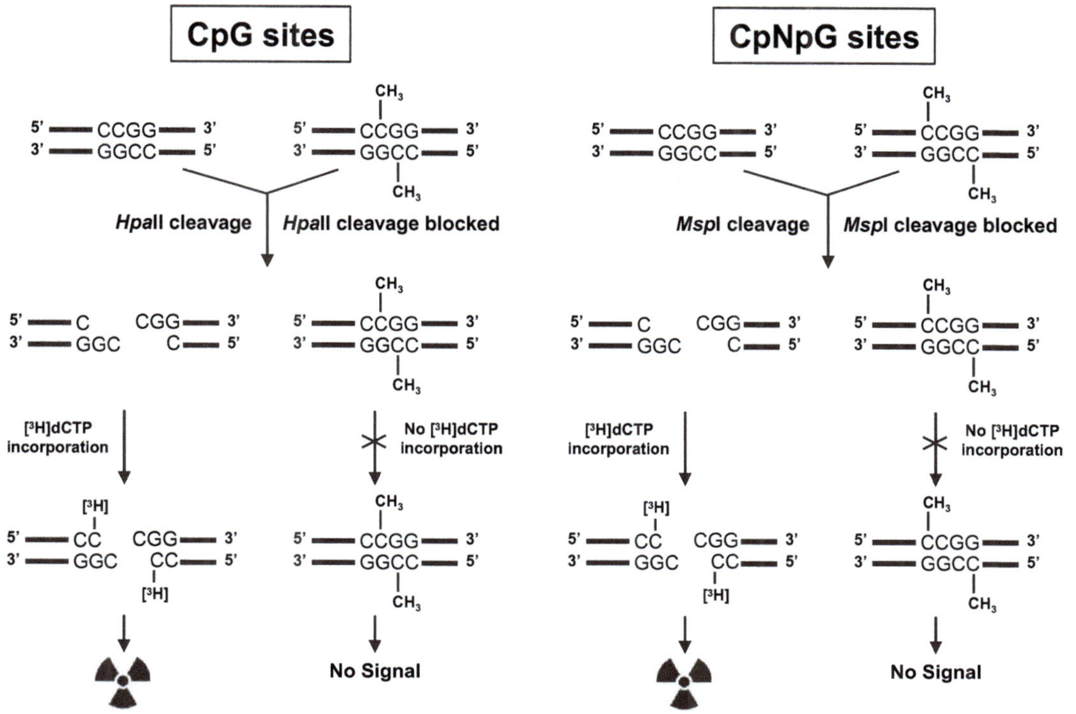

Fig. 1 The mechanism of methylation pattern detection using the cytosine-extension assay. Digestion of plant genomic DNA with the methylation-sensitive restriction endonucleases (here *Hpa*II and *Msp*I) creates 5′ guanine overhangs. The single-nucleotide primer extension reaction incorporates [^{3}H]dCTP nucleotides into the digested DNA. DNA methylation at a restriction site blocks cleavage, thereby preventing [^{3}H]dCTP incorporation

2 Materials

2.1 Restriction Enzyme Digestion

1. Restriction enzymes such as *Hpa*II, *Msp*I, and others (*see* Table 1) with suitable 10× reaction buffer.
2. Nuclease-free water.
3. 1.0 μg genomic DNA aliquots.

2.2 Single Nucleotide Extension Reaction

1. Nuclease-free water.
2. Methylation-sensitive restriction enzyme with suitable 10× reaction buffer.
3. Agarose, electrophoresis grade.
4. 1× TBE: 90 mM Tris–HCl, pH 8.0, 90 mM boric acid, 2 mM EDTA.
5. 6× DNA gel loading buffer.
6. 10× PCR buffer II without $MgCl_2$.
7. 25 mM $MgCl_2$.
8. AmpliTaq DNA polymerase.
9. [^{3}H]dCTP. **Caution: Radiation protection measures must be taken for handling 3H and all derived materials.** Store in a shielded container in a dedicated freezer at −20 °C.

Table 1
A list of the methylation-sensitive restriction endonucleases that can be used for the cytosine-extension assay

Restriction endonuclease	Restriction site (blocked by cytosine methylation)	Methylation pattern analysis
*Aci*I	C/CGG	Global methylation
*Age*I	A/CCGGT	Global methylation
*Asc*I	GG/CGCGCC	CpG islands
*BssH*II	G/CGCGC	CpG islands
*Bst*BI	TT/CGAA	Global methylation
*Hpa*II	C/CGG	Global methylation
*Hpy*CH4IV	A/CGT	Global methylation
*Mlu*I	A/CGCGT	Global and CpG islands
*Msp*I	C/CGG	Global methylation
*Nar*I	GG/CGCC	CpG islands

The most frequently used enzymes are *Hpa*II, *Msp*I, *Aci*I, and *BssH*II. If samples of animal DNA are analyzed, then *Hpa*II and *Msp*I can be used to measure a percentage of methylated CpG sites of the total number of restriction sites available. While *Hpa*II cleavage is blocked by methylation at the internal cytosine, its isoschizomer *Msp*I can cleave the same site regardless of cytosine methylation. In contrast, in plant DNA, *Msp*I cleavage is blocked, if the external cytosine at the restriction site is methylated. Hence, the combination of *Hpa*II and *Msp*I enzymes can be efficiently used for the analysis of plant DNA methylation to compare cytosine methylation in the CpG and CpHpG sequence context, respectively

10. Whatman DE-81 ion-exchange filters.
11. 500 mM Na-phosphate buffer, pH 7.0.
12. Scintillation vials.
13. A scintillation cocktail.
14. A Beckman LS 5000 CE liquid scintillation counter.

3 Methods

3.1 Restriction Enzyme Digestion

1. Using nuclease-free water, prepare two 1.0 μg genomic DNA aliquots. One aliquot is incubated with a methylation-sensitive endonuclease. The second DNA aliquot is incubated without restriction enzyme and serves as a background control (*see* **Notes 1–4**).
2. Set up digestion of the first aliquot in a final volume of 20 μl using a tenfold excess of restriction enzyme according to the manufacture's protocol. Use nuclease-free water in place of the enzyme for the second DNA aliquot. Incubate samples overnight at the temperature suggested by the manufacturer.

3.2 Single-Nucleotide Extension Reactions

1. Use 10 μl (0.5 μg DNA) of each digestion reaction for the single-nucleotide extension reaction.
2. Set up reactions in a final volume of 25 μl containing 0.5 μg DNA (10 μl), 1× PCR buffer II without $MgCl_2$, 1.0 mM $MgCl_2$, 0.5 units of AmpliTaq DNA polymerase, [^{3}H]dCTP (42.9 Ci/mmol) (*see* **Note 5**).
3. Incubate samples at 56 °C for 1 h, and then place samples on ice.
4. Apply 25 μl reactions to Whatman DE-81 ion-exchange filters. Air-dry filters (*see* **Note 6**).
5. Wash filters in 500 mM Na-phosphate buffer, pH 7.0 at room temperature for 10 min.
6. Repeat the washing step twice.
7. Air-dry filters and transfer them to scintillation vials containing 5 ml of a scintillation cocktail. Ensure that the filters are completely submersed into the scintillation cocktail.
8. Measure the incorporation of radioactive material in the samples in a liquid scintillation counter (e.g., Beckman LS 5000 CE liquid scintillation counter) using the settings suggested by the manufacturer. The readings taken from the enzyme-treated samples show the total radiolabel incorporation (RI_T) that negatively correlates with the number of methylated restriction sites. The readings taken from the samples incubated without a restriction enzyme show the background radiolabel incorporation (RI_B) that may reflect the quality and integrity of the input DNA.
9. Calculate the actual (due to restriction site demethylation) radiolabel incorporation (RI_A) using a formula $RI_A = RI_T - RI_B$, where RI_T and RI_B are the total and background radiolabel incorporation, respectively. Express the results as relative [^{3}H] dCTP incorporation/0.5 μg DNA. Alternatively, the results can be expressed as a percentage change in control samples (*see* **Note 4**). It should be noted that the methylation level inversely correlates with the amount of incorporated radionuclides.

4 Notes

1. The precise quantification of genomic DNA is essential for the assay. Samples can be quantified using either a spectrophotometer or NanoDrop. However, the equal sample loading must be confirmed using gel electrophoresis and, if necessary, should be adjusted accordingly.
2. It is important to ensure a purity of DNA preparation. We suggest using the ethanol-precipitated DNA for the assay. This prevents the interference of chemicals used during DNA extraction (SDS, EDTA, proteinase K, phenol, etc.) with the process of restriction digestion.

3. The choice of restriction enzyme determines the type of DNA methylation being analyzed: either global genome methylation or CpG islands (Table 1). The enzymes that have their recognition sites distributed randomly throughout the genome are suitable for global methylation analysis. The enzymes that have multiple CpGs in their recognition sequences are usually used to study methylation of CpG islands. Similarly, choosing the right enzyme makes it possible to selectively analyze methylation in both CpHpG and CpG sequence contexts. A good example is the enzyme pair (*Hpa*II and *Msp*I) that recognizes and cleaves the CCGG nucleotide sequence. In plants, the symmetrical CpG and CpHpG sites are the most common sites of methylation. Methylation of the external cytosine in CCGG representing CpHpG methylation prevents the digestion with *Msp*I and severely impairs the digestion with *Hpa*II [23]. Methylation of the internal cytosine in CCGG representing CG methylation does not influence the digestion with *Msp*I but prevents the digestion with *Hpa*II. Thus, the digestion with *Msp*I allows to evaluate the difference in methylation at symmetrical CpHpG sites between the selected samples.
4. A pair of isoschizomers comprised of one methylation sensitive and one methylation-insensitive enzyme can be used to determine the percentage of restriction sites available in the genome that contains the methylated cytosine residues. Three cytosine-extension reactions for each sample should be performed: a background control reaction (no enzyme added), a digestion reaction with a methylation-sensitive enzyme, and a digestion reaction with a methylation-insensitive enzyme. Once samples are corrected for background incorporation, the ratio of incorporations after methylation-sensitive and methylation-insensitive enzyme digestion will show the percentage of unmethylated restriction sites.
5. Reports demonstrate that [^{3}H]dCTP nucleotides can be efficiently substituted for biotinylated dCTP, thus eliminating the need to use radioactivity for the assay [24].
6. Using Whatman DE-81 ion-exchange filters is essential because they drastically reduce DNA contamination with unincorporated nucleotides [25].

References

1. Law JA, Jacobsen SE (2010) Establishing, maintaining and modifying DNA methylation patterns in plants and animals. Nat Rev Genet 11(3):204–220
2. Saze H, Tsugane K, Kanno T, Nishimura T (2012) DNA methylation in plants: relationship to small RNAs and histone modifications, and functions in transposon inactivation. Plant Cell Physiol 53(5):766–784
3. Shibuya K, Fukushima S, Takatsuji H (2009) RNA-directed DNA methylation induces transcriptional activation in plants. Proc Natl Acad Sci U S A 106(5):1660–1665
4. Bilichak A, Ilnystkyy Y, Hollunder J, Kovalchuk I (2012) The progeny of Arabidopsis thaliana plants exposed to salt exhibit changes in DNA methylation, histone modifications and gene expression. PLoS One 7(1):e30515

5. Boyko A, Blevins T, Yao Y, Golubov A, Bilichak A, Ilnytskyy Y, Hollunder J, Meins F Jr, Kovalchuk I (2010) Transgenerational adaptation of Arabidopsis to stress requires DNA methylation and the function of Dicer-like proteins. PLoS One 5(3):e9514
6. Cuozzo C, Porcellini A, Angrisano T, Morano A, Lee B, Di Pardo A, Messina S, Iuliano R, Fusco A, Santillo MR, Muller MT, Chiariotti L, Gottesman ME, Avvedimento EV (2007) DNA damage, homology-directed repair, and DNA methylation. PLoS Genet 3(7):e110
7. Schar P, Fritsch O (2011) DNA repair and the control of DNA methylation. Prog Drug Res 67:51–68
8. Kato M, Miura A, Bender J, Jacobsen SE, Kakutani T (2003) Role of CG and non-CG methylation in immobilization of transposons in Arabidopsis. Curr Biol 13(5):421–426
9. Bassing CH, Swat W, Alt FW (2002) The mechanism and regulation of chromosomal V(D)J recombination. Cell 109(Suppl):S45–S55
10. Bender J (1998) Cytosine methylation of repeated sequences in eukaryotes: the role of DNA pairing. Trends Biochem Sci 23(7):252–256
11. Lan J, Hua S, He X, Zhang Y (2010) DNA methyltransferases and methyl-binding proteins of mammals. Acta Biochim Biophys Sin Shanghai 42(4):243–252
12. Zhu JK (2009) Active DNA demethylation mediated by DNA glycosylases. Annu Rev Genet 43:143–166
13. Lippman Z, Gendrel AV, Black M, Vaughn MW, Dedhia N, McCombie WR, Lavine K, Mittal V, May B, Kasschau KD, Carrington JC, Doerge RW, Colot V, Martienssen R (2004) Role of transposable elements in heterochromatin and epigenetic control. Nature 430(6998):471–476
14. Vaughn MW, Tanurdzic M, Lippman Z, Jiang H, Carrasquillo R, Rabinowicz PD, Dedhia N, McCombie WR, Agier N, Bulski A, Colot V, Doerge RW, Martienssen RA (2007) Epigenetic natural variation in Arabidopsis thaliana. PLoS Biol 5(7):e174
15. Wang X, Elling AA, Li X, Li N, Peng Z, He G, Sun H, Qi Y, Liu XS, Deng XW (2009) Genome-wide and organ-specific landscapes of epigenetic modifications and their relationships to mRNA and small RNA transcriptomes in maize. Plant Cell 21(4):1053–1069
16. Zilberman D, Gehring M, Tran RK, Ballinger T, Henikoff S (2007) Genome-wide analysis of Arabidopsis thaliana DNA methylation uncovers an interdependence between methylation and transcription. Nat Genet 39(1):61–69
17. Cokus SJ, Feng S, Zhang X, Chen Z, Merriman B, Haudenschild CD, Pradhan S, Nelson SF, Pellegrini M, Jacobsen SE (2008) Shotgun bisulphite sequencing of the Arabidopsis genome reveals DNA methylation patterning. Nature 452(7184):215–219
18. Lister R, O'Malley RC, Tonti-Filippini J, Gregory BD, Berry CC, Millar AH, Ecker JR (2008) Highly integrated single-base resolution maps of the epigenome in Arabidopsis. Cell 133(3):523–536
19. Gehring M, Bubb KL, Henikoff S (2009) Extensive demethylation of repetitive elements during seed development underlies gene imprinting. Science 324(5933):1447–1451
20. Zemach A, Kim MY, Silva P, Rodrigues JA, Dotson B, Brooks MD, Zilberman D (2010) Local DNA hypomethylation activates genes in rice endosperm. Proc Natl Acad Sci U S A 107(43):18729–18734
21. Dowen RH, Pelizzola M, Schmitz RJ, Lister R, Dowen JM, Nery JR, Dixon JE, Ecker JR (2012) Widespread dynamic DNA methylation in response to biotic stress. Proc Natl Acad Sci U S A 109(32):E2183–E2191
22. Pogribny I, Yi P, James SJ (1999) A sensitive new method for rapid detection of abnormal methylation patterns in global DNA and within CpG islands. Biochem Biophys Res Commun 262(3):624–628
23. McClelland M, Nelson M, Raschke E (1994) Effect of site-specific modification on restriction endonucleases and DNA modification methyltransferases. Nucleic Acids Res 22(17):3640–3659
24. Fujiwara H, Ito M (2002) Nonisotopic cytosine extension assay: a highly sensitive method to evaluate CpG island methylation in the whole genome. Anal Biochem 307(2):386–389
25. Basnakian AG, James SJ (1996) Quantification of 3′OH DNA breaks by random oligonucleotide-primed synthesis (ROPS) assay. DNA Cell Biol 15(3):255–262

Chapter 7

In Situ Analysis of DNA Methylation in Plants

Palak Kathiria and Igor Kovalchuk

Abstract

Epigenetic regulation in the plant genome is associated with the determination of expression patterns of various genes. Methylation of DNA at cytosine residues is one of the mechanisms of epigenetic regulation and has been a subject of various studies. Various techniques have been developed to analyze DNA methylation, most of which involve isolation of chromatin from cells and further in vitro studies. Limited techniques are available for in situ study of DNA methylation in plants. Here, we present such an in situ method for DNA methylation analysis which has high sensitivity and good reproducibility.

Key words DNA methylation, Epigenetic regulation, In situ analysis, Immunohistochemistry

1 Introduction

The phenotypic characters and the development of an organism are determined by two essential components—the genetic composition and regulation of an organism. Different cellular processes are involved in the regulation of gene expression at the transcriptional, posttranscriptional, translational, or posttranslation level. At transcription stage, both genetic (e.g., transcription factors) and epigenetic (e.g., chromatin structure) factors are at play to determine the expression level of a gene. Hence, to understand any biological process, it becomes inevitable to study the role of epigenetic processes involved.

In conjunction with histone modifications, DNA methylation is an important determinant of chromatin structure [1]. DNA methylation is represented in a form of a methyl group added to cytosine residues of DNA. Cytosines may be present at symmetrical sites, such as CG and CNG sequences, as well as at nonsymmetrical sites abbreviated as CNN [2]. In plants, modifications of cytosine methylation have been correlated with the altered gene expression patterns during plant development and upon exposure to environmental stresses [1, 3, 4]. For example, transgenerational

Igor Kovalchuck (ed.), *Plant Epigenetics: Methods and Protocols*, Methods in Molecular Biology, vol. 1456,
DOI 10.1007/978-1-4899-7708-3_7, © Springer Science+Business Media New York 2017

priming against pathogens has been implicated to act via changes in DNA methylation [5–7].

Changes in DNA methylation can be analyzed by various techniques. Among them are in vitro techniques such as bisulfite-conversion-based PCRs, methylation-sensitive restriction fragment length polymorphism (RFLP) analysis, and chromatin immunoprecipitation (ChIP) assays [8]. These techniques require either chromatin or DNA isolation; hence, the in situ analysis of DNA methylation is not possible. The in situ technique has been used successfully for various studies, including studies on plants [9–11]. Here, we present an improved technique based on the immunological detection that allows for in situ analysis of DNA methylation in plants. The advantage of this technique is that the distribution of euchromatic and heterochromatin regions can be analyzed. The patterns are similar to those in previous reports documenting the distribution of these regions in the nucleus of animal tissues [12, 13].

Previously, the technique has been used for DNA methylation studies on tobacco and *Arabidopsis*. Experiments based on this technique were conducted on 5-week-old tobacco plants. However, the technique is applicable to other plant species, with tissue-specific modifications being required. The initial steps of the technique include the fixation and sectioning of plants. The tissue sections are treated to remove all RNA and proteins. The DNA is denatured for optimal recognition by antibodies. At this stage, an anti-5 MeC antibody is used for immunolabeling. The chromophore-conjugated secondary antibody is used, and the DNA is counterstained with DAPI. The analysis is carried out using confocal microscopy (Fig. 1).

2 Materials

2.1 Slide Coating with APES

1. 3, Aminopropyltriethoxysilane (APES).
2. 100% ethanol.
3. Acetone.
4. A 60 °C oven.
5. A slide holder.

2.2 Tissue Fixation and Cryosectioning

1. Fixative: a 4% paraformaldehyde (PFA) solution in 1× phosphate buffer saline (PBS) or 3:1 ethanol, an acetic acid solution (*see* **Note 1**). To prepare a 4% PFA solution, heat 90 mL of distilled water to 60 °C. Add 4 g of PFA powder. Dissolve the powder by adding the 1 N NaOH solution and adjust pH to 11. After the powder has completely dissolved, add 1 N HCl to adjust pH to 7.5. Bring the solution to room temperature and add 10 mL of 10× PBS. **Caution:** The PFA solution should be prepared in a well-ventilated fume hood because PFA is a hazardous chemical when inhaled.

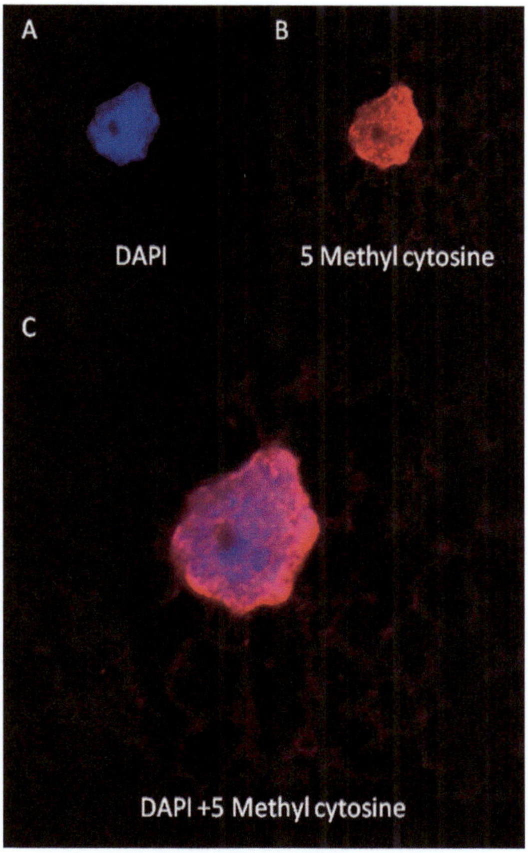

Fig. 1 In situ analysis of DNA methylation: (**a**) The plant nucleus showing DAPI-stained DNA in *blue*. (**b**) The same plant nucleus with 5-MC in *red*. (**c**) The superimposed image of DAPI and 5-MC. The euchromatic regions in the nucleus can be seen as more *blue* stained due to the relative scarceness of DNA methylation. The heterochromatic regions reveal high-level expression of *red* which indicates high-level methylation

2. A 30 % sucrose solution in distilled water.
3. Tissue-Tek® OCT (optimal cutting temperature) (Sakura Finetek, Netherland) solution for mounting cryopreserved specimens.
4. APES-coated glass slides.
5. Dry ice.
6. A cryomicrotome.

2.3 Immunodetection of DNA Methylation

1. 1× PBS and PBST: To prepare 10× PBS, dissolve NaCl (80 g), KCl (2 g), Na_2HPO_4 (14.4 g), and KH_2PO_4 (2.4 g) in 90 mL of water. Adjust pH to 7.5 and add water to make the total volume of 1 L. To make PBST, add Tween 20 up to 0.05 % in 1× PBS.

2. 2× SSC and 4× SSC: To prepare 20× SSC, dissolve NaCl (175.3 g) and sodium citrate (88.2 g) in 900 mL of water. Set pH to 7.0. Make a volume of 1 L. Dilute accordingly in ddH_2O to achieve 2× and 4× solutions.
3. 100, 80, 60, 40, and 20 % ethanol.
4. 100 μg/mL *RNAse A* in 2× SSC solution.
5. 100 μg/mL of *Protease K* solution or pepsin in 100 mM HCl solution.
6. 50 % formamide in 4× SSC solution. **Caution: Formamide is toxic. Please do all manipulations in a fume hood.**
7. Blocking buffer: 5 % BSA in 1× PBS solution (*see* **Note 2**).
8. The primary antibody solution: An anti-5′methylcytosine antibody diluted in blocking buffer at 1:200 dilutions.
9. The secondary antibody solution: Diluted in blocking buffer at 1:500 dilutions (*see* **Note 3**).
10. Antifade solution: To prepare an antifade solution, dissolve 50 mg of *p*-phenylenediamine (Sigma-Aldrich) in 5 mL of 1× PBS solution. Set pH to 9.0. Add 45 mL of glycerol. Mix well, aliquot in 1 mL tubes, and store at −80 °C.
11. Counter stain: A 1 μg/mL DAPI solution in water.
12. A hot plate.
13. A thermometer.
14. Beakers.
15. A confocal microscope.
16. Slide holders or glass/plastic Coplin jars.

3 Methods

3.1 Slide Coating with APES

Tissue retention on slides is one of the major problems in tissue section analysis. It can be enhanced by using various cross-linking agents. The suitability of a cross-linker is dependent on the target tissue type and has to be experimentally determined. Coating with APES is one of the commonly used methods, and it is described here. Many pre-coated slides are also available from various suppliers.

1. Add 2 mL of APES to 100 mL of acetone to prepare a 2 % solution in a beaker (*see* **Note 4**).
2. Add a 2 % solution of APES to a glass slide holder. Arrange the slides in the slide holder in such a way that the entire glass surface of the slide is exposed to the solution. Care must be taken that the slides do not touch each other. Incubate the slides for 2 min. During incubation, APES will react with the glass surface.

3. Carefully take out the slides using forceps and rinse them well in a beaker containing 100% ethanol to remove all the unreacted APES.
4. Air-dry the slides in a dust-free ventilated area such as a clean fume hood. After air-drying, incubate the slides at 60 °C for at least 3 h. The baking process creates the additional cross-linking between the glass and APES molecules.
5. Return the slides to the original box they came with and store at 4 °C until further use. It can be stored up to 2–3 months.

3.2 Tissue Fixation and Cryosectioning

1. Harvest plant tissues from healthy plants and prepare them for fixation. If mesophyll tissues are to be analyzed, dissect leaves into 1 × 1 cm pieces because larger pieces are harder to fix. Submerge the tissues in a fixative solution and vacuum-infiltrate them for 20–30 min. In the case of an ethanol–acetic acid fixative, store plants for 24 h or longer in the fixative without vacuum infiltration. Care must be taken not to prolong fixation for more than needed because over-fixation may cause tissue turning brittle. It negatively impacts tissue integrity and immuno-reaction.
2. After fixation, rinse the tissue once with 1× PBS solution to remove excess 4% PFA. Submerge the tissue in a 30% sucrose solution with vacuum infiltration for 10 min (avoid the vacuum step for delicate tissues). The 30% sucrose solution acts as a cryoprotectant and reduces injuries of cells due to freezing.
3. Store the tissue at 4 °C until it sinks to the bottom. At this stage, replace the tissue with a 1:1 30% sucrose/OCT solution. Incubate the tissue at 60 °C for 2–3 h and then overnight at room temperature. This allows ample time for the sucrose/OCT solution to infiltrate the tissue.
4. The tissue is ready for cryosectioning. For sectioning, the hardened tissue is required. To make it, carefully place the tissue in a 100% OCT solution and, then, on dry ice. At this stage, the OCT solution solidifies. Cryosectioning is carried out to obtain 10 μm thick sections of tissue using a cryomicrotome. The appropriate temperature for cryosectioning is tissue dependent. The temperature range of −10 to −15 °C is a good starting point. Sectioning at either too low or high temperature leads to tissue disruption that negatively impacts assay outcome.
5. Place the sections on APES coated slides and allow them to air-dry. The slides with sections can be stored at −80 °C for long periods.

3.3 Immunodetection of DNA Methylation

1. Thaw the slides with tissue sections at room temperature. Bake the slides at 60 °C for 20 min to induce additional cross-links between the tissue and slides.

2. Fix the tissue with 4 % paraformaldehyde for 10 min. Wash the slides two times with 1× PBS for 5 min each in a Coplin jar or a slide holder. At this point, tissue integrity should be evaluated under a microscope. If the tissue does not appear intact, it will be difficult to get results by following further steps.
3. The presence of RNA and proteins in cells may hinder an effective penetration of antibodies and antigen recognition. Hence, to enhance the antigen/antibody reaction, RNA and protein removal is required that will also remove chromatin proteins and unwind DNA, thus allowing antibodies to have more access to DNA. Incubate the sections for 1 h at 37 °C in 100 μg/mL RNAse A in 2× SSC solution to remove RNA from the tissue. Subsequently, treat the sections with a 100 μg/mL Protease K solution in 100 mL HCl solution for 30 min at room temperature (*see* **Note 5**). Then, wash two times with 1× PBS for 3 min each by changing the solution in the Coplin jar.
4. Dehydrate the tissue in progressively higher concentrations of a 20, 40, 60, 80, and 100 % ethanol solution (incubation time, 10 min). Take the slides out from the Coplin jar and remove the excess of 100 % ethanol by dripping the solution. Keep the sections horizontally on a bench top to air-dry for 10 min.
5. At this stage, DNA is denatured for optimal recognition by the antibody. Submerge the dried slides in a 50 % formamide in 4× SSC solution preheated to 80 °C in a beaker. Take the beaker off from the hotplate and place at room temperature to allow it to cool down. Wash the slides in two changes of 1× PBS for 5 min each time in the Coplin jar without agitation.
6. Block the sections using an appropriate blocking solution (*see* **Note 2**) for 1–2 h. The blocking step is required to eliminate nonspecific binding between the antibody and other cellular components.
7. Apply 200 μL of the primary antibody solution to each slide. Cover the slides with a piece of Parafilm to prevent evaporation and drying of the antibody solution during prolonged incubations. Incubate the slides from 5 h to overnight at 4 °C without agitation (*see* **Note 6**).
8. To remove excess antibodies from the slides, wash the slides in three changes of PBST solution for 15 min each time in the Coplin jar without agitation.
9. Apply the secondary antibody solution to the slides in the volume of 200 μL per slide. Then, cover the slides with a piece of Parafilm and incubate at room temperature for 3 h.
10. To remove the unreacted secondary antibodies, wash the slides with PBST solution three times (each wash –15 min) in the Coplin jar.

11. Apply DAPI counterstain to the sections for 10 min and destain in 1× PBS for 10 min.
12. Mount the sections in antifade solution and apply coverslips to the slides. Store the slides in dark at 4 °C. The antifade solution prevents photobleaching caused by strong light sources.

4 Notes

1. The PFA powder remains insoluble in water until the pH is adjusted to 7.0 with 1 M NaOH. The solution should be prepared fresh for optimal tissue fixation. As an option, the solution can stored at −20 °C approximately for a month. Never heat the solution above 60 °C.
2. The composition of blocking buffer can vary. It is better to use the serum of the animal in which the secondary antibody was raised as a blocking reagent. For example, if the secondary antibody is goat anti-rabbit, use 5 % goat serum instead of 10 % BSA solution for blocking.
3. The dilution of the secondary antibody has to be determined experimentally. It may vary, depending on the tissue type used. A 1:500 dilution can be used as an initial reference point.
4. The APES solution has to be diluted just before the use. Once prepared, it can be reused to coat a large number of slides and stored for 24 h at room temperature.
5. The Protease K solution makes the tissue more delicate to handle. Hence, over-digestion of soft tissues may lead to an excessive tissue damage. The time of incubation at 37 °C has to be standardized according to the tissue type used.
6. The time of incubation with the primary antibody depends on the type and thickness of tissue. An overnight incubation generally gives better results. To avoid drying of the solution during incubation, slides can be covered with a piece of paraplast.

References

1. Vaillant I, Paszkowski J (2007) Role of histone and DNA methylation in gene regulation. Curr Opin Plant Biol 10(5):528–533
2. Cao X, Jacobsen SE (2002) Locus-specific control of asymmetric and CpNpG methylation by the DRM and CMT3 methyltransferase genes. Proc Natl Acad Sci U S A 99(Suppl 4):16491–16498
3. Mathieu O, Reinders J, Caikovski M, Smathajitt C, Paszkowski J (2007) Transgenerational stability of the Arabidopsis epigenome is coordinated by CG methylation. Cell 130(5): 851–862
4. Widman N, Jacobsen SE, Pellegrini M (2009) Determining the conservation of DNA methylation in Arabidopsis. Epigenetics 4(2): 119–124
5. Kathiria P, Sidler C, Golubov A, Kalischuk M, Kawchuk LM, Kovalchuk I (2010) Tobacco mosaic virus infection results in an increase in recombination frequency and resistance to viral, bacterial, and fungal pathogens in the

progeny of infected tobacco plants. Plant Physiol 153(4):1859–1870
6. Luna E, Bruce TJ, Roberts MR, Flors V, Ton J (2012) Next-generation systemic acquired resistance. Plant Physiol 158(2):844–853
7. Slaughter A, Daniel X, Flors V, Luna E, Hohn B, Mauch-Mani B (2012) Descendants of primed Arabidopsis plants exhibit resistance to biotic stress. Plant Physiol 158(2): 835–843
8. DeAngelis JT, Farrington WJ, Tollefsbol TO (2008) An overview of epigenetic assays. Mol Biotechnol 38(2):179–183
9. Mayer W, Niveleau A, Walter J, Fundele R, Haaf T (2000) Embryogenesis: demethylation of the zygotic paternal genome. Nature 403(6769):501–502
10. Naumann K, Fischer A, Hofmann I, Krauss V, Phalke S, Irmler K, Hause G, Aurich AC, Dorn R, Jenuwein T, Reuter G (2005) Pivotal role of AtSUVH2 in heterochromatic histone methylation and gene silencing in Arabidopsis. EMBO J 24(7):1418–1429
11. Oakeley EJ, Podesta A, Jost JP (1997) Developmental changes in DNA methylation of the two tobacco pollen nuclei during maturation. Proc Natl Acad Sci U S A 94(21):11721–11725
12. Manak JR, Wen H, Van T, Andrejka L, Lipsick JS (2007) Loss of Drosophila Myb interrupts the progression of chromosome condensation. Nat Cell Biol 9(5):581–587
13. Zink D, Fischer AH, Nickerson JA (2004) Nuclear structure in cancer cells. Nat Rev Cancer 4(9):677–687

Chapter 8

Analysis of DNA Hydroxymethylation Using Colorimetric Assay

Andrey Golubov and Igor Kovalchuk

Abstract

Hydroxymethylcytosine (hmC or 5-hmC) is a nitrogen base occurring as a result of cytosine methylation followed by replacing a methyl group with a hydroxyl group through active oxidation. 5-hmC is considered to be one of the forms of epigenetic modification and is suggested as an intermediate step in a semi-active loss of DNA methylation mark. 5-hmC plays an important role in the epigenetic regulation of gene expression in animals, although its role in plants remains controversial. Here, we present a colorimetric method of quantification of 5-hmC using *Brassica rapa* DNA.

Key words Global genome methylation, Cytosine hydroxymethylation, 5-hmC, Colorimetric assay

1 Introduction

DNA methylation is a chemical modification of the nitrogenous base by the addition of a methyl group. In plants, such base is predominantly a cytosine, and a methyl group is added to the 5-carbon of the ring, resulting in 5-mC. In plants, 5-mCs occur in the symmetric (CG and CNG) and nonsymmetric (CNN) contexts, and it is an essential component of gene expression regulation, chromatin structure, the maintenance of genome stability, and possibly even DNA repair [1]. As such, cytosine methylation is essential for various physiological processes and responses to stress. It is worth noting that a methylation mark is not a rigid modification as it can be removed in either an active or a passive manner [2]. Methylation marks can be lost passively through DNA replication and DNA repair/resynthesis processes when a stretch of DNA with methylated cytosines is excised and replaced with the DNA containing unmethylated cytosines [3]. Methylation marks can also be removed actively through the activity of glycosylases that act upon methylated cytosines such as REPRESSOR OF SILENCING 1 (ROS1), DEMETER (DME), and DME-like proteins DML2 and DML3 [4]. Demethylation is a

Igor Kovalchuck (ed.), *Plant Epigenetics: Methods and Protocols*, Methods in Molecular Biology, vol. 1456,
DOI 10.1007/978-1-4899-7708-3_8, © Springer Science+Business Media New York 2017

reversal of the methylation status essential for developmental processes, including the germination and transition to flowering as well as a response to stress. One additional way how methylation marks are removed is the oxidation of methylated cytosines by the ten-eleven translocation (TET) family of enzymes. Oxidation occurs in several steps, with the first oxidation product being 5-hydroxymethylcytosine (5-hmC) which is then oxidized to 5-formylcytosine (5-fC) and finally to 5-carboxylcytosine (5-caC) [5, 6]. Both 5fC and 5caC can be actively removed from the genome by thymine DNA glycosylase (TDG) [6]. 5-hmC can be removed from the genome via the TET-mediated oxidation process or through a passive loss during replication/DNA repair processes because the maintenance DNA methyltransferase Dnmtl methylates hemihydroxymethylated CpGs with a much lower efficiency compared to hemi-methylated CpGs [7]. In addition, using in vitro study, it was shown that Dnmt3a and Dnmt3b might convert 5-hmC to the unmodified cytosine directly under oxidative conditions [8]; it remains to be shown how robust is the dehydroxymethylase activity of Dnmt3 proteins.

A biological meaning of 5-hmC, 5-fC, and 5-caC is not entirely clear, although it is likely that they are just the result of intermediate steps of an active demethylation process involving thymine DNA glycosylase. It was recently proposed that 5-fC might influence DNA conformation through the alteration of the structure of the DNA double helix; specifically, 5-fC changes the geometry of grooves and base pairs associated with the modified base, leading to helical underwinding [9]. 5hmC appears to play a more important role; it was discovered that 5-hmC was an abundant modification of cytosine Purkinje neurons and embryonic stem cells in mice [10]. Recent research shows age-specific changes in 5hmC and 5fC abundance in the human and mouse brain. Whereas the level of 5-hmC gradually increased with age, 5fC rapidly declines. Therefore, the authors suggested that this is an indication that 5-hmC is likely a stable epigenetic mark, whereas 5-fC is an intermediate product of an active demethylation process [11]. Differential levels of 5-hmC were demonstrated in placenta [12], in cancer cell lines [13], in T cells [14], and in response to stress [15] or Vitamin C [16].

The presence of 5-hmC, 5-fC, and 5-caC in plants was also demonstrated. Tang et al. [17] used a chemical derivatization strategy combined with a liquid chromatography–electrospray ionization tandem mass spectrometry (LC/ESI-MS/MS) method to identify 5-fC (called 5-fodC in the paper) and 5-caC (called 5-cadC) and showed the existence of 5-fodC and 5-cadC in genomic DNA of various plant tissues. In addition, the authors found that exposure to drought and salinity changed the level of 5-fC and 5-caC in plant genomes [17]. It is possible that these intermediates have a functional role in stress, but it is also likely that stress activates an alternative (to ROS1) active demethylation process. This study demonstrates that plants are likely to have mechanisms of active demethylation similar to animals. The study by Li et al. [15] also showed the presence of 5-hmC in plants [15]. Yu et al. [18] used a TET-assisted bisulfite sequencing (TAB-seq) approach [18]; TAB-seq requires the activity β-glucosyltransferase (βGT) that conjugates glucose to 5-hmC but not to 5-mC, resulting in β-glucosyl-5-hydroxymethylcytosine (5-gmC). 5-gmC is protected from oxidation by recombinant TET1, whereas 5-mC is oxidized to 5-caC, which is then converted to carboxyl uracil using bisulfite mutagenesis. Bisulfite also converts unmodified cytosine to uracil, but it spares 5-gmC, thus allowing to distinguish 5-hmC from 5-mC and 5-C.

At the same time, another study by Erdmann et al. [19] shows that the presence of the first oxidative product of the 5-mC, 5-hmC, is likely to be negligible in plants [19]. They failed to observe 5-hmC by using thin-layer chromatography (TLC), enzymatic radiolabeling, and mass spectrometry. Their detection limit of TLC was between 0.1 and 0.5 % of total cytosines, which was much lower than the level detected in cortex tissues (>1.0 %), but on the border line with the level detected in embryonic stem cells (~0.2 %). Enzymatic labeling represents a sensitive method that is based on the ability of the βGT enzyme to add a radioactively labeled glucose moiety to 5-hmC at the hydroxyl position. In contrast, methods using the antibody-based detection, such as dot blot, ChIP–chip, and ELISA, demonstrated the presence of low levels of 5-hmC, in the range of 0.07–0.17 % of total cytosines. Also, these experiments showed a potential cross-reactivity of anti-5hmC antibodies with 5-mC, although the signal from 5-mC was substantially weaker [19]. A similarly low level of 5-hmC—0.07 %—was found by other authors who used antibodies and dot blot for the detection [20]. Even if 5-hmC marks exist in plants, it is not clear how they are generated because plants are not known to have TET homologues.

Here, we used the genomic DNA prepared from leaves of heat-stressed *Brassica napus* plants to analyze the level of 5-hmC methylation. Previously, we have demonstrated that exposure of *B. napus* plants to heat stress early during the development resulted in changes in ncRNA expression most pronounced in unexposed pollen and endosperm of exposed plants as well as in changes in mRNA expression in the exposed leaves and the unexposed developing embryo and endosperm [21]. Curiously, the progeny of stressed plants also exhibited pronounced changes in ncRNA expression. We have not analyzed changes in DNA methylation in the exposed leaf samples, but previous studies have demonstrated an increase in DNA methylation (5-mC) in response to heat stress in several different plants species/experimental setups [22–24]. Here, we demonstrate a slight increase in the level of 5-hmC, which is likely due to an overall increase in 5-mC in response to heat stress.

2 Materials

2.1 Preparation of Positive Control (HC5)

1. The MethylFlash Hydroxymethylated DNA Quantification 48-Assay Kit, Colorimetric (Epigentek, USA).
2. HC5 positive control, 20 ng/μl (20 μg/ml), 6 μl.
3. 1× TE buffer or nuclease-free water.

2.2 DNA Binding

1. The MethylFlash Hydroxymethylated DNA Quantification 48-Assay Kit, Colorimetric (Epigentek, USA).
2. Nuclease-free water.
3. Sample DNA (200 ng/μl).
4. Parafilm M.

2.3 Hydroxymethylated DNA Capture

1. The MethylFlash Hydroxymethylated DNA Quantification 48-Assay Kit, Colorimetric (Epigentek, USA).
2. Nuclease-free water.
3. Parafilm M.

2.4 Signal Detection

1. The MethylFlash Hydroxymethylated DNA Quantification 48-Assay Kit, Colorimetric (Epigentek, USA).
2. Parafilm M.
3. Aluminum foil.

3 Methods

3.1 Preparation of Positive Control (HC5)

Prepare five different HC5 concentrations with the 20 ng/μl (20 μg/ml) HC5 stock solution (*see* **Note 1**) according to Table 1.

3.2 DNA Binding

1. Calculate the number of strips you need for your experiment: one strip contains eight wells; you need at least two wells per each sample, two wells for each positive control, two wells for Negative Control I (HC3, 20 μg/ml), and two wells for the Negative Control II (HC4, 20 μg/ml). Detach extra strips (*see* **Note 2**) that you do not need and place them back into a plastic bag, seal the bag tightly, and store it at 4 °C.
2. Add 80 μl of Binding Solution (HC2) to each well of the strips.
3. Add 1 μl of your samples (200 ng/μl) and the corresponding negative and positive controls into the designated strip wells with 80 μl of Binding Solution (HC2) according to Table 2 (the number of sample wells depends on the number of your samples), mix them thoroughly by pipetting.
4. Cover strip wells with Parafilm M and incubate them at 37 °C for 90 min. While waiting, dilute 10× Wash Buffer (HC1): add 13 ml of 10× HC1 to 117 ml of nuclease-free water.

Table 1
Preparation of a positive control of five different concentrations

Final concentration (ng/µl)	HC5 (µl)	1× TE or water (µl)
10	3	3
5	1	3
2	1	9
1	1	19
0.5	Mix 10 µl of 1 ng/µl diluted HC5 (from the row above) with 10 µl of 1× TE or water	

Table 2
Examples of loading of samples into wells

Well #	Strip 1	Strip 2	Strip 3
A	HC3, repeat 1	HC3, repeat 2	Sample 1, repeat 1
B	HC4, repeat 1	HC4, repeat 2	Sample 1, repeat 2
C	HC5, 0.5 ng/µl, repeat 1	HC5, 0.5 ng/µl, repeat 2	Sample 2, repeat 1
D	HC5, 1 ng/µl, repeat 1	HC5, 1 ng/µl, repeat 2	Sample 2, repeat 2
E	HC5, 2 ng/µl, repeat 1	HC5, 2 ng/µl, repeat 2	Sample 3, repeat 1
F	HC5, 5 ng/µl, repeat 1	HC5, 5 ng/µl, repeat 2	Sample 3, repeat 2
G	HC5, 10 ng/µl, repeat 1	HC5, 10 ng/µl, repeat 2	Sample 4, repeat 1
H	Sample 5, repeat 1	Sample 5, repeat 2	Sample 4, repeat 2

5. Remove the solution from each well (81 µl) using a multichannel pipette (change tips every time).
6. Wash wells three times with 1× Wash Buffer HC1, 150 µl by pipetting with a multichannel pipette.

3.3 Hydroxymethylated DNA Capture

1. Dilute Capture Antibody HC6 (1000 µg/ml) 1:1000 with 1× Wash Buffer HC1. You will need 50 µl per well plus 10% (in our example, 50 µl × 24 wells = 1200 µl, +10% = 1320 µl).
2. Add 50 µl of the diluted Capture Antibody HC6 to each well, cover strip wells with Parafilm M, and incubate them at room temperature (22 °C) for 60 min. While waiting, dilute Detection Antibody HC7 (400 µg/ml) 1:1000 with 1× Wash Buffer HC1. You will need 50 µl per well plus 10% (in our example, 50 µl × 24 wells = 1200 µl, +10% = 1320 µl).

3. Remove the diluted Capture Antibody HC6 solution from each well using a multichannel pipette, and wash wells three times with 150 μl (per well) of 1× Wash Buffer HC1.
4. Add 50 μl of the diluted Detection Antibody HC7 to each well, cover strip wells with Parafilm M and incubate them at room temperature (22 °C) for 30 min. While waiting, dilute Enhancer Solution HC8 1:5000 with 1× Wash Buffer HC1. You will need 50 μl per well plus 10% (in our example, 50 μl × 24 wells = 1200 μl, + 10% = 1320 μl).
5. Remove the diluted Detection Antibody HC7 from each well using a multichannel pipette.
6. Wash wells four times with 1× Wash Buffer HC1, 150 μl by pipetting with a multichannel pipette.
7. Add 50 μl of the diluted Enhancer Solution HC8 to each well, cover strip wells with Parafilm M, and incubate them at room temperature (22 °C) for 30 min.
8. Remove the diluted Enhancer Solution HC8 from each well using a multichannel pipette.
9. Wash wells five times with 1× Wash Buffer HC1, 150 μl by pipetting with a multichannel pipette.

3.4 Signal Detection

1. Add 100 μl of developer solution HC9 to each well, cover strip wells with Parafilm M and aluminum foil (*see* **Note 3**), and incubate them at room temperature (22 °C) for 1–10 min. Incubation time depends on the speed of color development: the solution in the wells with positive controls should turn medium blue.
2. Add 100 μl of stop solution HC10 to each well; mix the solution in the wells by pipetting 8–10 times. The color will change to yellow.
3. Read the absorbance on a microplate reader at 450 nm within 2–15 min.

3.5 Calculation: Absolute Quantification

1. Plot OD values versus the concentration of positive controls HC5 and generate a standard curve using Microsoft Excel or any other similar software.
2. Calculate the slope of a standard curve using a linear regression function.
3. Calculate the amount of 5-hmC in your samples using the following formula:

   5-hmC (ng) = (sample OD − HC4 OD)/slope − 5 (*see* **Note 4**).
4. Calculate the percentage of 5-hmC in your samples using the following formula:

   5-hmC (%) = 5-hmC (ng, from the formula above)/200 (the amount of input DNA in ng) × 100% (*see* **Note 5**).

Table 3
The level of 5-hmC in leaves of *B. rapa* plants in ng and in percentage

Absolute quantification		
Sample	5-hmC (ng)	5-hmC (%)
C1	0.5623	0.2812
C2	0.5444	0.2722
T1	0.5569	0.2785
T2	0.6479	0.3240

Plants were exposed to 42 °C for 3 h each day for 7 consecutive days at the age of 2 weeks, and the leaf tissues were collected at the end of the last exposure. C1 and C2 are control samples, whereas T1 and T2 are treated samples. "5-hmC (ng)" shows the amount of 5-hmC in 200 ng of DNA, whereas "5-hmC (%)" shows the percentage of total cytosine

Table 4
Statistical analysis of the differences in the levels of 5-hmC in leaves of *B. rapa* plants

Sample	**Two-tailed *p* value**
C1/T1	0.889
C1/T2	0.158
C2/T1	0.620
C2/T2	0.066
C/T	0.180

Statistical treatment of data from Table 3. Only C2 to T2 showed a significant difference ($p < 0.1$)

4 Notes

1. Make sure that your pipettes are calibrated and you have good low retention tips: pipetting errors will severely jeopardize final calculations.
2. Be careful with strips, they are very fragile and can be easily damaged.
3. This reaction is light sensitive, so make sure it is well covered with aluminum foil.
4. "5" is a relative factor that normalizes 20 % 5-hmC in the positive control to 100 %.
5. In a particular experiment performed by us, we used four samples—two samples from nonexposed *B. rapa* leaves and two

samples from heat-exposed *B. rapa* leaves. The analysis showed that the level of 5-hmC was on average ~0.3 % of total cytosine (Table 3). This is well in range of the levels observed before. Heat exposure resulted in a slight increase in the level of 5-hmC, although the increase was not significant (Tables 3 and 4).

References

1. He XJ, Chen T, Zhu JK (2011) Regulation and function of DNA methylation in plants and animals. Cell Res 21(3):442–465
2. Piccolo FM, Fisher AG (2014) Getting rid of DNA methylation. Trends Cell Biol 24(2): 136–143
3. Wossidlo M, Arand J, Sebastiano V, Lepikhov K, Boiani M, Reinhardt R, Schöler H, Walter J (2010) Dynamic link of DNA demethylation, DNA strand breaks and repair in mouse zygotes. EMBO J 29(11):1877–1888
4. Lei M, Zhang H, Julian R, Tang K, Xie S, Zhu JK (2015) Regulatory link between DNA methylation and active demethylation in Arabidopsis. Proc Natl Acad Sci U S A 112(11):3553–3557
5. Ito S, Shen L, Dai Q, Wu SC, Collins LB, Swenberg JA, He C, Zhang Y (2011) Tet proteins can convert 5-methylcytosine to 5-formylcytosine and 5-carboxylcytosine. Science 333(6047):1300–1303
6. He YF, Li BZ, Li Z, Liu P, Wang Y, Tang Q, Ding J, Jia Y, Chen Z, Li L, Sun Y, Li X, Dai Q, Song CX, Zhang K, He C, Xu GL (2011) Tet-mediated formation of 5-carboxylcytosine and its excision by TDG in mammalian DNA. Science 333(6047):1303–1307
7. Hashimoto H, Liu Y, Upadhyay AK, Chang Y, Howerton SB, Vertino PM, Zhang X, Cheng X (2012) Recognition and potential mechanisms for replication and erasure of cytosine hydroxymethylation. Nucleic Acids Res 40(11): 4841–4849
8. Chen CC, Wang KY, Shen CK (2012) The mammalian de novo DNA methyltransferases DNMT3A and DNMT3B are also DNA 5-hydroxymethylcytosine dehydroxymethylases. J Biol Chem 287(40):33116–33121
9. Raiber EA, Murat P, Chirgadze DY, Beraldi D, Luisi BF, Balasubramanian S (2015) 5-Formylcytosine alters the structure of the DNA double helix. Nat Struct Mol Biol 22(1):44–49
10. Kriaucionis S, Heintz N (2009) The nuclear DNA base 5-hydroxymethylcytosine is present in Purkinje neurons and the brain. Science 324(5929):929–930
11. Wagner M, Steinbacher J, Kraus TF, Michalakis S, Hackner B, Pfaffeneder T, Perera A, Müller M, Giese A, Kretzschmar HA, Carell T (2015) Age-dependent levels of 5-methyl-, 5-hydroxymethyl-, and 5-formylcytosine in human and mouse brain tissues. Angew Chem Int Ed Engl. doi:10.1002/anie.201502722
12. Fogarty NM, Burton GJ, Ferguson-Smith AC (2015) Different epigenetic states define syncytiotrophoblast and cytotrophoblast nuclei in the trophoblast of the human placenta. Placenta 36(8):796–802
13. Kraus TF, Kolck G, Greiner A, Schierl K, Guibourt V, Kretzschmar HA (2015) Loss of 5-hydroxymethylcytosine and intratumoral heterogeneity as an epigenomic hallmark of glioblastoma. Tumour Biol 36(11):8439–8446
14. Ichiyama K, Chen T, Wang X, Yan X, Kim BS, Tanaka S et al (2015) The methylcytosine dioxygenase Tet2 promotes DNA demethylation and activation of cytokine gene expression in T cells. Immunity 42(4):613–626
15. Li S, Papale LA, Kintner DB, Sabat G, Barrett-Wilt GA, Cengiz P, Alisch RS (2015) Hippocampal increase of 5-hmC in the glucocorticoid receptor gene following acute stress. Behav Brain Res 286:236–240
16. Young JI, Züchner S, Wang G (2015) Regulation of the epigenome by vitamin C. Annu Rev Nutr 35:545–564
17. Tang Y, Xiong J, Jiang HP, Zheng SJ, Feng YQ, Yuan BF (2014) Determination of oxidation products of 5-methylcytosine in plants by chemical derivatization coupled with liquid chromatography/tandem mass spectrometry analysis. Anal Chem 86(15):7764–7772
18. Yu M, Hon GC, Szulwach KE, Song CX, Zhang L, Kim A, Li X, Dai Q, Shen Y, Park B et al (2012) Base-resolution analysis of 5-hydroxymethylcytosine in the mammalian genome. Cell 149:1368–1380
19. Erdmann RM, Souza AL, Clish CB, Gehring M (2014) 5-Hydroxymethylcytosine is not present in appreciable quantities in Arabidopsis DNA. G3 (Bethesda) 5(1):1–8
20. Yao Q, Song CX, He C, Kumaran D, Dunn JJ (2012) Heterologous expression and purification of Arabidopsis thaliana VIM1 protein: in vitro evidence for its inability to recognize

hydroxymethylcytosine, a rare base in Arabidopsis DNA. Protein Expr Purif 83:104–111

21. Bilichak A, Ilnytskyy Y, Wóycicki R, Kepeshchuk N, Fogen D, Kovalchuk I (2015) The elucidation of stress memory inheritance in *Brassica rapa* plants. Front Plant Sci 6:5
22. Correia B, Valledor L, Meijon M, Rodriguez JL, Dias MC, Santos C et al (2013) Is the interplay between epigenetic markers related to the acclimation of cork oak plants to high temperatures? PLoS One 8:e53543
23. Naydenov M, Baev V, Apostolova E, Gospodinova N, Sablok G, Gozmanova M et al (2015) High-temperature effect on genes engaged in DNA methylation and affected by DNA methylation in Arabidopsis. Plant Physiol Biochem 87:102–108
24. Gao G, Li J, Li H, Li F, Xu K, Yan G et al (2014) Comparison of the heat stress induced variations in DNA methylation between heat-tolerant and heat-sensitive rape seed seedlings. Breed Sci 64:125–133

Chapter 9

Analysis of DNA Cytosine Methylation Patterns Using Methylation-Sensitive Amplification Polymorphism (MSAP)

María Ángeles Guevara, Nuria de María, Enrique Sáez-Laguna, María Dolores Vélez, María Teresa Cervera, and José Antonio Cabezas

Abstract

Different molecular techniques have been developed to study either the global level of methylated cytosines or methylation at specific gene sequences. One of them is the methylation-sensitive amplified polymorphism technique (MSAP) which is a modification of amplified fragment length polymorphism (AFLP). It has been used to study methylation of anonymous CCGG sequences in different fungi, plants, and animal species. The main variation of this technique resides on the use of isoschizomers with different methylation sensitivity (such as *Hpa*II and *Msp*I) as a frequent-cutter restriction enzyme. For each sample, MSAP analysis is performed using both *Eco*RI/*Hpa*II- and *Eco*RI/*Msp*I-digested samples. A comparative analysis between *Eco*RI/*Hpa*II and *Eco*RI/*Msp*I fragment patterns allows the identification of two types of polymorphisms: (1) methylation-insensitive polymorphisms that show common *Eco*RI/*Hpa*II and *Eco*RI/*Msp*I patterns but are detected as polymorphic amplified fragments among samples and (2) methylation-sensitive polymorphisms which are associated with the amplified fragments that differ in their presence or absence or in their intensity between *Eco*RI/*Hpa*II and *Eco*RI/*Msp*I patterns. This chapter describes a detailed protocol of this technique and discusses the modifications that can be applied to adjust the technology to different species of interest.

Key words MSAP, AFLP-based technique, Isoschizomers, Cytosine methylation, Anonymous CCGG sites, Methylation pattern

1 Introduction

Nuclear plant DNA is highly methylated and contains 5-methylcytosine. Methylation of cytosine residues occurs predominantly at symmetrical CG and CNG sequences (where N is any nucleotide) and provides a mechanism of gene control. Different techniques have been developed to study DNA methylation variations in the nuclear genomes. Some of these techniques are based on the use of restriction enzyme isoschizomers which recognize the same restriction site but display differential sensitivity to cytosine methylation. Tetracutter restriction enzymes, such as

Igor Kovalchuck (ed.), *Plant Epigenetics: Methods and Protocols*, Methods in Molecular Biology, vol. 1456,
DOI 10.1007/978-1-4899-7708-3_9,

*Hpa*II and *Msp*I, are isoschizomers frequently used to detect anonymous 5′-CCGG sites which flanking sequences are unknown and cytosines are differentially methylated. Both restriction enzymes recognize the sequence 5′-CCGG. *Hpa*II only cleaves sites that are hemi-methylated at the external cytosine (5′-^{hm}C-CGG), whereas *Msp*I cleaves 5′-C^{m}CGG but not hemi- or fully methylated 5′-^{m}CCGG sequences.

Amplified fragment length polymorphism (AFLP) is a polymerase chain reaction (PCR)-based technique that allows a fast and relatively inexpensive analysis of a large number of marker fragments for any organism, without prior knowledge of its genomic sequence. It is based on the selective amplification of anonymous DNA fragments obtained after digestion of total DNA with two restriction enzymes (a hexacutter, i.e., *Eco*RI, and a tetracutter, i.e., *Mse*I) and ligation of oligonucleotide adapters [1]. Methylation-sensitive amplification polymorphism (MSAP) is an AFLP-derived technique adapted to study cytosine methylation by using restriction enzyme isoschizomers instead of the frequent cutter enzyme. Therefore, in this technique, two simultaneous analyses will be carried out, using *Eco*RI and either *Hpa*II or *Msp*I to digest each sample. After *Eco*RI and *Hpa*II–*Msp*I adapter ligation, DNA fragments are subjected to two successive PCR amplification steps: pre-amplification and selective amplification. In both amplifications, *Eco*RI and *Hpa*II–*Msp*I primers that contain the adapter, the restriction site, and several selective nucleotides are used to amplify either *Eco*RI/*Hpa*II or *Eco*RI/*Msp*I DNA fragments. The use of two PCR steps ensures an optimal reduction of DNA fragment complexity to end up with an adequate number of them to be visualized and scored after separation on denaturing polyacrylamide gels (Fig. 1) or capillary electrophoresis (*see* **Note 1**).

Different labels may be used to detect MSAP fragments. In most cases, only one of the primers (*Eco*RI primer) is 5′-labeled and used in the selective amplification (the second PCR reaction). In this chapter, we will describe a detailed protocol that can be applied to any DNA fragment detection system (*see* **Note 1**).

The final step of the analysis is the scoring of DNA fragment profiles (Fig. 2). A comparative analysis between *Eco*RI/*Hpa*II and *Eco*RI/*Msp*I MSAP fragment patterns reveals a genetic variability associated with methylation-insensitive polymorphism and methylation-sensitive polymorphism. Insensitive polymorphisms will show common *Eco*RI/*Hpa*II and *Eco*RI/*Msp*I patterns of polymorphic amplified fragments among samples, while sensitive polymorphisms will show the epigenetic variability through amplified fragments that differ either in their presence or absence or in their intensity between *Eco*RI/*Hpa*II and *Eco*RI/*Msp*I patterns of the same sample (Fig. 3). Thus, methylation of the internal cytosine would lead to the appearance of

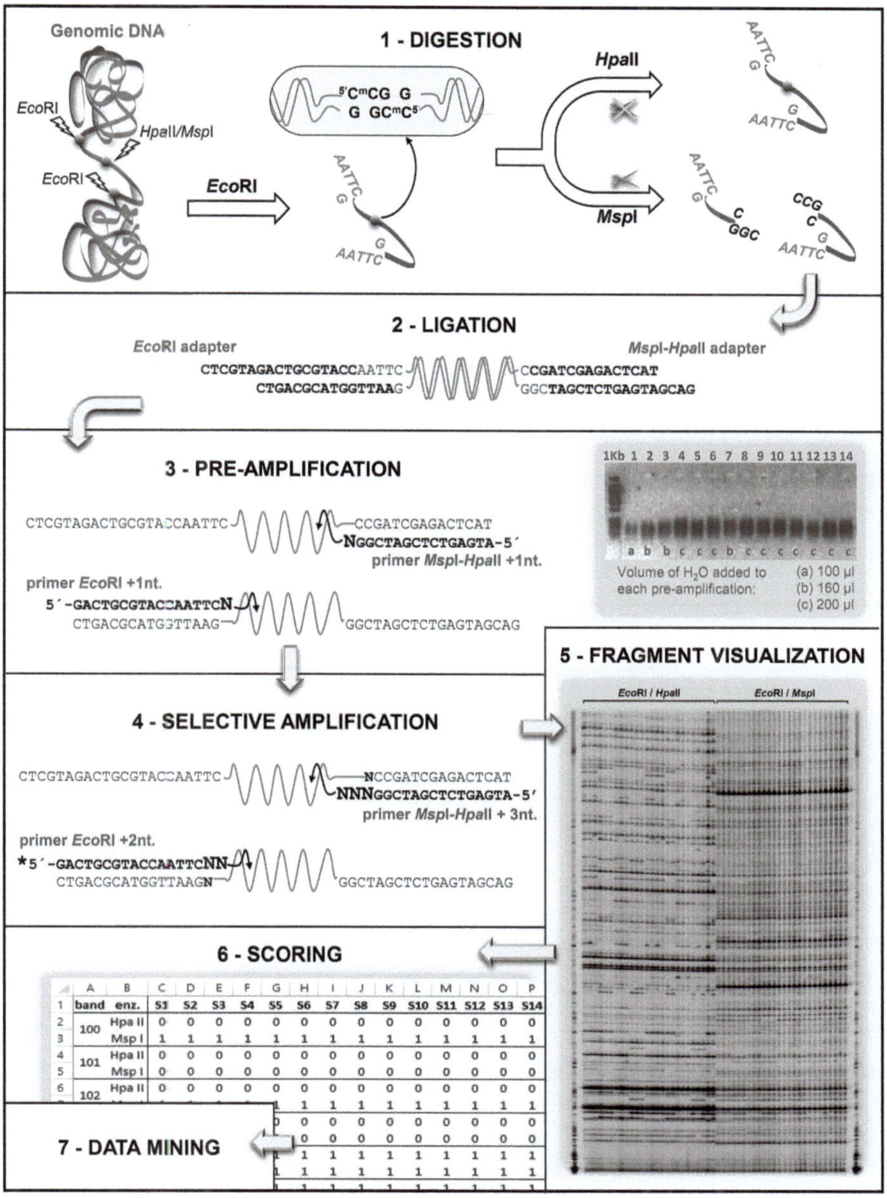

Fig. 1 A schematic representation of the MSAP technique. (*1*) Digestion of genomic DNA with *Eco*RI/*Hpa*II or *Eco*RI/*Msp*I restriction enzymes. (*2*) Ligation of *Eco*RI and *Hpa*II–*Msp*I double-stranded adapters to the ends of restriction fragments. (*3*) Pre-amplification of one sixteenth of restriction fragments using one primer complementary to each adapter, with one selective nucleotide (+1/+1) at the 3′ end (N stands for A, T, C, or G residues). After the amplification, each sample is diluted according to the intensity of the smear of DNA fragments observed in 0.8 % agarose gel. (*4*) The selective amplification using primers with +2/+3 selective nucleotides at their 3′ ends, maintaining the selective nucleotide used in the pre-amplification. Only *Eco*RI primers are labeled at their 5′ ends. The *arrows* indicate the direction of DNA polymerization and * means a labeled primer. (*5*) Gel electrophoresis of the amplified restriction fragments in a Li-Cor 4300 DNA Analysis System using the 16 % Long Ranger polyacrylamide 50 % gel solution. (*6*) Scoring of the DNA fingerprints obtained with each analysis (either *Eco*RI/*Hpa*II or *Eco*RI/*Msp*I). (*7*) The interpretation of the results (data mining)

<table>
<tr><th rowspan="2"></th><th rowspan="2" colspan="3">Banding pattern (EcoRI/HpaII)/(EcoRI/MspI)</th><th rowspan="2">Interpretation</th><th rowspan="2">Methylation scoring</th><th rowspan="2">Non-methylation scoring</th><th colspan="3">Mixed scoring</th></tr>
<tr><th>Non met.</th><th>Hemi & int. Cyt met.</th><th>Hemi & ext. Cyt met.</th></tr>
<tr><td>Status I ➞</td><td>$^{5'}$CCGG
GGCC$_{5'}$</td><td>Non methylation</td><td>1//1</td><td>Non-methylated cytosines</td><td>0</td><td>1</td><td>1</td><td>1|0</td><td>1|0</td></tr>
<tr><td>Status II ➞</td><td>$^{5'}$C^{m}CG G
G GC^{m}C$_{5'}$
$^{5'}$C^{m}CGG
G GCC$_{5'}$</td><td>Full & hemi-methylation of internal cytosine</td><td>0//1</td><td>Fully methylated & hemi-methylated internal cytosines</td><td>1</td><td>0</td><td>0</td><td>1</td><td>0</td></tr>
<tr><td>Status III ➞</td><td>$^{5'm}$CCGG
GGCC$_{5}$</td><td>Hemi-methylation of external cytosine</td><td>1//0</td><td>Hemi-methylated external cytosine</td><td>1</td><td>NA|0</td><td>0</td><td>0</td><td>1</td></tr>
<tr><td rowspan="3">Status IV ➞</td><td>$^{5'm}$C^{m}CG G
G GC^{m}C$^{m}_{5'}$</td><td>Full & hemi-methylation of both cytosines</td><td rowspan="3">0//0</td><td rowspan="3">No information</td><td rowspan="3">0|NA</td><td rowspan="3">0</td><td rowspan="3">0</td><td rowspan="3">0</td><td rowspan="3">0</td></tr>
<tr><td>$^{5'}$CCTG
GGNC$_{5'}$</td><td>Mutation (unknow)</td></tr>
<tr><td>$^{5'm}$CCGG
GGCC$^{m}_{5'}$</td><td>Full-methylation of external cytosine</td></tr>
</table>

Fig. 2 Strategies used for the scoring and interpretation of different banding patterns that can be obtained with the MSAP technique (NA means missing data)

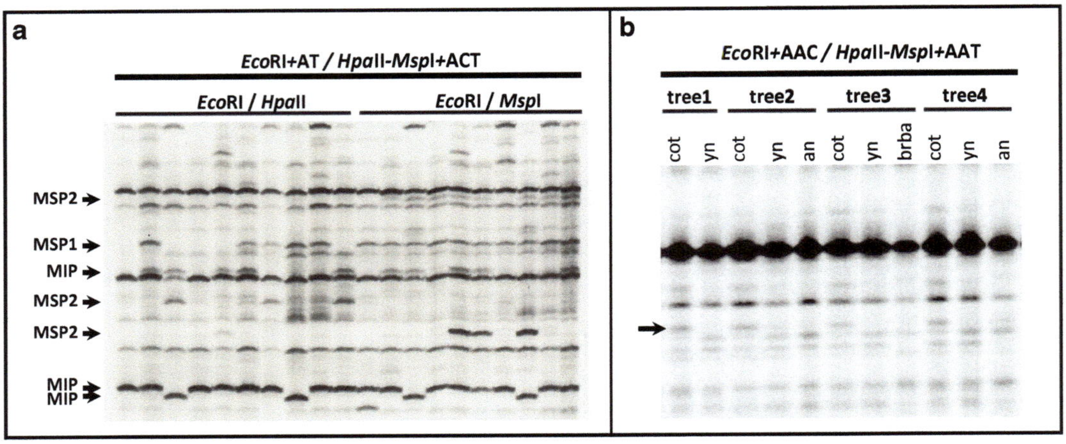

Fig. 3 An example of different MSAP patterns. (**a**) Details of fingerprints of ten ecotypes of *Arabidopsis thaliana* (*Eco*RI + AT/*Hpa*II–*Msp*I + ACT) pointing out methylation-insensitive (MIP) polymorphisms and methylation-sensitive polymorphisms found with both isoschizomers (MSP1) or with only one of them (MSP2). (**b**) Methylation-sensitive polymorphisms throughout the development identified in *Pinus pinea* (*Eco*RI +AAC/ *Hpa*II–*Msp*I +AAT): the *arrow* indicates a fragment not detected with *Hpa*II in any sample (data not shown) and detected with *Msp*I only in cotyledons, but not in young (yn) or adult needles (an)

amplified fragments in *Eco*RI/*Msp*I but not in *Eco*RI/*Hpa*II profiles. Indeed, hemi-methylation of CCGG site, in which the external cytosine is methylated only in one strand, would lead to the appearance of fragments in *Eco*RI/*Hpa*II but not in the *Eco*RI/*Msp*I profiles.

Different technologies have been developed to approach the large-scale ^{5m}C methylation analysis of plant genomes (for reviews, see Laird [2] and references therein). MSAP, initially developed by Reyna-López et al. [3], is such a cost-effective technology that it has been very frequently used for studying ^{5m}C methylation in a large number of samples. MSAPs have been recently used to study natural epigenetic variation at varietal, population, and species level [4–8] as well as to associate or map the location of differential methylation involved in the regulation of complex traits [9], including adaptive responses [10–13]. MSAP markers have also been used to analyze DNA methylation patterns associated with plant development [14–16], heterosis, and polyploidization [17–21] and to evaluate genetic integrity of micropropagated plants [22–25].

2 Materials

2.1 Equipment and Supplies

1. 1.5 and 2.0 mL microtubes.
2. PCR microtubes or plates and seals.
3. A thermocycler (PCR machine).
4. An agarose gel electrophoresis system.
5. Fragment separation and visualization support (*see* **Note 1**) such as the Li-Cor 4300 DNA Analysis System or ABI3130 for fluorochrome-labeled primers or a sequencing gel electrophoretic system with a gel dryer and a high-voltage power supply for radioactively labeled primers.
6. X-ray films, a phosphorimager device (if manual systems are used).
7. An incubator.
8. A vortex.

2.2 Buffers and Reagents

1. Restriction enzymes: *Eco*RI, *Hpa*II, and *Msp*I (New England Biolabs).
2. 10× *Hpa*II restriction buffer: 100 mM Bis-Tris Propane–HCl, 100 mM $MgCl_2$, 10 mM DTT, pH 7.0 (Buffer 1, New England Biolabs).
3. Digestion and ligation buffer (10× RL buffer): 100 mM Tris-HAc, 100 mM MgAc, 500 mM KAc, 50 mM DTT, and 500 ng/μL BSA, pH 7.5.
4. A double-stranded *Eco*RI-adapter (5 pmol/μL) (*see* **Note 2**). It is made of two primers: 5′-CTCGTAGACTGCGTACC and 5′-AATTGGTACGCAGTC.
5. A double-stranded *Hpa*II–*Msp*I adapter (50 pmol/μL) (*see* **Note 2**). It consists of the combination of primers 5′-GACGATGAGTCTCGAT and 5′ CGATCGAGACTCAT.
6. ATP 10 mM (Roche) (*see* **Note 3**).
7. ATP polynucleotide ligase (Promega) (*see* **Note 4**).

8. *Eco*RI primer +1 (50 ng/μL): 5′-GACTGCGTACCAATTCN.
9. *Eco*RI primer +3 (12 ng/μL): 5′-GACTGCGTACCAATTC NNN.
10. *Hpa*II–*Msp*I primer +1 (50 ng/μL): 5′-GATGAGTCTCGATC GGN.
11. *Hpa*II–*Msp*I primer +3 (50 ng/μL): 5′-GATGAGTCTCGAT CGGNNN.
12. dNTPs 10 mM (a mix of dATP, dTTP, dCTP, and dGTP).
13. Taq DNA polymerase (Invitrogen, 5 U/μL).
14. 10× PCR buffer (Invitrogen): 200 mM Tris–HCl, pH 8.4, 500 mM KCl.
15. $MgCl_2$ 25 mM.
16. Agarose gels: 0.8% agarose, 1× TBE, 0.03 μg/mL ethidium bromide (**Caution:** ethidium bromide is a mutagenic reagent; nitrile gloves and a laboratory coat should be worn when handled).
17. Denaturing 8% polyacrylamide gels are made of 16% Long Ranger polyacrylamide 50% gel solution (Cambrex Bio Science Rockland), 7.0 M urea, and 1× TBE. A total of 24 mL gel solution is used to prepare the gel on a 25 × 25 cm plate with 0.25 mm thick spacers (*see* **Note 5**). **Caution:** The acrylamide solution is carcinogenic, mutagenic, teratogenic, and neurotoxic; the use of nitrile gloves is required.
18. TBE 10× (pH 8): 1 M Tris base, 1 M boric acid, and 0.5 M EDTA (pH 8.0) (*see* **Note 6**).
19. *N*,*N*,*N*′,*N*′-tetramethylethylenediamine (TEMED): for 24 mL of gel, add 15 μL TEMED.
20. Ammonium persulfate (APS, 100 mg/mL): for 24 mL of gel, add 150 μL APS (*see* **Note 7**).
21. Formamide buffer: 98% formamide (deionized and filtered), 10 mM EDTA, pH 8.0, and 0.06% bromophenol blue.
22. A DNA ladder (*see* **Note 8**).
23. Ethanol.
24. NaOAc 3 M pH 5.2.

3 Methods

Just as in other AFLP-based technologies, the MSAP protocol (Fig. 1) consists of four major steps: the digestion of the genomic DNA and ligation of adapters, the pre-amplification of digested-ligated fragments, a selective amplification of pre-amplified fragments, and fragment detection and scoring.

3.1 DNA Extraction

High-quality DNA is required to avoid the inhibition of endonuclease activities. This can be achieved using manual protocols such as Dellaporta et al. [26] or commercial kits.

3.2 Digestion–Ligation

This step involves two digestions of the genomic DNA with two different restriction enzymes and the ligation of double-stranded adapters to the sticky ends generated (Fig. 1). The adapter and restriction site sequences will serve as primer binding sites for the subsequent amplification steps. Here, a complete digestion of the DNA is crucial to prevent the later amplification of uncut fragments (*see* **Note 9**). The complete digestion is achieved by the use of high-quality DNA and an excess of restriction enzyme.

3.2.1 Digestion

Two different MSAP analyses have to be performed using *Eco*RI and either *Hpa*II or *Msp*I to digest each sample. DNA digestion with *Eco*RI and *Hpa*II cannot be performed simultaneously because each restriction enzyme has different restriction buffer requirements. Thus, 250–500 ng of genomic DNA (*see* **Note 10**) is incubated in a final volume of 25 μL with 6 U *Hpa*II and Buffer 1 for 2 h at 37 °C. After digestion, DNA is precipitated by adding 0.1 volumes of sodium acetate (NaOAc 3 M, pH 5.2) and 2.5 volumes of ethanol. The samples must be incubated overnight at −20 °C after a thorough manual mixing and a subsequent spin in a centrifuge to recover droplets. The digested product is centrifuged at 15,500 × *g* for 15 min at 4 °C to discard the supernatant, and then it is washed with 1 mL ethanol (70 %). After washing, the pellet is dried at room temperature for 10–15 min and resuspended in 24 μL of dH_2O. The resuspended DNA is then digested with 10 U of *Eco*RI incubated for 2 h at 37 °C in a final volume of 35 μL with 1× RL buffer. For DNA digestion with *Eco*RI/*Msp*I, both restriction enzymes can be used together. The reaction is carried out in a final volume of 35 μL with 1× RL buffer, 10 U *Eco*RI, 10 U *Msp*I, and 250–500 ng of the genomic DNA for 3 h at 37 °C.

3.2.2 Ligation

Two different adapters (one for the *Eco*RI sticky ends and one for the *Hpa*II–*Msp*I sticky ends) are ligated to DNA fragments after digestions. Five μL of a mix containing 5 pmol *Eco*RI adapter, 50 pmol *Hpa*II–*Msp*I adapter, 8 mM ATP, 1× RL buffer, and 2 U T4 DNA ligase is added to each final digestion (*see* **Note 4**). The ligation is incubated at 37 °C for 3 h and then at 4 °C overnight (*see* **Note 9**).

3.2.3 Dilution

Digested–ligated DNA fragments are diluted fivefold with sterile dH_2O and stored at −20 °C.

DNA digestion generates thousands of fragments. The complexity of this fragment population is reduced by two successive PCR reactions using primers with an increased number of selective nucleotides at their 3′ ends in order to accurately visualize a single subset of diluted–ligated DNA fragments at the end of each analysis (*see* **Note 10**). The first PCR reaction (pre-amplification) is performed using a single selective nucleotide at the 3′ end of both *Eco*RI and *Hpa*II–*Msp*I primers. The second PCR (selective amplification) is carried out using more than one selective nucleotide at the 3′ end of both *Eco*RI and *Hpa*II–*Msp*I primers, depending on the genome size.

3.3 Pre-amplification

A pre-amplification PCR reaction uses primers that are complementary to the *Eco*RI and *Hpa*II–*Msp*I adapters with an additional selective 3′ nucleotide (i.e., *Eco*RI + A and *Hpa*II–*Msp*I + C), thus selecting 1/16 of the digested fragments. The PCR reaction is performed in a 20 μL volume of 1× PCR buffer, 0.4 mM Cl_2Mg, 0.2 mM of each dNTP, 30 ng of each primer (*Eco*RI + 1 and *Hpa*II–*Msp*I + 1), 0.4 U Taq DNA polymerase, and 5 μL of digested–diluted fragments (*see* **Note 11**). PCR amplifications are carried out using 16–28 cycles (*see* **Note 10**), each cycle consisting of 30 s at 94 °C, 1 min at 60 °C, and 1 min at 72 °C.

In order to verify the efficiency of pre-amplification, 2 μL of final products are electrophoresed on a 0.8 % agarose gel and separated in a short run (10–15 min) to visually compare intensities among the amplified samples (Fig. 1). The longer runs lead to very faint smears, thus hampering the comparisons. The pre-amplified DNA fragments are diluted at least fivefold up to tenfold with dH_2O to approximately even concentrations, depending on the intensity of the smears visualized in the agarose gels. The diluted pre-amplification can be stored at −20 °C for more than 1 year.

3.4 Selective Amplification

The selective amplification consists of a PCR reaction with primers that are complementary to the *Eco*RI and *Hpa*II–*Msp*I adapters with two or three selective nucleotides at their 3′ ends, thus selecting a subset of diluted pre-amplified fragments (*see* **Note 10**). It is important to point out that the selective nucleotide used in the pre-amplification has to be maintained in the selective amplification. For the selective amplification, only *Eco*RI primers are labeled at their 5′ ends with fluorescent or radioactive tags, depending on the detection method used (*see* **Notes 1** and **12**).

The selective PCR reaction is performed in a 10 μL volume of 1× PCR buffer, 0.1 mM of each dNTP, 3 ng of the labeled-*Eco*RI primer (*see* **Note 11**), 15 ng of the *Hpa*II–*Msp*I primer, 0.2 U Taq DNA polymerase, and 2.5 μL of diluted pre-amplified DNA. The PCR reaction is carried out using classical AFLP cycling parameters [1]: 1 cycle of 30 s at 95 °C, 30 s at 65 °C, and 1 min at 72 °C followed by 13 cycles in which the annealing temperature decreases 0.7 °C per cycle, followed by 23 cycles of 30 s at 95 °C, 30 s at 56 °C, and 1 min at 72 °C (*see* **Note 13**).

3.5 Fragment Detection and Scoring

The final step of MSAP technique is the separation and visualization of amplified fragments followed by data interpretation. MSAP products can be separated and scored using a variety of systems (*see* **Note 1**). Polyacrylamide gel electrophoresis (a conventional or an automated sequencer) and capillary electrophoresis provide the maximum resolution of banding patterns.

At the end of selective PCR, samples are denatured by adding an equal volume of formamide buffer and heating for 2 min at 94 °C followed by quick cooling on ice.

Before loading samples, polyacrylamide gels have to be pre-run for 15 min to warm up the gel using the same settings as for the run. These settings depend on gel size and thickness and the electrophoresis system used. With a Li-Cor 4300 DNA Analysis System, these settings are: 1500 V, 35 W, 35 mA, and 45 °C. After pre-running the gel, remove urea precipitate or pieces of gel with a syringe before loading. A total of 0.8–2 μL of each sample is loaded on Li-Cor or conventional polyacrylamide gels, respectively. For capillary electrophoresis, after heating for 5 min at 95 °C for denaturalization, an aliquot of a mix containing 1 μL of the diluted sample (diluted 10–100 times depending on the observed on a 2% agarose gel), 14 μL formamide, and 0.11 μL of a molecular weight marker (i.e., LIZ500) is loaded.

Fragments can be scored visually or by using different scoring software developed for AFLPs [28], such as 1 when the fragment is present, and 0 when it is absent. Progressive fragment appearance or disappearance can also be illustrated in a table indicating the number and percentage of methylation-sensitive fragments showing a specific pattern.

4 Data Interpretation

The direct comparison of *Eco*RI/*Msp*I and *Eco*RI/*Hpa*II profiles for each sample allows obtaining the information about the cytosine methylation state at the CCGG sequences analyzed. Figure 2 summarizes the four different situations that can be observed, and Fig. 3 illustrates several examples. The analyzed range of MSAP is between 40 and 400 nt; therefore the probability of containing an internal CCGG restriction site is relatively low. The fragments present in both profiles, *Eco*RI/*Msp*I and *Eco*RI/*Hpa*II, are associated with non-methylated sites. The fragments present only in the *Eco*RI/*Msp*I profiles are associated with the hemi- or fully methylated internal cytosines (5′-C^{hm}CGG or 5′-C^{m}CGG sites). The fragments present only in the *Eco*RI/*Hpa*II profiles are associated with the hemi-methylated external cytosines (5′-^{hm}CCGG sites). The absence of fragments in both profiles indicates an uninformative state that can be related with several methylation situations or polymorphisms at the restriction site of the given samples.

In addition, different approaches have been used to integrate the presence/absence of fragments in both *Eco*RI/*Msp*I and

*Eco*RI/*Hpa*II matrices into one or several binary matrices that can be used for downstream analyses. These integrated matrices can be grouped based on three scoring types (Fig. 2, *see* ref. 28 for further details): methylation scoring [5, 29, 30], non-methylation scoring [10, 28], and mixed scoring [28, 31].

- *Methylation scoring*: The amplified fragments differing in the presence/absence or intensity between *Eco*RI/*Msp*I and *Eco*RI/*Hpa*II profiles (status II and III on Fig. 2) are scored as 1. The fragments present in both *Eco*RI/*Msp*I and *Eco*RI/*Hpa*II profiles (status I) are scored as 0. The absence of fragments in both profiles (status IV) can be scored as 0 or missing data.
- *Non-methylation scoring*: Only the amplified fragments present in both analyses (non-methylated sites, status I) are scored as 1. The fragments present only in *Eco*RI/*Hpa*II profiles (hemi-methylated at the external cytosine, status III) are excluded or scored as 0.
- *Mixed scoring*: MSAP raw data are transformed into three matrices that correspond to: non-methylated fragments, fragments that are hemi-methylated or fully methylated at the internal cytosine, and fragments that are hemi-methylated at the external cytosine. The fragments present in both profiles (status I) are scored as 1 for the three sets of matrices or, in an alternative interpretation, only for the non-methylated set; the fragments present only in *Eco*RI/*Msp*I profiles (status II) are scored as 1 only for the fragments that are hemi-methylated or fully methylated at the internal cytosine set. The fragments present only in *Eco*RI/*Hpa*II profiles (status III) are scored as 1 only for the fragments that are hemi-methylated at the external cytosine set.

The optimal MSAP scoring and data interpretation approaches can vary depending on the characteristics of the studied species and the type of analysis. Although the results obtained with the three approaches are largely consistent, some of the descriptive indices can vary, including diversity and differentiation, and have to be considered when comparing results obtained by different authors [28].

In addition, although MSAP has not been initially developed as a quantitative technique, methylation-sensitive fragments showing a different intensity are usually observed. These changes may be due to differences in cytosine methylation status among different cells analyzed in each sample [32–35]. Moreover, there have been found changes in MSAP profiles associated with developmental stages [14, 35] or stress responses [8] in which a decrease in intensity or the disappearance of fragments indicates methylation, while the appearance or an increase in intensity of fragments indicates demethylation.

5 Notes

1. Different detection systems can be selected to visualize MSAP fragments. These detection systems are based on the use of labeled or non-labeled primers (such as the silver staining method) and the fragment separation support. When using labeled primers, only *Eco*RI primers are labeled. A different chemistry can be used, including radioactive isotopes and fluorescent dyes (such as IRD 700 or IRD 800 from LI-COR; FAM, HEX, ROX, TAMRA, and TET from Applied Biosystems; Cy from Amersham Biosciences; or Yakima Yellow from Epoch Biosciences). The visualization of amplified products can be achieved using manual or automatic fragment analysers based on gel electrophoresis or capillary electrophoresis. It is important to point out that if radioactive MSAP analyses are carried out using gel-based electrophoresis, ^{33}P-labeled primers provide a better resolution of amplified products than ^{32}P-labeled primers. After the completion of electrophoresis, radioactive gels can be directly dried without fixation and exposed to X-ray film for 24–72 h at room temperature.
2. Double-stranded *Eco*RI and *Hpa*II–*Msp*I adapters are made of 17 and 15 and 14 and 16 base pair primers, respectively. When the adapter primer pairs are mixed for the first time, they should be heated at 65 °C for 5 min to denature for improving the annealing of two strands of each adapter stock. Then, allow cooling slowly to renature completely. The adapters can be stored at −20 °C. When non-phosphorylated adapters are used, a single strand of each adapter is ligated to the DNA. The recessed 3′ ends of the template are filled-in by the Taq polymerase in the presence of dNTPs during the first cycle.
3. 10 mM ATP aliquots must be prepared and stored at −20 °C. Do not refreeze the rest of the aliquot that has not been used.
4. In order to ensure the addition of small volumes to the ligation mix, a highly concentrated ATP polynucleotide ligase has to be used (>6 U/μL).
5. MSAP reaction products are analyzed on 4.5–8 % acrylamide gels. The detection of radiolabeled products is performed using the conventional gel electrophoresis systems and 4.5 % denaturing polyacrylamide gels (acrylamide/bisacrylamide, 19:1) containing 7.5 M urea and 1× TBE. If a LI-COR automated DNA sequencer is used, the 16 % Long Ranger 50 % Gel solution containing 7.0 M urea and 1× TBE is prepared. Once the urea is dissolved, the solution is filtrated and maintained at 4 °C in the dark. The gels should be casted at least for 2 h before using to ensure sufficient time for gel polymerization and may be stored for 24 h at 4 °C.

6. 10× TBE: Dissolve 108 g Tris base, 55 g boric acid, and 40 mL EDTA (pH 8.0) in 700 mL distilled water, stir to dissolve, and finally add distilled water to bring up the total volume to 1 L. If only dry ingredients are used, boric acid should be added last after EDTA is dissolved.
7. To prepare the 100 mg/mL APS solution, it is important to be sure that APS powder is dry. APS solutions can be used for a maximum of a week, but they are not stable at room temperature and should be stored at 4 °C or at −20 °C.
8. Different commercial DNA ladders may be used for MSAP analysis in gel electrophoresis: IRD-labeled Li-Cor ladders, ABI size standards (Applied Biosystems), the labeled 30–330 bp DNA Ladder (Life Technologies), and the labeled 100-bp ladder (Gibco Life Technologies). Home-made DNA ladders made of combinations of labeled DNA fragments of known sizes may also be used.
9. The adapter design avoids the reconstruction of restriction sites (Fig. 2). Thus, the presence of restriction enzymes in the ligation step results in almost complete adapter-fragment ligation because primer concatemers which may be generated by ligation are restricted.
10. Several parameters need to be adjusted to analyze different plant species depending on their genome sizes that range from 0.50 to 40 pg/2C [36]: (1) the amount of template DNA, ranging from 250 ng for small genomes, such as *Arabidopsis*, to 500 ng for large genomes such as conifer which are 170-fold larger than the Arabidopsis genome, (2) the number of cycles used in the pre-amplification ranges from 16 for small genomes to 28 for large genomes, and (3) the number of selective nucleotides used in both PCR steps. The use of one, two, or three selective nucleotides at the 3′ end of one of the primers (i.e., *Eco*RI + 1, *Eco*RI + 2, or *Eco*RI + 3, respectively) reduces the number of amplified fragments by factors of 4, 16, and 64, respectively. The use of a level of selection +2/+2 (i.e., *Eco*RI + AC/*Hpa*II + CG) will decrease the number of amplified fragments to 1/256. When it is necessary, the intermediate levels of selection can be achieved by combining two *Eco*RI primers that share some of the selective nucleotides (i.e., 2 *Eco*RI(+AC,+AG)/*Hpa*II + CG results in a level of selection of 1/128). The protocol for selective amplification commonly ranges from *Eco*RI + 2/*Hpa*II–*Msp*I + 3 (for genome sizes that are smaller than 0.60 pg/2C) to 2 *EcoRI* + 3/*Hpa*II–*Msp*I + 3 (for genome sizes ranging between 0.60 and 1.00 pg/2C) or *Eco*RI + 3/*Hpa*II–*Msp*I + 3 (for genome sizes over 1.00 pg/2C). This is always preceded by *Eco*RI + 1/*Hpa*II–*Msp*I + 1 pre-amplification. For extremely large genomes, such as in conifer species (with 20–38 pg/2C), higher selection levels must be

used for the pre-amplification (i.e., *Eco*RI + 1/*Hpa*II–*Msp*I + 2) if *Eco*RI + 3/*Hpa*II–*Msp*I + 4 is used for the selective amplification.

11. MSAP reaction mixes should be prepared for a minimum of ten different samples to minimize discrepancies due to inaccurate pipetting of small volumes.
12. The mobility of the two DNA fragment strands is slightly different. Since only one of two primers is labeled, the comparison of MSAP profiles should be carried out using the same labeled primer.
13. The start of PCR at a very high annealing temperature allows the optimal primer selectivity. By gradually decreasing the annealing temperature, we could increase the efficiency of primer binding.

References

1. Vos P, Hogers R, Bleeker M, Reijans M, van de Lee T, Hornes M, Frijters A, Pot J, Peleman J, Kuiper M, Zabeau M (1995) AFLP: a new technique for DNA fingerprinting. Nucleic Acids Res 23:4407–4414
2. Laird PW (2010) Principles and challenges of genome wide DNA methylation analysis. Nat Rev Genet 11:191–203
3. Reyna-López GE, Simpson J, Ruiz-Herrera J (1997) Differences in DNA methylation patterns are detectable during the dimorphic transition of fungi by amplification of restriction polymorphisms. Mol Gen Genet 253:703–710
4. Li Y, Shan X, Liu X, Hu L, Guo W, Liu B (2008) Utility of the methylation-sensitive amplified polymorphism (MSAP) marker for detection of DNA methylation polymorphism and epigenetic population structure in a wild barley species (*Hordeum brevisubulatum*). Ecol Res 23:927–930
5. Herrera CM, Bazaga P (2010) Epigenetic differentiation and relationship to adaptive genetic divergence in discrete populations of the violet *Viola cazorlensis*. New Phytol 187:867–876
6. Herrera CM, Bazaga P (2011) Untangling individual variation in natural populations: ecological, genetic and epigenetic correlates of long-term inequality in herbivory. Mol Ecol 20:1675–1688
7. Ocaña J, Walter B, Schellenbaum P (2013) Stable MSAP markers for the distinction of *Vitis vinifera* cv Pinot noir clones. Mol Biotechnol 55:236–248
8. Sáez-Laguna E, Guevara MA, Díaz LM, Sánchez-Gómez D, Collada C, Aranda I, Cervera MT (2014) Epigenetic variability in the genetically uniform forest tree species *Pinus pinea* L. PLoS One 9:e103145
9. Long Y, Xia W, Li R, Wang J, Shao M, Feng J, King GJ, Meng J (2011) Epigenetic QTL mapping in *Brassica napus*. Genetics 189:1093–1102
10. Lira-Medeiros CF, Parisod C, Fernandes RA, Mata CS, Cardoso MA, Ferreira PCG (2010) Epigenetic variation in mangrove plants occurring in contrasting natural environment. PLoS One 5(4):e10326
11. Wang WS, Pan YJ, Zhao XQ, Dwivedi D, Zhu LH, Ali J, Fu BY, Li ZK (2011) Drought-induced site-specific DNA methylation and its association with drought tolerance in rice (*Oryza sativa* L.). J Exp Bot 62:1951–1960
12. Karan R, DeLeon T, Biradar H, Subudhi PK (2012) Salt stress induced variation in DNA methylation pattern and its influence on gene expression in contrasting rice genotypes. PLoS One 7:e40203
13. Herrera CM, Bazaga P (2013) Epigenetic correlates of plant phenotypic plasticity: DNA methylation differs between prickly and nonprickly leaves in heterophyllous *Ilex aquifolium* (Aquifoliaceae) trees. Bot J Linn Soc 171:441–452
14. Ruiz-García L, Cervera MT, Martinez-Zapater JM (2005) DNA methylation increases throughout Arabidopsis development. Planta 222:301–306

15. Meng FR, Li YC, Yin J et al (2012) Analysis of DNA methylation during the germination of wheat seeds. Biol Plantarum 56:269–275
16. Osabe K, Clement JD, Bedon F, Pettolino FA, Ziolkowski L, Llewellyn DJ, Finnegan EJ, Wilson IW (2014) Genetic and DNA methylation changes in cotton (Gossypium) genotypes and tissues. PLoS One 9:e86049
17. Salmon A, Ainouche ML, Wendel JF (2005) Genetic and epigenetic consequences of recent hybridization and polyploidy in Spartina (Poaceae). Mol Ecol 14:1163–1175
18. Zhao Y, Yu S, Xing C, Fan S, Song M (2008) Analysis of DNA methylation in cotton hybrids and their parents. Mol Biol 42:169–178
19. Hegarty MJ, Batstone T, Barker GL, Edwards KJ, Abbott RJ, Hiscock SJ (2011) Nonadditive changes to cytosine methylation as a consequence of hybridization and genome duplication in Senecio (Asteraceae). Mol Ecol 20:105–113
20. Li A, Hu BQ, Xue ZY, Chen L, Wang WX, Song WQ, Chen CB, Wang CG (2011) DNA methylation in genomes of several annual herbaceous and woody perennial plants of varying ploidy as detected by MSAP. Plant Mol Biol Report 29:784–793
21. Rodriguez MP, Cervigni GDL, Quarin CL, Ortiz JPA (2012) Frequencies and variation in cytosine methylation patterns in diploid and tetraploid cytotypes of *Paspalum notatum*. Biol Plantarum 56:276–282
22. Hanai LR, Floh EIS, Fungaro MHP, Anta-Catarina C, de Paula FM, Viana AM, Vieira MLC (2010) Methylation patterns revealed by MSAP profiling in genetically stable somatic embryogenic cultures of *Ocotea catharinensis* (Lauraceae). In Vitro Cell Dev Biol Plant 46:368–377
23. Bobadilla Landey R, Cenci A, Georget F, Bertrand B, Camayo G, Dechamp E, Herrera JC, Santoni S, Lashermes S, Simpson J, Etienne H (2013) High genetic and epigenetic stability in *Coffea arabica* plants derived from embryogenic suspensions and secondary embryogenesis as revealed by AFLP, MSAP and the phenotypic variation rate. PLoS One: e56372.
24. Tiwari JK, Chandel P, Gupta S, Gopal J, Singh BP, Bhardwaj V (2013) Analysis of genetic stability of in vitro propagated potato microtubers using DNA markers. Physiol Mol Biol Plants 19:587–595
25. Rathore MS, Mastan SG, Agarwal PK (2015) Evaluation of DNA methylation using methylation-sensitive amplification polymorphism in plant tissues grown in vivo and in vitro. Plant Growth Regul 75:11–19
26. Dellaporta SL, Wood J, Hicks JB (1985) Maize DNA miniprep. In: Malberg R, Messing J, Sussex I (eds) Molecular biology of plants. Cold Spring Harbor Laboratory, Cold Spring Harbor, NY, pp 36–37
27. Meudt HM, Clarke AC (2007) Almost forgotten or latest practice? AFLP applications, analyses and advances. Trends Plant Sci 12:106–117
28. Schulz B, Eckstein RL, Durka W (2013) Scoring and analysis of methylation-sensitive amplification polymorphisms for epigenetic population studies. Mol Ecol Resour 13: 642–653
29. Salmon A, Clotault J, Jenczewski E, Chable V, Manzanares-Dauleux MJ (2008) *Brassica oleracea* displays a high level of DNA methylation polymorphism. Plant Sci 174:61–70
30. Vergeer P, Wagemaker N, Ouborg NJ (2012) Evidence for an epigenetic role in inbreeding depression. Biol Lett 8:798–801
31. Paun O, Bateman RM, Fay MF et al (2010) Stable epigenetic effects impact adaptation in allopolyploid orchids (*Dactylorhiza: Orchidaceae*). Mol Biol Evol 27:2465–2473
32. Cervera MT, Ruiz-García L, Martínez-Zapater JM (2002) Analysis of DNA methylation-sensitive AFLP markers. Mol Genet Genomics 268:543–552
33. Xiong LZ, Xu CG, Saghai Maroof MA, Zhang Q (1999) Patterns of cytosine methylation in an elite rice hybrid and its parental lines, detected by a methylation-sensitive amplification polymorphism technique. Mol Gen Genet 261:439–446
34. Peraza-Echeverria S, Herrera-Valencia VA, Kay A (2001) Detection of DNA methylation changes in micropropagated banana plants using methylation-sensitive amplification polymorphism (MSAP). Plant Sci 161:359–367
35. Candaele J, Demuynck K, Mosoti D, Beemster GTS, Inzé D, Nelissen H (2014) Differential methylation during maize leaf growth targets developmentally regulated genes. Plant Physiol 164:1350–1364
36. Cervera MT, Remington D, Frigerio JM, Storme V, Ivens B, Boerjan W, Plomion C (2000) Improved AFLP analysis of tree species. Can J For Res 30:1608–1616

Chapter 10

Differentially Methylated Region-Representational Difference Analysis (DMR-RDA): A Powerful Method to Identify DMRs in Uncharacterized Genomes

Pavlina Sasheva and Ueli Grossniklaus

Abstract

Over the last years, it has become increasingly clear that environmental influences can affect the epigenomic landscape and that some epigenetic variants can have heritable, phenotypic effects. While there are a variety of methods to perform genome-wide analyses of DNA methylation in model organisms, this is still a challenging task for non-model organisms without a reference genome. Differentially methylated region-representational difference analysis (DMR-RDA) is a sensitive and powerful PCR-based technique that isolates DNA fragments that are differentially methylated between two otherwise identical genomes. The technique does not require special equipment and is independent of prior knowledge about the genome. It is even applicable to genomes that have high complexity and a large size, being the method of choice for the analysis of plant non-model systems.

Key words Complex genomes, DNA methylation, Epigenetics, Genome comparison, Genome representation, Non-model plants, Polymerase chain reaction, Subtractive hybridization

1 Introduction

Most current methods used for epigenetic profiling in model systems or in the context of human disease are based on microarrays or next-generation sequencing (NGS) and are not well suited for species for which no genome information is available. However, before genome information was available, representational difference analysis (RDA) was used for the targeted identification of differences between complex genomes, e.g., between the normal and cancerous tissue of a patient [1]. RDA was also adapted to identify differences between transcriptomes [2] and DNA methylation patterns [3], although the latter have rarely been used because it was superseded by other techniques. Most methods for the genome-wide analysis of DNA methylation are microarray or NGS-based and thus require previous genomic information [4]. Therefore, these methods have only been widely applied to

Igor Kovalchuck (ed.), *Plant Epigenetics: Methods and Protocols*, Methods in Molecular Biology, vol. 1456,
DOI 10.1007/978-1-4899-7708-3_10, © Springer Science+Business Media New York 2017

humans and genetic model systems. In contrast, methylation-sensitive amplified polymorphism (MSAP) analyses [5] do not require genomic information and allow a comparison of how different two genomes are at the level of DNA methylation. Much more powerful, however, is RDA in combination with methylation-sensitive restriction enzymes, as it will directly identify the genomic regions that are differentially methylated between two samples. As the method is highly sensitive and relatively cheap, and requires neither special equipment nor previous genomic knowledge, it is the method of choice for non-model systems with often complex genomes.

Differentially methylated region-representational difference analysis (DMR-RDA) is a genome-based method that identifies differences in the methylomes between identical genomes. The technique does not require prior genomic information and aims at isolating specific DNA regions that are differentially methylated between the analyzed genomes, while avoiding the scanning though repetitive elements and DNA stretches that share the same cytosine methylation patterns. Since DNA methylation is a common epigenetic mark found in plants in three different contexts (CG, CHG, and CHH, where H is A, T, or C), analyzing whole genomes of uncharacterized species by NGS may become a time-consuming, bioinformatically highly challenging, and expensive approach. The advantage of DMR-RDA lies in its sensitivity and robustness, promising to be the method of choice for the analysis of (epi)genomic differences in plant non-model systems.

DMR-RDA is based on the classical RDA protocol developed by Lisitsyn and colleagues (1993) to identify differences between genomes [6]. The technique creates representations of the two genomes to be compared, which then undergo a subtractive hybridization followed by a kinetic enrichment of the differential products. Two names are assigned to the genome pools: the DRIVER is used to subtract the fragments of interest from the TESTER by rendering the sequences common to the two populations inert for subsequent analyses.

In order to establish a robust and effective DMR-RDA protocol, we used two lines of *Arabidopsis thaliana* accession Zürich that are genetically identical but have a differentially methylated transgene coding for hygromycin resistance (called C-insert). The insertion in the hygromycin sensitive line (C2S) is heavily methylated in the CG and CHG contexts, and hence silent, while the resistant line (C2R) has unchanged methylation status of the C-insert [7].

DMR-RDA uses the combination of two restriction enzymes. The methylation-sensitive restriction cleavage of the genome occurs in the first place, determining the specificity of the technique toward extracting fragments from the non-methylated fraction of the genome. The enzyme will digest only the non-methylated parts (mostly coding sequences of the genome) into sizes that are suitable for amplification. On the other hand, if the cytosine in the restriction site is methylated, the enzyme will not cut, yielding long fragments that usually derive from the repetitive fraction of the genome (Fig. 1a).

A universal adaptor is then ligated to both ends of the fragmented DNA, but before the first amplification event occurs, a second restriction cleavage takes place. A methylation-dependent restriction enzyme is used to digest the DNA that is methylated. By introducing this step, we eliminate the possibility of linear amplification of the methylated fraction of the genome, as well as the amplification of DRIVER-derived sequences that are unevenly methylated within the cell populations used for the analysis (Fig. 1b).

The newly created amplicons transform the problem from distinguishing between methylated and non-methylated DNA fragments into the classical RDA paradigm of amplifying fragments that are present only in the TESTER DNA pool (Fig. 1c). The successful output of the analysis is set at this stage, as even a small fraction of DRIVER-derived fragments may decrease the efficiency [3].

Subsequently, the conditions to selectively enrich the differential products are set by exchanging the first adaptor with a second one, thus creating primer-binding sites that are specific for the TESTER (Fig. 1d). To be able to specifically enrich only the differences between the genomes, the TESTER is mixed with an excess amount of DRIVER, denatured, and allowed to reanneal under high stringency conditions. During this step, the DNA TESTER fragments that have counterparts in the DRIVER will form TESTER:DRIVER duplexes, which will be amplified only in a linear fashion. The self-reannealed TESTER duplexes are the true differential products and will be amplified exponentially, while the DRIVER-derived duplexes will not be amplified at all (Fig. 1e).

After the first round, a second round takes place to further eliminate fragments that are identical between the two genomes. The first round product receives a third set of adaptors and undergoes hybridization against an excess amount of DRIVER. The product of this round refers to as second round differential product.

2 Materials

2.1 Equipment

1. Autoclave.
2. Laminar flow cabinet.
3. Benchtop centrifuge.
4. Refrigerated centrifuge.

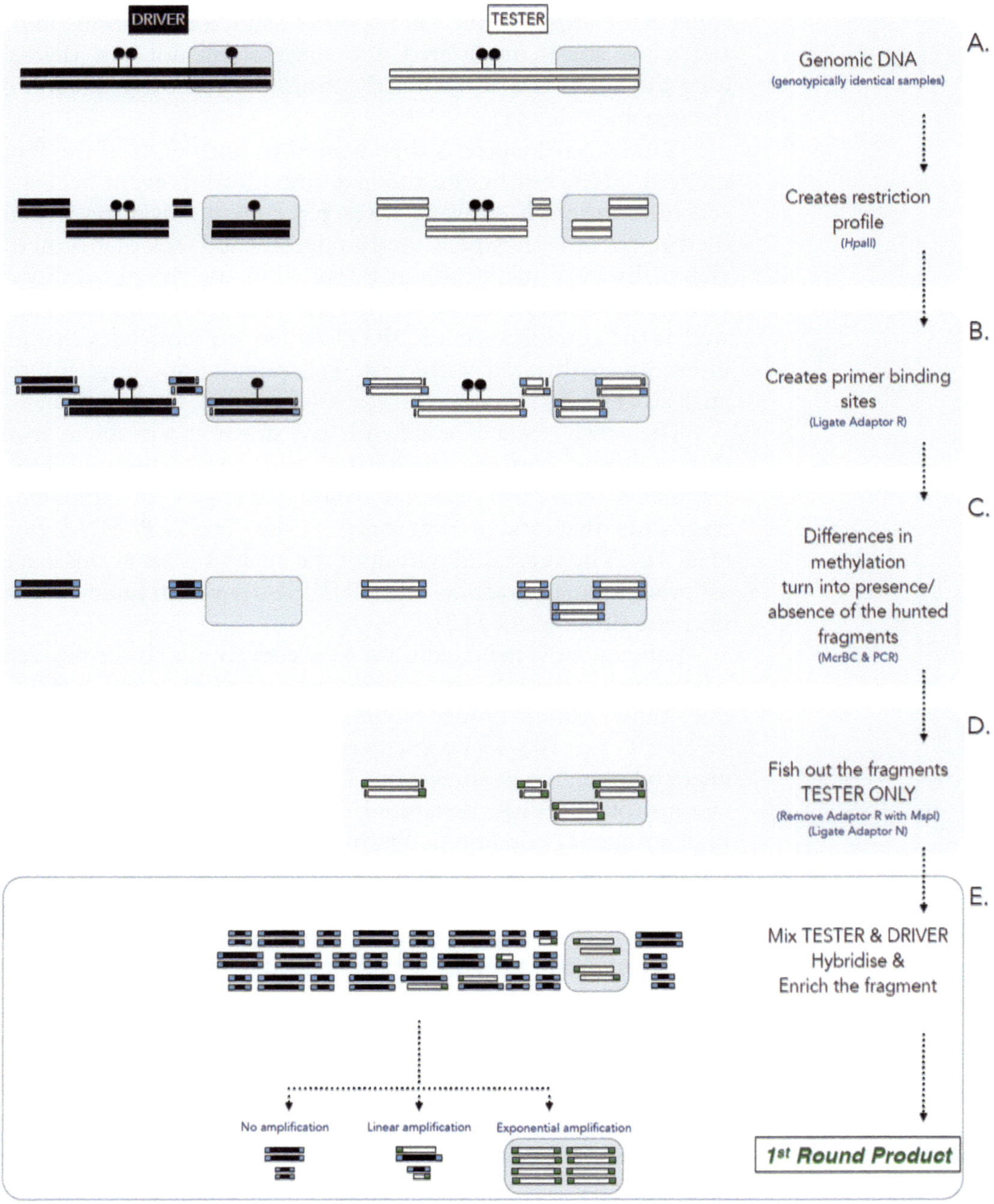

Fig. 1 Schematic presentation of the DMR-RDA protocol. The two DNA pools that are genotypically identical, but contain differences in the DNA methylation landscape, are TESTER (*open rectangle*) and DRIVER (*closed rectangle*). The sites of DNA methylation are denoted with a "lollipop" structure on the top of the DNA. The filled areas (**a–d**) denote an active step, which is performed separately for each of the individual DNA pools. In the framed section (**e**), the TESTER and DRIVER are combined prior to the hybridization. (**a**) Two genotypically identical DNA pools that have differences in the DNA methylation are digested separately, to create restriction profiles using the methylation-sensitive restriction enzyme *Hpa*II. (**b**) Universal adaptor (*blue squares*) is ligated to the *CG* overhang at the 5′ end of the fragmented DNA, creating primer-binding sites. (**c**) At this step, the amplicons are created. The combination of methylation-dependent McrBC digest before the PCR amplification

5. Benchtop incubator.
6. Dry heating block.
7. Thermal cycler.
8. Hybridization incubator.
9. UV-cross linker.
10. Agarose gel electrophoresis equipment.
11. Gel documentation system.
12. Chemiluminescence imaging system.
13. Fluorometer for nucleic acid and protein concentration measurement.
14. Spectrophotometer for nucleic acid and protein concentration measurement.

2.2 Reagents

1. CI: Chloroform:isoamyl alcohol 24:1 (v/v).
2. Glycogen (20 mg/ml).
3. 10 mM spermidine.
4. 5 M NaCl.
5. 3 M NaOAc.
6. 1× HYB dilution buffer: 0.5 M NaCl, 8 mM Tris-HCl, and 0.8 mM EDTA.
7. 1× TE: 10 mM Tris-HCl (pH 8.0) and 1 mM EDTA.
8. 1× TBE: 100 mM Tris, 100 mM boric acid, and 2 mM EDTA, pH 8.3.
9. 10× GTP (NEB).
10. 10× PCR reaction buffer for Taq DNA polymerase, containing $MgCl_2$ (Sigma-Aldrich).
11. 10× NEBuffer 2 (NEB).
12. 10× NEBuffer 1 (NEB).
13. 10× T4 DNA ligase reaction buffer, containing ATP (NEB).

Fig. 1 (continued) converts the differences between the two DNA pools from non-methylated (TESTER) vs methylated (DRIVER) DNA stretches into presence (TESTER) vs absence (DRIVER) of the fragment(s) of interest. Amplification serves a double function: (1) to lower the complexity of the two genomes and (2) to enrich the fraction of single-copy DNA. (**d**) Only TESTER DNA receives new adaptor pair, thus setting the conditions for enrichment of the DNA fragments unique to the TESTER. This is achieved by digesting with *Msp*I (an isoschizomer of *Hpa*II), which removes the adaptor R (*blue squares*), followed by a ligation of the second adaptor N (*green squares*). (**e**) Excess amount of the DRIVER is mixed with small amount of the TESTER DNA for competitive hybridization under stringent conditions. The duplexes between the TESTER DNA fragments are the true differential products, which are exponentially enriched because only these hybrids have primer-binding sites at the both ends. The TESTER:DRIVER duplexes are linearly amplified and their presence will be further suppressed after the second hybridization

14. 12× EEN: 120 mM EPPS (4-(2-hydroxyethyl)-1-piperazine propanesulfonic acid, pH 8.0), 12 mM EDTA, and 2 M NaCl.
15. 25 mM dNTPs.
16. 100× BSA (NEB).
17. 100% DMSO.

2.3 Enzymes

1. *Hpa*II (NEB).
2. *Msp*I (NEB).
3. Mung bean nuclease (NEB).
4. Taq DNA polymerase (Sigma-Aldrich).
5. T4 DNA ligase (NEB).

2.4 Oligonucleotides

1. R24: 5′-AGCACTCTCCAGCCTCTCACCGAC-3′
2. R11: 5′-**CG**GTCGGTGAG-3′
3. J24: 5′-ACCGACGTCGACTATCCATGAAAC-3′
4. J11: 5′-**CG**GTTTCATGG-3′
5. N24: 5′-AGGCAACTGTGCTATCCGAGGGAC-3′
6. N11: 5′-**CG**GTCCCTCGG-3′

3 Methods

3.1 General Considerations

1. Before starting the protocol, test the adaptor pairs for nonspecific amplification of the DNA from the plant to be used for DMR-RDA.
2. Calculate the total amount of the DRIVER that will be needed for the two hybridization events in order to calculate the volume of the PCR reaction on Day 3 (Subheading 3.4, **step 2**). Keep in mind that the yield may differ between species.
3. Never repeat an adaptor pair: the primer-binding site generated by the adaptor ligation cannot be removed completely. This will create a bias for the second/third hybridization, because amplification of fragments that have retained the adaptor pair 1/2 from the previous cycles cannot be avoided.

3.2 Day 1

1. *DNA Isolation.* Isolate clean, genomic DNA using a method of choice. Check the DNA quality by running a 0.8% agarose gel (you should see clear bands at or higher than 12 kb and no RNA on the gel). Measure the DNA concentration (Qubit) and estimate the purity (NanoDrop).
2. *Hpa*II *Digest.* This step creates the restriction profile of the two DNA pools, as set by the methylation-sensitive restriction digest with *Hpa*II. Digest 1 μg clean DNA in a total volume

of 100 µl containing 1× NEBuffer 1, 1 mM spermidine, and 10 U *Hpa*II (10 U/µl, NEB). Incubate O/N at 37 °C (*see* **Note 1**).

3. Prepare the adaptors as 100 µM solutions, aliquot, and store at –20 °C. The primers R24, N24, and J24 are used for PCR reactions that are performed later in the protocol.

3.3 Day 2

1. Add 2 U *Hpa*II to the reaction and incubate for 1 h more. Purify the digestion product with CI and precipitate with 1/10 Vol 3 M NaOAc and 2.5 Vol ice-cold absolute EtOH in the presence of 20 µg glycogen (*see* **Note 2**). Wash the pellet twice with 80% ice-cold EtOH and resuspend in 20 µl 1× TE. Measure the DNA concentration (Qubit) and confirm the DNA was digested by electrophorezing an aliquot on 1.2% agarose gel (Fig. 2).
2. *Ligation of Adaptor R.* This step creates primer-binding sites by ligating universal adaptors. Ligate 100 pmol of the first adaptor pair R (R24-R11) to 100 ng of the digested DNA in the presence of 1× ligase buffer (30 µl reaction volume). Heat for 5 min at 55 °C and cool down to 12 °C (1 °C/min, takes approx. 50 min). This allows the 12-mers and 24-mers to anneal to themselves and to the CG overhangs of the digested DNA. Add 80 units T4 DNA ligase (400 U/µl, NEB), gently mix, and incubate O/N at 12 °C (*see* **Note 3**).

3.4 Day 3

1. *Digestion with* McrBC (*see* **Note 4**). Inactivate the ligase according to the manufacturer's instructions. Adjust the reaction mixture to 2× NEBuffer 2, 1× BSA, and 1× GTP and add 6 U McrBC (10 U/µl, NEB) (*see* **Note 5**). Incubate for 7 h at 37 °C. Heat-inactivate the enzyme according to the manufacturer's instructions and adjust the DNA concentration to 1 ng/µl before the amplification step.
2. *PCR Amplification.* Prepare a PCR reaction mix with a final volume of 400 µl containing 1× PCR reaction buffer, 7% DMSO, 200 µM dNTPs, 1 µM R24, 20 U Taq DNA Pol (5 U/µl), and 40 ng of the DNA. Incubate for 5 min at 72 °C (to fill in the ends) and perform 28 cycles of PCR amplification (1 min at 94 °C, 1 min at 70 °C, 2 min at 72 °C), followed by a final extension of 10 min at 72 °C (*see* **Note 6**). Measure the DNA concentration (Qubit) and electrophorese an aliquot of the amplicons on a 1.2% agarose gel (Fig. 3, lanes 4–9).

3.5 Day 4

1. *Purification of TESTER and DRIVER.* Purify the amplicons using spin columns (MN), elute in 1:1 elution buffer/sterile ddH_2O and quantify the DNA yield. The concentration of the DRIVER DNA should be around 600 ng/µl—if necessary, use a Speed Vac to concentrate.

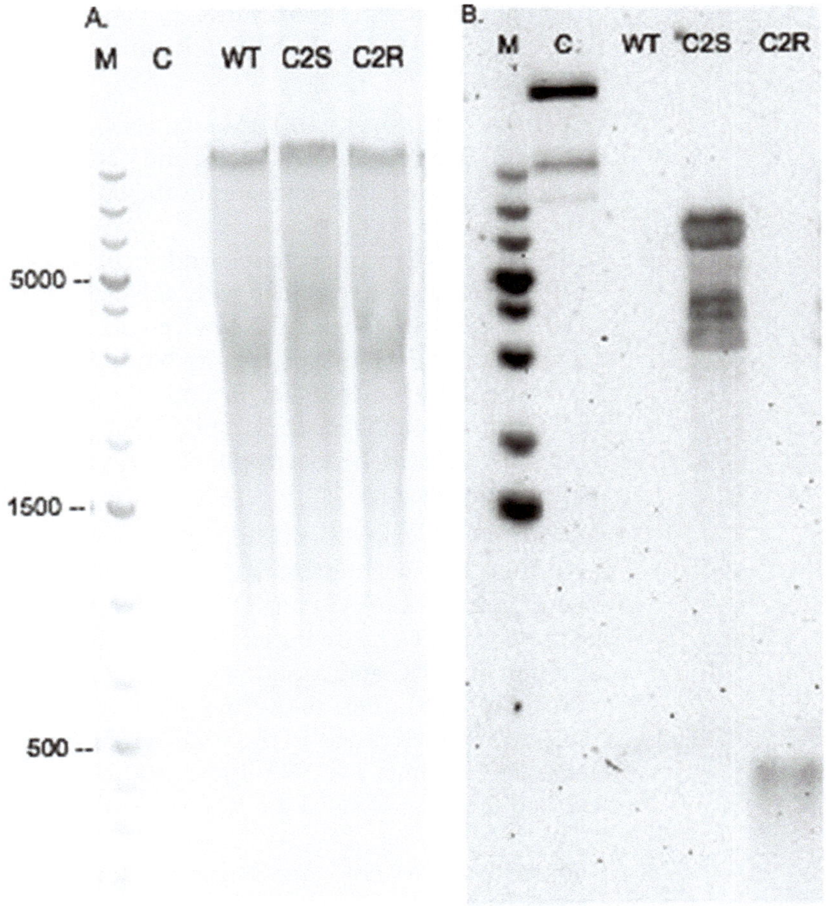

Fig. 2 EtBr stained 0.8 % agarose gel electrophoresis (**a**) and Southern blot analysis (**b**) of plant samples (WT, C2S, and C2R) digested with *Hpa*II, blotted and hybridized with a probe spanning a 1 kb region of the C-insert coding for hygromycin resistance. M, DNA marker; C, 20 pg pMDC107 harboring gene for hygromycin resistance; WT, 1 μg of *A. thaliana* accession Zürich WT; C2S, 1 μg of *A. thaliana* accession Zürich C2S, containing a methylated C-insert; C2R, 1 μg of *A. thaliana* accession Zürich C2R, containing an un-methylated C-insert

2. *TESTER ONLY: Remove Adaptor R*. This step sets the conditions to kinetically enrich unique TESTER DNA fragments. Remove the first adaptor pair by digesting 1 μg of the TESTER DNA in a 50 μl reaction volume containing 1× NEBuffer 2 and 20 U *Msp*I (20 U/μl, NEB). Purify the digested TESTER DNA using spin columns (MN), elute in 1:1 elution buffer/sterile ddH_2O (*see* **Note 7**), and quantify the DNA yield.
3. *TESTER ONLY: Ligate Adaptor N*. Ligate the second adaptor pair N24-N11 as described above (Day 2: Subheading 3.3, **step 2**).

3.6 Day 5

1. *Purify TESTER*. Purify the ligated TESTER DNA using spin columns (MN), elute in 1:1 elution buffer/sterile ddH_2O, and quantify the DNA yield. The concentration of the TESTER

DNA should be around 30 ng/μl—if necessary, adjust the DNA concentration.

2. *First Round Hybridization.* In a 1.5-ml microcentrifuge tube pipet 1 μl of the TESTER DNA, 5 μl of the DRIVER DNA, and 2 μl 12× EEN buffer. Spin briefly to collect the DNA at the bottom of the tube and cover gently with 20 μl of mineral oil. Denature the DNA by heating for 10 min at 95 °C, decrease the temperature to 68 °C (the slower the better), and leave for at least 16 h. This will allow the TESTER and DRIVER DNA to reanneal, forming three types of duplex DNA: TESTER:TESTER, TESTER:DRIVER, and DRIVER:DRIVER (*see* **Note 8**).

3.7 Day 6

1. *PCR 1.* The sequence of events in this step should be fast but accurate. Prepare PCR mix (100 μl) consisting of 1× PCR reaction buffer, 7% DMSO, 200 μM dNTPs, and 5 U Taq DNA pol (5 U/μl, Sigma-Aldrich); aliquot, label, and keep on ice until needed. Heat the HYB dilution buffer to 68 °C and add 92 μl to the hybridized DNA. Heat the PCR mix to 68 °C, remove the mineral oil from the tube containing the diluted hybridized DNA, and add 10 μl DNA (equivalent to 3 ng TESTER DNA) to the PCR mix (*see* **Note 9**). Incubate 5 min at 72 °C. This allows the Taq DNA polymerase to fill in the 3′ ends of the TESTER:TESTER and TESTER:DRIVER duplexes, using the ligated adaptor N24 as a template. Add the N24 primer (0.5 μM final) to the PCR reaction in the first 94 °C step and perform 10 cycles of the selective amplification as follows: 1 min at 94 °C, 1 min at 70 °C, and 2 min at 72 °C.
2. *Mung Bean Nuclease (MBN) Treatment* (*see* **Note 10**). Take the amplicon and adjust to 1× MBN buffer and 10 U MBN (10 U/μl, NEB). Incubate for 30 min at 30 °C. Purify the amplicon using spin columns (MN) and elute in 50 μl 1:1 elution buffer/sterile ddH$_2$O (*see* **Note 11**).
3. *PCR 2 and First Round Difference Product* (Fig. 3, lanes 10–15). Prepare PCR mix 2 (100 μl) containing 1× PCR reaction buffer, 7% DMSO, 200 μM dNTPs, 1 μM N24, 5 μl DNA, and 10 U Taq DNA pol (5 U/μl, Sigma-Aldrich). Perform 25 cycles of amplification as follows: 1 min at 94 °C, 1 min at 70 °C, and 2 min at 72 °C, with a final extension of 5 min at 72 °C. Run the first round difference product on a 1.5% agarose gel and estimate the DNA concentration; it should be enough to perform the next hybridization round (Days 7–9) (*see* **Note 12**).

3.8 Day 7

1. *Purify First Round Product.* Purify the first round difference product using spin columns (MN) and elute in 50 μl 1:1 elution buffer/sterile ddH$_2$O. Estimate the DNA concentration.
2. *Msp*I *Digest.* Digest 1 μg of the first round difference product as described in Day 4 (Subheading 3.5, **step 2**).

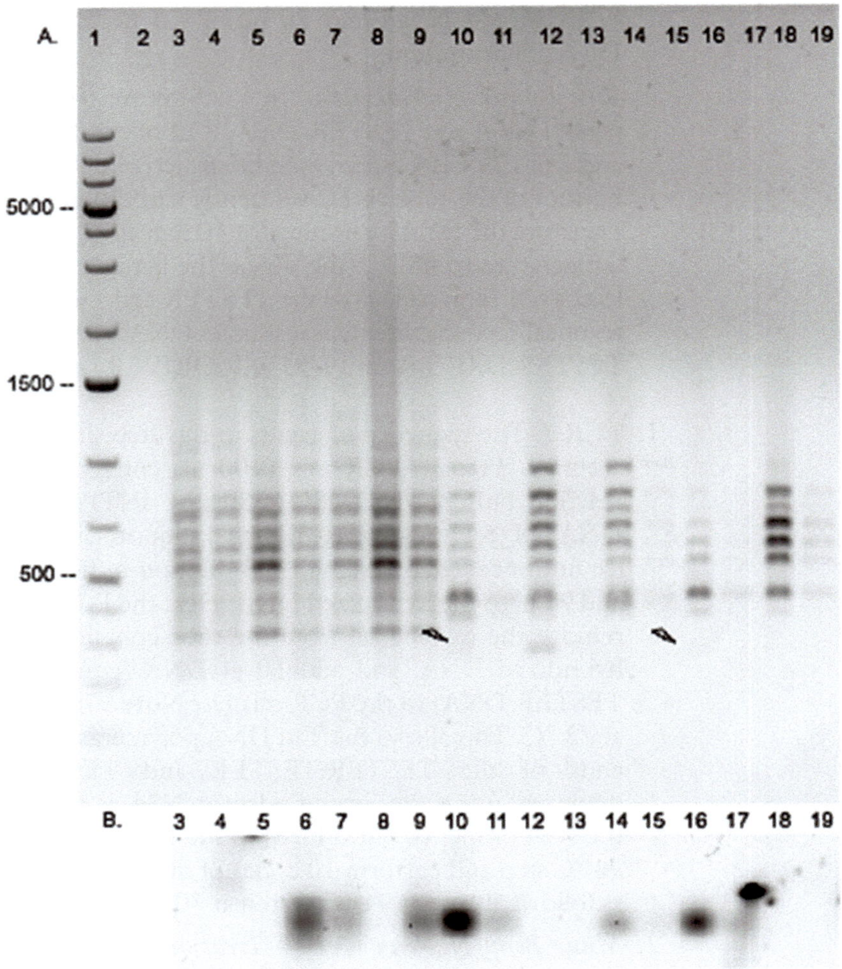

Fig. 3 EtBr stained 0.8 % agarose gel electrophoresis (**a**) and Southern blot analysis (**b**) of the amplicons (3–9) and the hybridized difference products (10–19), blotted and hybridized with a probe spanning a 1 kb region of the C-insert coding for hygromycin resistance. *1*, DNA marker; *2*, 20 pg pMDC107 harboring the gene for hygromycin resistance; *3*, 500 ng of *A. thaliana* accession Zürich WT amplicon; *4*, 500 ng of C2S amplicon without McrBC treatment; *5* and *8*, 500 ng of C2S amplicon with McrBC treatment; *6*, 500 ng of C2R amplicon without McrBC treatment; *7* and *9*, 500 ng of C2R amplicon with McrBC treatment; *10* and *11*, first round difference product (500 ng and 50 ng); *12* and *13*, first round difference product diluted 20 times instead of mung bean nuclease (MBN) treatment, Day 6 (*2.*) (500 ng and 50 ng); *14* and *15*, first round difference product amplified without addition of DMSO (500 ng and 50 ng); *16* and *17*, second round difference product (500 ng and 50 ng); *18* and *19*, second round difference product diluted 20 times instead of MBN treatment, Day 9 (*2.*) (500 ng and 50 ng)

3. *Ligate Adaptor J.* Ligate the third adaptor pair J24-J11 to the first round difference product only, as described in Day 2 (Subheading 3.3, **step 2**).

3.9 Day 8

1. *Step 1. Purify the Ligate.* Inactivate the ligase according to the manufacturer's instructions. Purify the ligate using spin columns (MN) and elute in 50 μl 1:1 elution buffer/sterile ddH$_2$O. Estimate the DNA concentration; it should be around 15 ng/μl.

2. *Second Round Hybridization.* Prepare the second competitive hybridization as described in Day 5 (Subheading 3.6, **step 2**) but increase the ratio of DRIVER:TESTER to approx. 160:1.

3.10 Day 9

1. *PCR 1.* Follow description from Day 6 (Subheading 3.7, **step 1**). Add the J24 primer (0.5 μM final) to the PCR reaction in the first 94 °C step and perform 10 cycles of the selective amplification as follows: 1 min at 94 °C, 1 min at 69 °C, and 2 min at 72 °C.
2. *MBN Treatment.* Follow the description from Day 6 (Subheading 3.7, **step 2**). (Fig. 3, lanes 16–19).
3. *PCR 2 and Second Round Difference Product.* Follow the description from Day 6 (Subheading 3.7, **step 3**); however, use a different annealing temperature of 69 °C in the PCR program. The amplicon is the second round difference product. In our example, the difference product was not further enriched, but some bands in the high molecular weight range disappeared (Fig. 3, lanes 16–19).

4 Notes

1. Overnight digest with an excess amount of enzyme is a standard approach for genomic DNA digestion. *Hpa*II is a restriction endonuclease type IIE: it creates a loop between 2 restriction sites and uses one as an effector and cuts the other [8]. The enzyme has slow cleavage kinetics, which is increased by the addition of the polyamine spermidine [9]. *Hpa*II is a four base pair cutter (C^CGG), which is blocked if the internal cytosine is methylated (C^^{me}CGG)m such that it can be used to assess CG methylation.
2. Glycogen allows the precipitation of DNA fragments as small as 8 bp. It does not interfere with downstream applications if it is at concentrations of less than 8 μg/μl, and the pellet at this concentration is denser, preventing DNA loss. The volume of the precipitation reaction should not exceed 1 ml, because the glycogen concentration will be ineffective.
3. The ligation is performed at 12 °C for more stringent conditions (recommended 12–16 °C), but still favorable for the ligase activity.
4. This step results in the elimination of nonspecific cleavage of *Hpa*II or uneven methylation in the gene pool. The digestion will not affect the non-methylated part of the genome, which received adaptors on the both ends and is the target of this method. It is expected that the long DNA stretches that derive from the methylated part of the genome will be too long to be amplified in the next step, although random fragments may be amplified in a linear fashion.

5. Heating has double function—to inactivate the ligase and to deplete the ATP from the reaction mixture, since the ATP competes with the GTP for the McrBC binding site and inhibits the enzyme.
6. R11 is used to recognize the CG overhang and to serve as annealing matrix for the R24 primer. The ligase, however, cannot catalyze covalent bonding between the 3′-OH end of the DNA fragment and the un-phosphorylated 5′-end of R11. Upon heating (the first 72 °C step), R11 dissociates from the DNA and leaves the R24 sequence as the only primer-binding site.
7. *Msp*I is isoschizomer of *Hpa*II, a restriction enzyme type IIP, and is active as a monomer. It recognizes C^CGG and leaves the same CG overhang as *Hpa*II [10]. The DNA is directly purified using spin columns, because the enzyme cannot be heat-inactivated.
8. The ratio of DRIVER to TESTER is 100:1 for the first hybridization and 160:1 for the second hybridization. The DRIVER is added in excess amount to counteract self-reannealing of the TESTER-derived fragments that are common to both amplicons. While the TESTER:DRIVER duplexes will undergo only a linear amplification, the duplexes formed from DRIVER-derived fragments will not be amplified at all. The true difference products are the TESTER:TESTER duplexes, which will be exponentially amplified.
9. Do not let the temperature drop below 68 °C because nonspecific hybridization may take place.
10. The MBN will digest the single-stranded primer-binding sites protruding from the 5′ end of the TESTER:DRIVER duplexes. This step decreases the amount of PCR products that are not true difference products. The first round difference product cannot be enriched sufficiently if DMSO is omitted, or the product is diluted 20 times between the two PCR events (Fig. 3, lanes 10–15).
11. Do not heat-inactivate the enzyme. This may lead to undesirable DNA degradation occurring before the enzyme is inactivated.
12. During a second hybridization round, the background should be decreased and the difference products should be visible as faint bands (Fig. 3, lanes 16–19).

Acknowledgments

We thank Ortrun Mittelsten Scheid (Gregor Mendel Institute, Vienna) for providing the *Arabidopsis* lines that served as the basis to develop this protocol, and the members of the Grossniklaus lab for helpfup discussions. This work was supported by the University of Zürich and Sciex-NMS[ch] Fellowship 12.222 to P. Sasheva from the Department of Pharmacognosy, Medical University of Sofia, Sofia, Bulgaria.

References

1. Lisitsyn NA (1995) Representational difference analysis: finding the differences between genomes. Trends Genet 11:303–307
2. Wallrapp C, Gress TM (2001) Isolation of differentially expressed genes by representational difference analysis. Methods Mol Biol 175:279–294
3. Kaneda A, Takai D, Kaminishi M, Okochi E, Ushijima T (2003) Methylation-sensitive representational difference analysis and its application to cancer research. Ann N Y Acad Sci 983:131–141
4. Zuo T, Tycko B, Liu TM, Lin JJ, Huang TH (2009) Methods in DNA methylation profiling. Epigenomics 1:331–345
5. Reyna-Lopez GE, Simpson J, Ruiz-Hererra J (1997) Differences in DNA methylation patterns are detectable during the dimorphic transition of fungi by amplification of restriction polymorphism. Mol Gen Genet 253: 703–710
6. Lisitsyn N, Lisitsyn N, Wigler M (1993) Cloning the differences between two complex genomes. Science 259:946–951
7. Mittelsten Scheid O, Afsar K, Paszkowski J (2003) Formation of stable epialleles and their paramutation-like interaction in tetraploid *Arabidopsis thaliana*. Nat Genet 34:450–454
8. Gemmen GJ, Millin R, Smith DE (2006) DNA looping by two-site restriction endonucleases: heterogeneous probability distributions for loop size and unbinding force. Nucleic Acids Res 34:2864–2877
9. Oller AR, Broek WV, Conrad M, Topal MD (1991) Ability of DNA and spermidine to affect the activity of restriction endonucleases from several bacterial species. Biochemistry 30:2543–2549
10. Xu QS, Roberts RJ, Guo H-C (2005) Two crystal forms of the restriction enzyme *Msp*I-DNA complex show the same novel structure. Protein Sci 14:2590–2600

Chapter 11

Analysis of Small RNA Populations Using Hybridization to DNA Tiling Arrays

Martine Boccara, Alexis Sarazin, Bernard Billoud, Agnes Bulski, Louise Chapell, David Baulcombe, and Vincent Colot

Abstract

Epigenetic response to stress in plants involves changes in DNA methylation, histone modifications, and expression of small noncoding RNAs (sRNA). Here we present the method of analysis of differential expression of sRNA populations using DNA tiling arrays. sRNA extracted from *Arabidopsis thaliana* plants exposed to pathogen elicitor or control plants were reverse-transcribed into cDNAs, and subsequently hybridized after labeling to a custom-made DNA tiling array covering Arabidopsis chromosome 4. We first designed a control experiment with eight cDNA clones corresponding to sequences located on chromosome 4 and obtained robust and specific hybridization signals. Furthermore, hybridization signals along chromosome 4 were in good agreement with sRNA abundance as previously determined by massive parallel sequence signature (MPSS) in the case of untreated plants, but differed substantially after stress treatment. These results demonstrate the utility of hybridization to DNA tiling arrays to detect major changes in sRNA abundance.

Key words Small RNA, cDNA libraries, Cy-dye indirect labeling, Hypersensitive response, Microarray, DNA tiling array, Harpin protein

1 Introduction

Plants contain two predominant classes of small RNAs [1, 2]. Most sequenced sRNAs are 24 nucleotide (nt) short interfering RNAs (siRNAs) mapping to transposable elements and other repeated sequences. These siRNAs are thought to direct methylation of DNA and establishment of repressive histone marks over repeated sequences of the genome. The other most abundant classes of small RNAs correspond to microRNAs (miRNA). MiRNAs are 21 nt long; they are involved in several developmental processes, and in some cases accumulate in response to various biotic and abiotic stresses [3–7]. We are interested in the study of sRNA populations during the plant hypersensitive response (HR), a form of programmed cell death that occurs at the site of infection, when

Igor Kovalchuck (ed.), *Plant Epigenetics: Methods and Protocols*, Methods in Molecular Biology, vol. 1456,
DOI 10.1007/978-1-4899-7708-3_11,

plants are challenged by pathogens [8]. We used the harpin protein from *Erwinia amylovora*, an elicitor of HR in several plant species [9], to infiltrate *Arabidopsis thaliana* leaves. Small RNAs were extracted from these leaves as well as control leaves infiltrated with buffer to produce two cDNA libraries [10–12]. The procedure of cloning sRNAs implies that these sRNAs are 5′P and not degradation products from conventional ribonucleases which release 5′OH. The sRNAs were ligated sequentially to 5′ and 3′ RNA/DNA chimeric oligonucleotide adapters with T4 RNA ligase and reverse transcribed. Here, we describe a method to label such cDNA libraries and to hybridize them to a custom-made DNA tiling array covering Arabidopsis chromosome 4 [13, 14]. PCR amplification and purification are first required to obtain cDNAs. The labeling reaction can be divided into two steps: the first step involves the incorporation of amino-allyl-modified deoxynucleotide (AA-dUTP) into PCR-amplified cDNAs of sRNAs; the second step is the chemical coupling of amine-reactive Cy-Dye. Although this procedure is longer than direct labeling and is a more labor-intensive protocol, Cy3 or Cy5 is incorporated more evenly, and more Cy-Dye is incorporated into DNA. Results are presented which demonstrate the validity of our method to characterize small RNA populations and to identify major differences in sRNA abundance between populations.

2 Materials

2.1 RNA Extraction and sRNA Isolation (See Note 1*)*

1. Trizol (Invitrogen) or Tri-reagent (Sigma-Aldrich) corresponding to 4 M guanidinium isothiocyanate and acidic phenol (pH 4.3). *CAUTION: Phenol is toxic and corrosive.*
2. Chloroform. *CAUTION: Chloroform is toxic and a suspected carcinogen.*
3. Isopropanol.
4. 75 % Ethanol.
5. 15 % Denaturing polyacrylamide/urea gel mix: 21 g Urea (7 M), 2.5 mL 10× TBE (0.5×), 18.75 mL 40 % 19:1 acrylamide:bis-acrylamide (15 %), make up to 50 mL with MQH_2O (*see* **Note 2**). *CAUTION: Acrylamide monomer is a neurotoxin and a potential carcinogen. Wear gloves while handling acrylamide and clean any spillage thoroughly.*
6. 10 % Ammonium persulfate (freshly prepared).
7. TEMED.
8. Ethidium bromide (10 mg/mL stock), diluted 10,000× in 1× TBE buffer for gel staining. *CAUTION: Ethidium bromide is an intercalating agent, a mutagen, and thought to be carcinogenic. Handle with care, wearing nitrile gloves.*

9. Formamide mix: 0.05 % Bromophenol blue, 0.05 % xylene cyanol in formamide.
10. 10× TBE: 890 mM Tris, 890 mM boric acid, 20 mM EDTA.
11. Oligonucleotide markers: 20 and 30 nucleotides (nt) in size.
12. 0.3 M NaCl.
13. Phenol-chloroform (buffer not added to keep acid the pH) (Sigma).
14. Absolute ethanol.

2.2 Adapter Ligation and Reverse Transcription

1. Chimeric DNA/RNA oligonucleotide adapters: 5′ adapter: ACGGAATTCCTCACTaaa and 3′ adapter: uuuCTATCCATGGACTGTidT (idT:inverted deoxythymidine) (lower case are RNA). The 3′ adapter is 5′ phosphorylated.
2. 50 % Dimethyl sulfoxide (DMSO).
3. 10× PAN ligation buffer: 0.5 M Tris–HCl pH 7.6, 0.1 M $MgCl_2$, 0.1 M ß-mercaptoethanol, 2 mM ATP, 1 mg/mL acetylated BSA.
4. Acetylated BSA (Sigma).
5. T4 RNA ligase (Roche).
6. Primers for reverse transcriptase and first PCR: Forward primer: 5′ CAG CCA ACG GAA TTC CTC ACT AAA 3′; reverse primer: 5′ CGA ACA TGT ACA GTC CAT GGA TAG 3′.
7. 100 mM dNTP Set (Promega).
8. RT mix: 20 μL 0.1 M DTT, 40 μL 5× first-strand buffer (both supplied with SuperScript II Reverse Transcriptase), 56 μL 2 mM dNTPs.
9. SuperScript II Reverse Transcriptase (200 U/μL) (Invitrogen).
10. Alkali mix: 150 mM KOH, 20 mM Tris base.

2.3 Amplification, Labeling, and Hybridization of cDNAs to Microarray

1. Taq polymerase (New England Biolabs).
2. 100 mM dNTP Set (Promega).
3. 20 nt low ladder 40 μg (Sigma).
4. 15 % Native polyacrylamide gel: 2.5 mL 10× TBE (0.5×), 18.75 mL, 40% 19:1 acrylamide:bis-acrylamide, 28.4 mL MQH_2O.
5. TE buffer pH 7.5.
6. Primers for labeling: sRNArev TGTACAGTCCATGGATA and sRNA for ACGGAATTCCTCACTAA.
7. (3-Aminoallyl)-2′deoxyuridine-5′-triphosphate (AA-dUTP) (Sigma): For a final concentration of 20 mM add 95.5 μL of TE pH 7.5 to a stock vial containing 1 mg of aa-dUTP. Gently vortex to mix and store at –20 °C.
8. Labeling mix (25×): dNTP (minus dTTP) with aa-dUTP: 2 μL dATP (final concentration, 10 mM), 2 μL dCTP (final

concentration, 10 mM), 2 μL dGTP (final concentration, 10 mM), and 10 μL aa-dUTP (final concentration, 10 mM), make up 20 μL with RNase-free H_2O, store at −20 °C.

9. QIAquick Nucleotide Removal Kit (Qiagen).
10. Sodium bi-carbonate buffer (Na_2HCO_3): 0.05 M, pH 9.0.
11. Cy-dye esters (Amersham-GE) (*see* **Note 3**).
12. 0.3 M Sodium acetate pH 5.2.
13. Acrylamide (2.5 μg/μL).
14. Yeast RNA (10 mg/mL in RNase-free H_2O) (Invitrogen).
15. 20× SSC (Sigma).
16. 10 or 20% SDS solution.
17. Formamide (Sigma).
18. Bovine serum albumin (BSA) 10% (filter solution before using it and store at −20 °C).
19. Pre-hybridization solution: 1× SSC, 0.1% SDS, 1% BSA.
20. 2× Hybridization buffer: 50% Formamide, 10× SSC, 0.2% SDS.
21. 22×60 mm Lifterslips (electron microscopy sciences).
22. Corning® hybridization chambers (Sigma).
23. pGEM®-T Easy Vector (pGEM®-T Easy Vector Systems, Promega Cat#A1360).

3 Methods

3.1 Small RNA Isolation, Adapter Ligation, and Reverse Transcription

3.1.1 RNA Extraction

1. Grind tissue under liquid nitrogen with pestle and mortar. Add 1 mL Trizol (per 50 to 100 μg tissues) and grind into slurry. Pipette into a 2 mL microfuge tube and incubate at room temperature for 3 min.
2. Add 0.2 mL of chloroform and shake vigorously by hand for 15 s. Leave at room temperature for 2–3 min.
3. Centrifuge at 10,000×*g* at 4 °C for 15 min. Transfer the aqueous phase to a 1.5 mL microfuge tube. Add 0.5 mL of isopropanol and incubate for 10 min at room temperature.
4. Centrifuge at 16,000×*g* at 4 °C for 20 min. Remove the supernatant, wash the pellet with 1 mL of 75% ethanol (vortexing), and centrifuge at 16,000×*g* at 4 °C for 5 min.
5. Remove supernatant and air-dry the pellet.
6. Resuspend the pellet in 20 μL of RNase-free H_2O, make up a 1/100 dilution, and quantify at spectrophotometer (*see* **Note 4**).

3.1.2 sRNA Isolation

1. Add to 50 mL (15×17 cm, 1.5 mm thick gel) denaturing 15 % polyacrylamide/urea mix, 350 μL of 10 % ammonium persulfate, and 17.5 μL TEMED; pour immediately; and let the gel set for 1 h.
2. Mix 200 μg total RNA with an equal volume of formamide mix, denature for 30 s at 90 °C, and place on ice.
3. Load the samples (*see* **Note 5**), run the denaturing 15 % polyacrylamide/urea gel in 0.5× TBE at 25 V/cm, and stop when the xylene cyanol has migrated to the middle of the gel.
4. Stain the gel in ethidium bromide (0.5 μg/mL), excise under UV (360 nm) a gel slice encompassing 20–30 nt (using the oligonucleotide markers), and determine the weight of the slice.
5. Cut the gel slice in small fragments, elute into 0.3 M NaCl (2–3 volumes v/w) at 4 °C overnight with agitation, extract once with phenol:chloroform, and precipitate the aqueous phase with 3 volumes of absolute ethanol at 20 °C for at least 2 h.
6. The pellet of sRNAs is collected after centrifugation (16,000 ×*g* 20 min 4 °C) and after drying resuspended in 20 μL RNase-free H_2O.

3.1.3 Adapter Ligation

1. Prepare a reaction mixture for ligation of 5′ adapter by combining the following components: 20 μL of gel-eluted sRNAs, 3 μL of 100 μM 5′ adapter, 15 μL 50 % DMSO, 5 μL 10× PAN ligation buffer for a final volume of 48 μL.
2. Denature for 30 s at 90 °C and place on ice.
3. Add 2 μL of T4 RNA ligase (40 U/μL) and incubate at 37 °C for 1 h.
4. Add an equal volume of formamide mix, denature for 30 s at 90 °C, place on ice, and load on a 15 % polyacrylamide/urea gel.
5. Run the denaturing 15 % polyacrylamide/urea gel in 0.5× TBE at 25 V/cm and stop when the xylene cyanol has migrated to the middle of the gel.
6. Excise a gel slice encompassing 39–43 nt (just above and including xylene cyanol loading dye and above the 30 nt marker).
7. Elute into 0.3 M NaCl at 4 °C overnight with agitation, extract once with phenol:chloroform, and precipitate the aqueous phase with 3 volumes of ethanol at –20 °C for at least 2 h (*see* **Note 6**).
8. The pellet of sRNAs is collected after centrifugation (16,000 ×*g* 20 min 4 °C) and drying, and resuspended in 19 μL RNase-free H_2O.

9. Prepare a reaction mixture for ligation of 3′ adapter by combining the following components: 19 μL sRNAs ligated to 5′adapter, 3.8 μL 100 μM 3′ adapter, 12 μL 50% DMSO, 4 μL 10× PAN ligation buffer. Mix all the reagents, denature for 30 s at 90 °C, and place on ice.
10. Add 1.2 μL of T4 RNA ligase (40 U/μL) and incubate at 37 °C for 1 h.
11. Add an equal volume of formamide mix, denature for 30 s at 90 °C, place on ice, and load on a 15% polyacrylamide/urea gel.
12. Run the denaturing 15% polyacrylamide/urea gel in 0.5× TBE at 25 V/cm and stop when the xylene cyanol has migrated to the middle of the gel.
13. Stain the gel in ethidium bromide, and excise under UV (360 nm) a gel slice encompassing 58 to 62 nt (just above—but *not* including—xylene cyanol loading dye).
14. Elute into 0.3 M NaCl at 4 °C overnight with agitation, extract once with phenol:chloroform, and precipitate the aqueous phase with 3 volumes of ethanol and 2 μL of 100 μM reverse primer at −20 °C for at least 2 h (*see* **Note 7**).
15. The pellet collected after centrifugation (16,000×*g* 20 min 4 °C) and after drying is resuspended in 11.1 μL RNase-free H_2O.

3.1.4 Reverse Transcription

1. Denature sRNAs ligated to 5′ and 3′ adapters 30 s at 90 °C and place on ice.
2. Add 17.4 μL of RT mix and incubate at 42 °C for 3 min.
3. Add 1.5 μL Superscript RT II (200 U/μL) and incubate at 42 °C for 30 min.
4. Hydrolyze the RNAs by adding 80 μL of alkali mix, incubate at 90 °C for 10 min, and place on ice.
5. Neutralize the solution by adding 80 μL of 150 mM HCl and check the pH value with pH paper (should be around 8–9). Store the cDNAs at −20 °C.

3.2 Amplification, Labeling, and Hybridization of cDNAs to a Tiling Microarray

3.2.1 First PCR Amplification of cDNA

1. 10 μL of cDNAs were amplified with 10 μL of 2 mM dNTP, 10 μL of 10× PCR buffer (provided with the Taq polymerase), 1 μL of 100 μM reverse primer, 1 μL of forward primer, and 2 μL of Taq polymerase (5 U/μL) for a final volume of 100 μL.
2. The program of cycling is 45 s at 94 °C, 1 min 25 at 50 °C, and 1 min at 72 °C for 25 cycles.
3. After amplification, the PCR products are run on a native 15% polyacrylamide gel alongside 10 μL of 20 nt ladder at 2 V/cm for 3 h.

4. Stain the gel in ethidium bromide and excise a gel slice encompassing 70–80 nt (*see* **Note 8**).
5. Elute into 0.3 M NaCl at 4 °C overnight and purify by phenol:chloroform extraction.
6. Precipitate the aqueous phase in 3 volumes of ethanol at –20 °C for at least 2 h.
7. The pellet collected after centrifugation (16,000 × *g* 20 min 4 °C) and drying is resuspended in 50 μL TE pH 7.5.

3.2.2 Amplification with AA-dUTP

1. For the second PCR, 1 μL of previously amplified DNAs is used in a reaction containing 1 μL labeling mix with aa-dUTP (25×), 2.5 μL 10× PCR buffer, 0.75 μL 100 μM sRNArev, 0.75 μL 100 μM sRNA for primers, and 0.2 μL Taq polymerase (5 U/μL), in a final volume of 25 μL.
2. The program of cycling after initial denaturation at 94 °C for 3 min is 30 s at 94 °C, 30 s at 55 °C, and 30 s at 72 °C for 30 cycles.
3. The PCR products are purified with QIAquick Nucleotide Removal Kit to remove unincorporated nucleotides and primers, according to the supplier's instructions. The samples can be kept at –20 °C.

3.2.3 Coupling with cy5-dye (See Note 9*)*

1. The amplified DNAs are dried in a SpeedVac® and are resuspended in 10 μL of sodium bicarbonate buffer 0.05 M (pH 9) at room temperature for 30 min (*see* **Note 10**).
2. Cy5-ester is provided as a dried product in five tubes (Cy5 Mono-Reactive Dye Pack (Amersham-GE). Resuspend a tube of dye ester in 8 μL of DMSO, distribute 1.5 μL in microfuge tube, and dry in a SpeedVac®. Tubes are stored at 4 °C in the dark.
3. 10 μL is transferred to a tube containing the dried dye and, after pipetting and brief centrifugation, is incubated at room temperature for 30 min in the dark.
4. The excess dye is eliminated by purification with QIAquick Nucleotide Removal Kit, according to the supplier's instructions. The recovered volume (after two times elution) is 60 μL in TE.
5. For each sample, measure absorbance at 260 nm and 650 nm (corresponding to the maximum absorbance of Cy5-dye).
6. For each sample: calculate the total μg of DNA using μg of DNA = [OD260 × 50 ng/μL × volume (μL)/1000] (1 OD260 = 50 ng/μL for DNA). Calculate the total picomoles of dye incorporation using pmol Cy5 = OD650 × volume (μL)/0.25. Calculate the frequency of incorporation = pmol Cy-dye

incorporated *324.5/ng DNA (324.5 average molar mass of dNTP) (*see* **Note 11**).

7. 30 pmol of labeled DNA in 100 µL TE pH = 7.5 is precipitated with 10 µL 0.3 M sodium acetate pH 5.2, 4 µL acrylamide (2.5 µg/µL) (*see* **Note 2**), 2 µL of yeast RNA (10 mg/mL), and 3 volumes of ethanol. Keep for 2 h at −20 °C, then centrifuge, and resuspend the pellet in 35 µL RNase-free H_2O.

3.2.4 *Prehybridization and Hybridization to Arabidopsis thaliana Chromosome 4 Tilling Array (See* Note 12*)*

1. Prepare 50 mL of prehybridization solution.
2. Prehybridize the array on slide at 42 °C for a minimum of 45 min.
3. Rinse in MilliQ water for 2 min and 1 min in isopropanol and centrifuge for 1 min at 800 × *g* to dry the array. Keep the slide out of light and use within 2 h.
4. Position slide in Corning® hybridization chamber with 22 × 60 mm Lifterslip covering array area.
5. 30 pmol of labeled DNAs in 35 µL of RNase-free H_2O are heated at 95 °C for 1 min, immediately mixed with 35 µL of 2× hybridization buffer preheated to 42 °C, and applied to slide.
6. Hybridize overnight at 42 °C (water bath) the Corning® hybridization chamber.

3.2.5 *Washing the Slides (See* Note 13*)*

1. First wash: 2× SSC, 0.1 % SDS at 42 °C, Lifterslip is removed during this step by a gentle hand agitation.
2. Second wash: 4 min in fresh preheated buffer (2× SSC, 0.1 % SDS) with agitation.
3. Third wash: 1× SSC at room temperature, for 4 min with agitation.
4. Fourth wash: 0.2× SSC at room temperature, for 4 min with agitation.
5. Fifth wash: 0.05× SSC at room temperature, for 4 min with agitation.
6. Spin 2 min at 800 g to dry the array (*see* **Note 14**).
7. Scan with the same PMT for red (635 nm) and green (532 nm) (around 600–650 V).

3.2.6 *Data Treatment*

1. Amplification, labeling, and hybridization were done in triplicate on the same cDNA preparation.
2. Hybridized probes were ranked according to the intensity of the hybridization signal (1 = highest signal) and the mean ranking was plotted as a function of the standard deviation computed from the three experiments.

Name	Sequence	Size(nt)	Annotation	Tm(;C)	ΔG(Kcal/mol)	Expected hybridized tiles (mean ranking, standard deviation)
1A08	TCGGACCAGGCTTCATTCCCC	21	miR 166	58	-29.9	**ta05f10 (3,3)**
1F11	CTTGGGTGGCTCAGGTGGCCTGAA	24	TE	63	-35.1	**ta57e01(10,5)**, ta65b11(4077,6983), **ta65b12 (33,19)**
2F01	GGGCCGGCCCTGGGCTAAAGTTG	23	TE	64	-35.4	**ta06d08 (4,2)**
TC05	TAGGTAGGACGTGTCGGCTG	20	Rib 26S	56	-27.9	**ta01a11(5,5)**
TD06	GATCCATTGGAGGGCAAG	18	Rib 18S	50	-22.5	ta01a07(9213,15888)
TF09	GCATGCCTGTCGGGACCCGA	20	Rib 26S	60	-30.7	**ta01a11(5,5)**
TH10	ATATACGGATTTGCATTGCATATA	24	Rib 5S	49	-26.9	**tc27a02(24,9), tc27a03 (42,25), tc27a04 (34,16)**
TG12	CATGGGTCAACTCAACTCAACTC	23	TE	55	-28.7	**tc18h11(5,3),tc23b04(3,2),tc23b03(19,8),ta60e08 (22,7) , ta60e07(8,4), ta36h11(30,13),ta74g01(20,7)**, ta98a06 (654,1017),ta98a07(420,700) ,**ta59f08(12,10)**

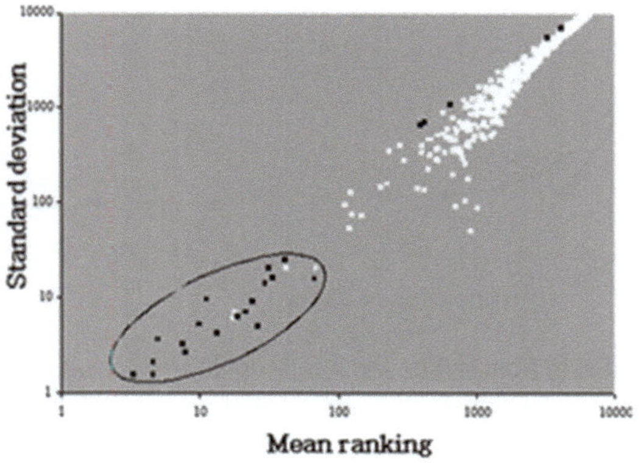

Fig. 1 Pilot hybridization to the chromosome 4 tilling array. The sRNAs used for this experiment are indicated, together with the ~1 kb DNA tiles they should hybridize to. Mean values of ranking and standard deviation are indicated in ***parentheses. Bold characters***: Tiles with highest hybridization rank and lowest standard deviation. ***Black squares***: Tiles expected to hybridize; ***white squares***: tiles not expected to hybridize according to an approximate matching approach using the eight cDNA sequences fused to the 5′ and 3′ adaptors as queries. Matches were considered whenever they covered 23 nucleotides or more, with less than 2 mismatches in any window of 12 consecutive nucleotides (reproduced from ref. 20) with permission from Elsevier Science)

3.3 Hybridization of Genomic Tiling Array: Validation Experiments

3.3.1 Hybridization with Known Sequences

The PCR products from the first amplification were ligated to pGEM®-T Easy Vector. Plasmids were prepared from clones and used for sequencing. From this sequencing we extracted eight cDNA clones corresponding to sequences located on chromosome 4. They were labeled and hybridized to the microarray in three independent experiments. Tiles containing the exact sequence of cDNAs were expected to rank highest and with the lowest standard deviation. Indeed, we observed a clear-cut separation between two populations of tiles, with those expected to hybridize exhibiting the highest mean ranking and lowest standard deviation (Fig. 1) (*see* **Note 15**).

3.3.2 Hybridization with sRNA Populations from Stressed and Unstressed Leaves

In a second step, the chromosome 4 tiling array was hybridized with labeled cDNAs derived from sRNAs that were extracted from buffer or harpin-infiltrated leaves. The experiment was repeated three times, and the same statistical procedure was applied as before to select tiles giving robust hybridization signals. After elimination of the overlapping tiles and tiles not located on chromosome 4, a set of 155 tiles was selected in this manner for buffer, while 164 tiles were obtained from the harpin-treated sample. The hybrid-

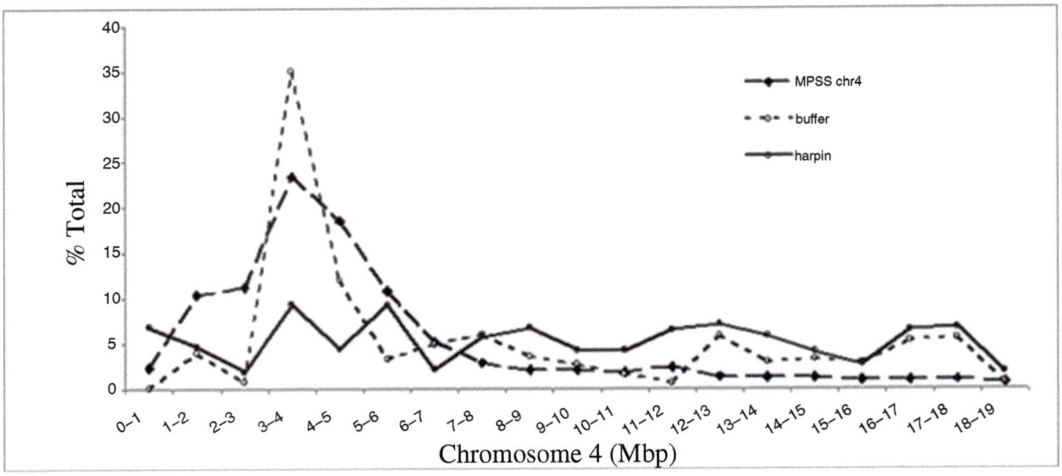

Fig. 2 Distribution of tiles hybridized to sRNAs from buffer and harpin-infiltrated leaves, and comparison with MPSS data. Values for hybridized tiles (***closed or opened round symbols***) and MPSS expression levels (***diamond symbols***) were computed in non-overlapping windows of 1 Mb along the Arabidopsis chromosome 4 sequence. Values are normalized to the total nucleotides number in each set, i.e., 100 % = 155 tiles from harpin-treated sample (***continuous line***), 100 % = 164 tiles from buffer-treated sample (***stripped line***), and 100 % = sum of MPSS expression levels (***dashed line***)

ized tiles from buffer-treated leaves were located mainly within pericentromeric regions in the 3–5 Mb interval (Fig. 2). Significantly, the number of hybridized tiles in each region is in good agreement with the accumulation level of sRNAs in the same region as determined by MPSS (massive parallel sequencing) [15, 16] (Fig. 2). In contrast, the distribution of hybridized tiles from the harpin-treated sample was uniform along the chromosome 4 (Fig. 2), suggesting major changes in the accumulation of small RNAs during stress.

3.3.3 Conclusions

Hybridization of labeled cDNAs derived from sRNAs to a DNA tiling microarray can lead to robust and meaningful hybridization signals. This method can be considered as cheap (provided that a tiling array is available) and can be useful to rapidly evaluate major differences between small RNAs accumulated in different conditions. The use of genomic oligonucleotide tiling arrays [17] should be very valuable to improve these analyses.

High-throughput sequencing is now the most used method to analyze small RNA molecules; this method which gives very deep sequencing has revealed bias in library construction. Two papers review this problem pointing the role of secondary structure and suggest new adaptors with internal randomized regions favoring the formation of structures favorable for ligation with RNA ligase [18, 19].

4 Notes

1. All solutions should be RNase free. RNA can be stored at –20 °C or below to minimize hydrolysis.
2. Acrylamide/bis-acrylamide 40% stock solution (19:1 ratio) (Sigma). Store at 4 °C.
3. Wrap all reaction tubes with foil and keep covered as much as possible in order to prevent photobleaching of the dyes. Any introduced water to the dye esters will result in a lower coupling efficiency due to the hydrolysis of the dye esters.
4. To resuspend RNA, we use RNase-free water; otherwise we used MilliQ water (resistivity> 5 MΩ.cm at 25 °C, with organic content < 30 ppb). The RNA pellet should not be over-dried; otherwise it would be difficult to resuspend. Heating at 55–60 °C may assist in resuspending the pellet.
5. When loading the gel, leave an empty place between samples to avoid cross contamination.
6. To precipitate the sRNAs after ligation steps, it is advisable to add glycogen (Invitrogen) at a final concentration of 1 μg/μL to make the pellet visible.
7. We precipitate in the presence of the reverse primer as a carrier to help recovery of the sRNAs ligated to adapters.
8. After amplification a double band is observed: a 50 nt corresponding to self-ligated adapters and a 70–80 nt to be collected. Although better resolution is obtained with polyacrylamide gel, the 3% agarose gel can be alternatively used.
9. For the labeling of sRNas, we used the cy-5 dye which appears red in scanning and allows the identification with no ambiguity of hybridized cDNAs over the background of probe DNAs (green).
10. Bicarbonate buffer changes composition over time; make a 1 M solution aliquot and store frozen.
11. 50–100 pmol of dye incorporation per sample and the frequency of incorporation in the range of 15–30 are optimal for hybridizations.
12. Our experiments were performed with a custom-made *Arabidopsis thaliana* chromosome 4 DNA tiling microarray. PCR amplification using selected primers at 1 kb interval was performed on bacterial artificial chromosome template (BAC) covering the chromosome IV of *Arabidopsis thaliana*. The generated fragments were printed on Ultragaps Coated Slides (Corning). All PCR products were checked on agarose gels, before printing onto glass slides.

13. Do not let slides dry. Transfer them as quickly as possible between each wash solution and between wash and centrifuge.
14. It is very important to do no more than two slides at a time and to proceed very quickly as any droplet that dries on the surface of the slide will leave spots.
15. Some tiles with an exact match to the labeled targets were absent from that group, due to one of the three experiments failing to provide significant hybridization over the tile in question. Conversely, some tiles from that group did not show any clear match with the set of labeled cDNAs used as targets, denoting some limitations in the computational prediction of hybridization patterns using small sequences as targets.

Acknowledgments

MB was supported by a Visiting Scientist Fellowship from INRA. VC and DB are members of the European Union Network of Excellence "The Epigenome."

References

1. Baulcombe D (2004) RNA silencing in plants. Nature 431:356–363
2. Brodersen P, Voinnet O (2006) The diversity of RNA silencing pathways in plants. Trends Genet 22:268–280
3. Baev V, Milev I, Naydenov M, Vachev T, Apostolova E, Mehterov N, Gozmanva M, Minkov G, Sablok G, Yahubyan G (2014) Insight into small RNA abundance and expression in high- and low-temperature stress response using deep sequencing in Arabidopsis. Plant Physiol Biochem 84:105–114
4. Sunkar R, Zhu JK (2004) Novel and stress-regulated MicroRNAs and other small RNAs from Arabidopsis. Plant Cell 16:2001–2019
5. Navarro L, Dunoyer P, Jay F, Arnold B, Dharmasiri N, Estelle M, Voinnet O, Jones JD (2006) A plant miRNA contributes to antibacterial resistance by repressing auxin signaling. Science 312:436–439
6. Sunkar R, Chinnusamy V, Zhu J, Zhu JK (2007) Small RNAs as big players in plant abiotic stress responses and nutrient deprivation. Trends Plant Sci 12:301–309
7. Bilichak A, Ilnytskyy Y, Wóycicki R, Kepeshchuk N, Fogen D, Kovalchuk I (2015) The elucidation of stress memory inheritance in *Brassica rapa* plants. Front Plant Sci 6:5
8. Greenberg JT, Yao N (2004) The role and regulation of programmed cell death in plant-pathogen interactions. Cell Microbiol 6:201–211
9. Wei ZM, Laby RJ, Zumoff CH, Bauer DW, He SY, Collmer A, Beer SV (1992) Harpin, elicitor of the hypersensitive response produced by the plant pathogen *Erwinia amylovora*. Science 257:85–88
10. Llave C, Kasschau KD, Rector MA, Carrington JC (2002) Endogenous and silencing associated small RNAs in plants. Plant Cell 14:1605–1619
11. Pfeffer S, Lagos-Quintana M, Tuschl T (2005) Cloning of small RNA molecules. Curr Protoc Mol Biol 26(Unit 26.4):26410–26418
12. Hafner M, Landgraf P, Ludwig J, Rice A, Ojo T, Lin C, Holoch D, Lim C, Tuschl T (2008) Identification of microRNAs and other small regulatory RNAs using cDNA library sequencing. Methods 44:3–12
13. Martienssen RA, Doerge RW, Colot V (2005) Epigenomic mapping in Arabidopsis using tiling microarrays. Chromosome Res 13:299–308
14. Vaughn MW, Tanurd IM, Lippman Z, Jiang H, Carrasquillo R, Rabinowicz PD, Dedhia N, McCombie WR, Agier N, Bulski A, Colot V,

Doerge RW, Martienssen RA (2007) Epigenetic Natural Variation in Arabidopsis thaliana. PLoS Biol 5, e174

15. Lu C, Tej SS, Luo S, Haudenschild CD, Meyers BC, Green PJ (2006) Elucidation of the small RNA component of the transcriptome. Science 309:1567–1569
16. Nakano M, Nobuta K, Vemaraju K, Tej SS, Skogen JW, Meyers BC (2006) Plant MPSS databases: signature-based transcriptional resources for analyses of mRNA and small RNA. Nucleic Acids Res 34:D731–D735
17. Thibaud-Nissen F, Wu H, Richmond T, Redman JC, Johnson C, Green R, Arias J, Town CD (2006) Development of Arabidopsis whole-genome microarrays and their application to the discovery of binding sites for the TGA2 transcription factor in salicylic acid-treated plants. Plant J 47:152–162
18. Sorefan K, Pais H, Hall AE, Kozomara A, Griffiths-Jones S, Moulton V, Dalmay T (2012) Reducing ligation bias of small RNAs in libraries for next generation sequencing. Silence 3:4
19. Fuchs RT, Sun Z, Zhuang F, Robb GB (2015) Bias in Ligation-Based Small RNA Sequencing Library Construction Is Determined by Adaptor and RNA Structure. PLoS One 10(5), e0126049
20. Boccara M, Sarazin A, Billoud B, Jolly V, Martienssen R, Baulcombe D, Colot V (2007) New approaches for the analysis of *Arabidopsis thaliana* small RNAs. Biochimie 89:1252–1256

Chapter 12

Northern Blotting Techniques for Small RNAs

Todd Blevins

Abstract

Cells have evolved intricate RNA-directed mechanisms that destroy viruses, silence transposons, and regulate gene expression. These nucleic acid surveillance and gene silencing mechanisms rely upon the selective base-pairing of ~19–25 nt small RNAs to complementary RNA targets. This chapter describes northern blot hybridization techniques for the detection of such small RNAs. Blots spiked with synthetic standards are used to illustrate the detection specificity and sensitivity of DNA oligonucleotide probes. Known endogenous small RNAs are then analyzed in samples prepared from several model plants, including *Arabidopsis thaliana*, *Nicotiana benthamiana*, *Oryza sativa*, *Zea mays*, and *Physcomitrella patens*, as well as from the animals *Drosophila melanogaster* and *Mus musculus*. Finally, the value of northern blotting for dissecting small RNA biogenesis is shown using an example of virus infection in *A. thaliana*.

Key words RNA silencing, Northern blot, RNA hybridization, Small RNA, siRNA, miRNA

1 Introduction

Detecting specific sequences of nucleic acids extracted from biological samples is an essential task in molecular biology. One standard approach is the electrophoretic separation of nucleic acids by length, their blotting to a membrane substrate, and a specific detection with radioactively labeled probes [1]. Distinct methods are used for the analysis of DNA and RNA by blot hybridization. DNA blot hybridization is called Southern blotting, an homage to its inventor, Edwin Mellor Southern [2, 3]. Analogous RNA blotting methods were developed and are now commonly referred to as "northern" blotting [4, 5].

Northern blot hybridization is primarily used to detect specific messenger RNAs (mRNAs) or other high-molecular-weight transcripts in RNA extracted from cells [1, 6]. However, the last two decades of research have uncovered novel populations of small regulatory RNAs in various eukaryotes [7–17]. Small RNAs afford sequence specificity to RNA silencing, a process by which Argonaute-family proteins repress gene expression. RNA silencing typically involves the following mechanism: stem-loop hairpin RNAs or long double-stranded RNAs (dsRNAs) are processed into

Igor Kovalchuck (ed.), *Plant Epigenetics: Methods and Protocols*, Methods in Molecular Biology, vol. 1456,
DOI 10.1007/978-1-4899-7708-3_12, © Springer Science+Business Media New York 2017

~19–25 nt small RNA duplexes by a Dicer RNase III endoribonuclease enzyme. A single small RNA strand then guides Argonaute-containing effector complexes to cleave specific RNA transcripts, block productive mRNA translation or direct repressive chromatin modifications to particular genomic loci [18–20].

The core mechanism of RNA silencing (a.k.a. RNA interference) and several functionally distinct small RNA pathways were elucidated, in part, by northern blot experiments with plants [7, 9, 21–29]. One class of small RNAs are microRNAs (miRNAs) which are excised from stem-loop hairpin structures in RNA transcripts; miRNAs can regulate the expression of mRNAs containing complementary miRNA-binding sites. Like the *let-7* miRNA archetype discovered in *C. elegans* [30–34], many miRNAs are highly conserved across either the animal or plant kingdoms, and they regulate development in these multicellular organisms [35–37]. However, less conserved, plant family-specific miRNAs have also been identified [38, 39]. Another class of small RNAs are small interfering RNAs (siRNAs), which are processed from perfect dsRNA substrates and can have multiple biological functions. These include antiviral defense, the regulation of developmental timing, and maintenance of silent chromatin [40]. The latter siRNAs guide histone modifications and DNA methylation to specific chromosomal regions that are themselves subject to heritable, potentially reversible states of gene expression, also known as "epigenetic memory" [41–44].

Although next-generation sequencing (NGS) can facilitate small RNA analyses (e.g., with an increased sensitivity, precision, and quantitative readout), northern blotting remains a key technique for dissecting small RNA biogenesis and function. NGS requires time-consuming and expensive investments in library construction, Illumina HiSeq runs, and bioinformatics support [45, 46]. By contrast, an individual scientist can complete a highly informative small RNA blot experiment in less than 5 days, at minimal cost and using conventional lab equipment. Studies in the field of RNA silencing typically combine NGS and northern blot approaches, privileging blot hybridization for assays that span numerous test conditions, immunoprecipitated protein fractions, or diverse genetic backgrounds [14, 47–50]+REF: Blevins T, Podicheti R, Mishra V, Marasco M, Tang H, Pikaard CS (2015). Identification of Pol IV and RDR2-dependent precursors of 24 nt siRNAs guiding de novo DNA methylation in Arabidopsis. Elife. 2015 4:e09591. doi: 10.7554/eLife.09591.

The blot hybridization techniques presented here are optimized for the detection of small RNAs in either plant or animal tissue and their analysis to a resolution of one nucleotide. A small RNA northern blot experiment consists of several independent techniques: (1) isolation of total RNA from fresh or frozen biological material, (2) size fractionation of the total RNA into high-molecular-weight (HMW) and low-molecular-weight (LMW) fractions, (3) denaturing polyacrylamide gel electrophoresis of LMW RNA (or total RNA) followed by its transfer to a nylon membrane, and (4) hybridization of membrane-bound RNA with a radioactively labeled DNA oligonucleotide probe to detect specific small RNAs. This last procedure can be repeated at least 10–15 times by stripping the membrane prior to each new hybridization. Most experiments require at least two rounds of hybridization: i.e., one round to detect a small RNA whose signal is expected to vary amongst samples, and a second to detect a loading control that should not vary significantly. The workflow diagram in Fig. 1 summarizes these steps.

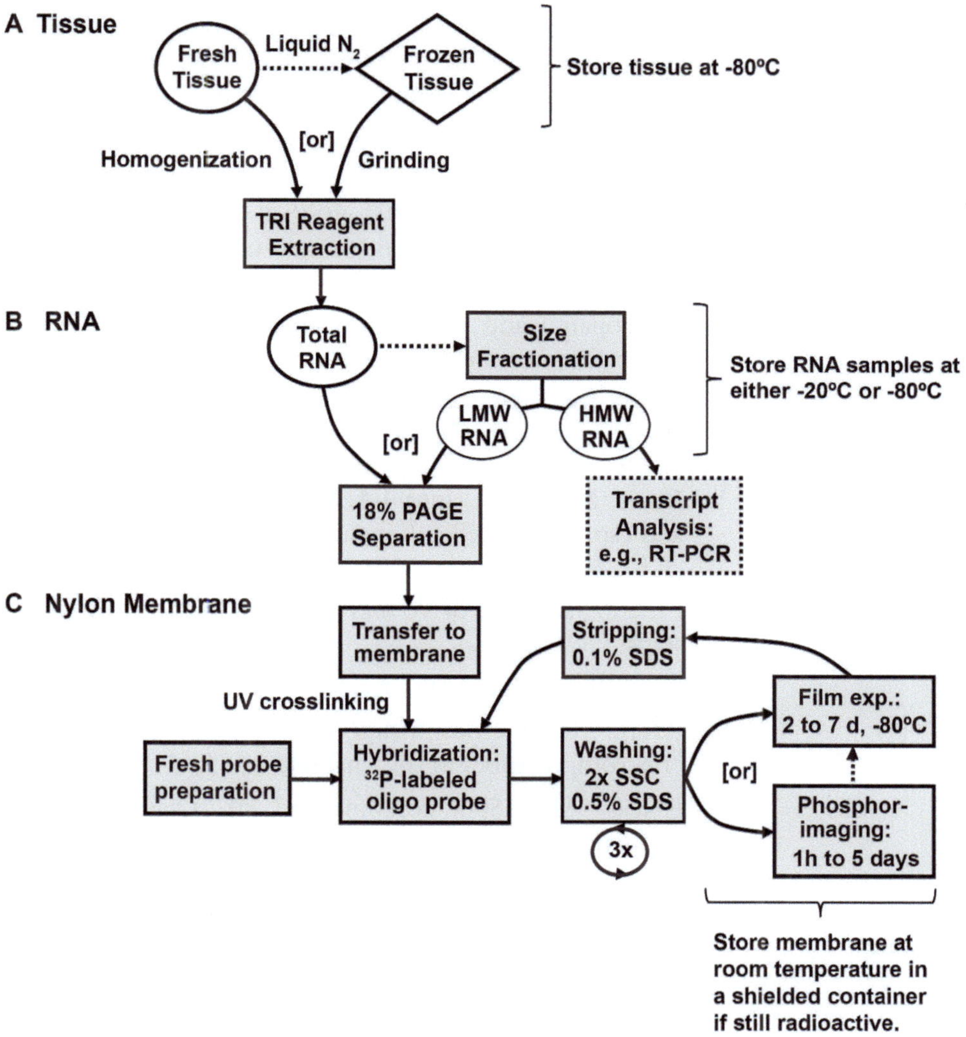

Fig. 1 A workflow diagram summarizing the small RNA northern blot procedure. Blot hybridization analysis of low-molecular-weight (LMW) RNA is broken into three major steps: (**a**) extraction of total RNA from fresh or frozen tissue; (**b**) size fractionation of total RNA to obtain the enriched LMW RNA (optional), followed by size separation of RNA using polyacrylamide gel electrophoresis (PAGE); and (**c**) transfer of separated RNA to a nylon membrane, UV cross-linking of the RNA to the membrane, hybridization with a radiolabeled probe, washing with high-salt/SDS buffer, and the detection of a radioactive signal using a phosphorimaging screen or photographic film. Subsequent hybridizations require stripping of the membrane, fresh probe preparation, washing, and detection. This last cycle of steps can be repeated at least 10–15 times without a substantial loss in signal. Unprocessed tissue samples should be stored at −80 °C, whereas RNA can be stored at either −20 °C or −80 °C; RNA cross-linked to a nylon membrane is stable at RT

2 Materials

2.1 RNA Isolation and Size Fractionation

1. A ceramic mortar (6–8 cm in diameter), fitted pestle, and small metal spatula, all precooled with liquid nitrogen; 100–200 mL liquid nitrogen per tissue sample.
2. Round-bottomed polypropylene tubes (13 mL) with tight-sealing screw caps (Sarstedt, Nümbrecht, Germany).
3. Standard centrifuge rotor types JA-20 (8-tube; Beckman Coulter, Fullerton, CA, USA) or SA-600 (12-tube; Sorvall, Asheville, NC, USA), with type 8441 rubber adapter inserts, ¾ inch diameter (Corning Inc., Corning, NY, USA).
4. TRI Reagent® (Molecular Research Center, Cincinnati, OH, USA) or TRIzol® (Invitrogen, Carlsbad, USA). *CAUTION: Phenol is toxic and corrosive.* Store at 4 °C.
5. Serological pipettes, 10 mL single-use, two per tissue sample (Sarstedt).
6. Isopropanol and chloroform kept cold on ice. *CAUTION: Chloroform is toxic and a suspected carcinogen.*
7. RNase-free bidistilled water, e.g., diethylpyrocarbonate (DEPC) treated. *CAUTION: DEPC is a suspected carcinogen* (*see* **Note 1**). Briefly, pipette 1 mL DEPC (Sigma-Aldrich, St. Louis, MO, USA) into 1 L bidistilled water in a glass bottle and cap. Shake well and incubate for 4 h at room temperature (RT). Autoclave to remove DEPC and store at RT.
8. Ethanol, absolute or 96 % as available, for use in RNA cleanup steps.
9. 75 % Ethanol (in DEPC-treated water) kept cold on ice.
10. RNase-free microcentrifuge tubes, 1.5 mL, referred to as "microfuge" tubes in this chapter.
11. RNeasy Mini Spin columns and kit (Qiagen, Venlo, The Netherlands).

2.2 Polyacrylamide Gel Preparation

1. Glass gel plates, 14 cm tall × 16 cm wide (or similar), with 1 mm thick spacers.
2. Gel comb (1 mm thick) with ~15 mm deep × 6 mm wide teeth (13–15 teeth/comb). Gel thickness and slot dimensions are important parameters that affect the resolution of the RNA during gel electrophoresis.
3. Four large binder clips, 2 in. size (sometimes called banker's or dog clips).
4. TBE buffer stock solution (10×): 0.9 M Tris-borate, 20 mM EDTA, pH 8.0 [1]. Store at RT.
5. Agarose (standard electrophoresis grade), dissolved in 1× TBE to 1 % concentration (w/v) using a microwave oven, and cooled to 65 °C before use.

6. Urea, ReagentPlus, ≥99.5 % (Sigma-Aldrich).
7. Acrylamide, 30 % solution (w/v), AccuGel 19:1 Acrylamide:Bis-Acrylamide, ultrapure sequencing grade (National Diagnostics, Atlanta, GA, USA). *CAUTION: Acrylamide monomer is a neurotoxin and a potential carcinogen.* Store at 4 °C.
8. Millipore Express Plus (0.22 μm) presterilized filter, 250 mL capacity container (Millipore, Billerica, MA, USA).
9. *N*, *N*, *N′*, *N′*-Tetramethylethylenediamine (TEMED) for electrophoresis, ~99 % (Sigma-Aldrich). Store at 4 °C.
10. Ammonium persulfate, APS (USB Corp., Cleveland, OH, USA), prepare as 10 % (w/v) in bidistilled water. Store 300 μL aliquots at −20 °C; these can be used for at least 3 months.
11. Vertical gel apparatus (gel rig) with upper and lower buffer reservoirs, for 14 × 16 cm polyacrylamide gels or close equivalent; power supply to generate 14–16 W or 450–550 V (protein minigel rigs are not recommended if a small RNA resolution of 1 nt is required).
12. Luer-lok syringes (2), 30 mL each, and 21G 1-½″ needle (BD, Franklin Lakes, NJ, USA).

2.3 Small RNA Sample Preparation, Electrophoresis, and Electroblotting

1. A spectrophotometer for the quantification of small volumes of RNA: e.g., NanoDrop™ ND-1000 (NanoDrop Technologies, Wilmington, DE, USA).
2. A Speed-Vac apparatus for drying down 20–40 μL volumes of aqueous RNA solution (Thermo Scientific, Waltham, USA).
3. RNA size markers (20–25 nt range): e.g., the "microRNA Marker" 17, 21, and 25 nt RNA oligo mixture (New England Biolabs, Ipswich, MA, USA), or custom-synthesized oligos (such as 21 and 24 nt RNA oligos shown in Fig. 2A).
4. RNA gel-loading buffer: 95 % (v/v) formamide, 0.025 % (w/v) bromophenol blue, 0.025 % (w/v) xylene cyanol FF, 5 mM EDTA, 0.025 % (w/v) SDS, pH 8.5 [1]. Store in 500 μL aliquots at −20 °C.
5. A thermoblock or water bath set to 95 °C.
6. Microcapillary pipette tips, 1–200 μL (United Laboratory Plastics, St. Louis, MO, USA).
7. Ethidium bromide (10 mg/mL stock), diluted 10,000× in 1× TBE buffer for gel staining. *CAUTION: Ethidium bromide is an intercalating agent, a mutagen, and thought to be carcinogenic. Handle with care, wearing nitrile gloves.*
8. An ultraviolet gel documentation system.
9. Whatman #1 filter paper (Whatman/GE Healthcare, Chalfont St. Giles, UK), or 3 MW paper (Midsci, St. Louis, MO, USA).

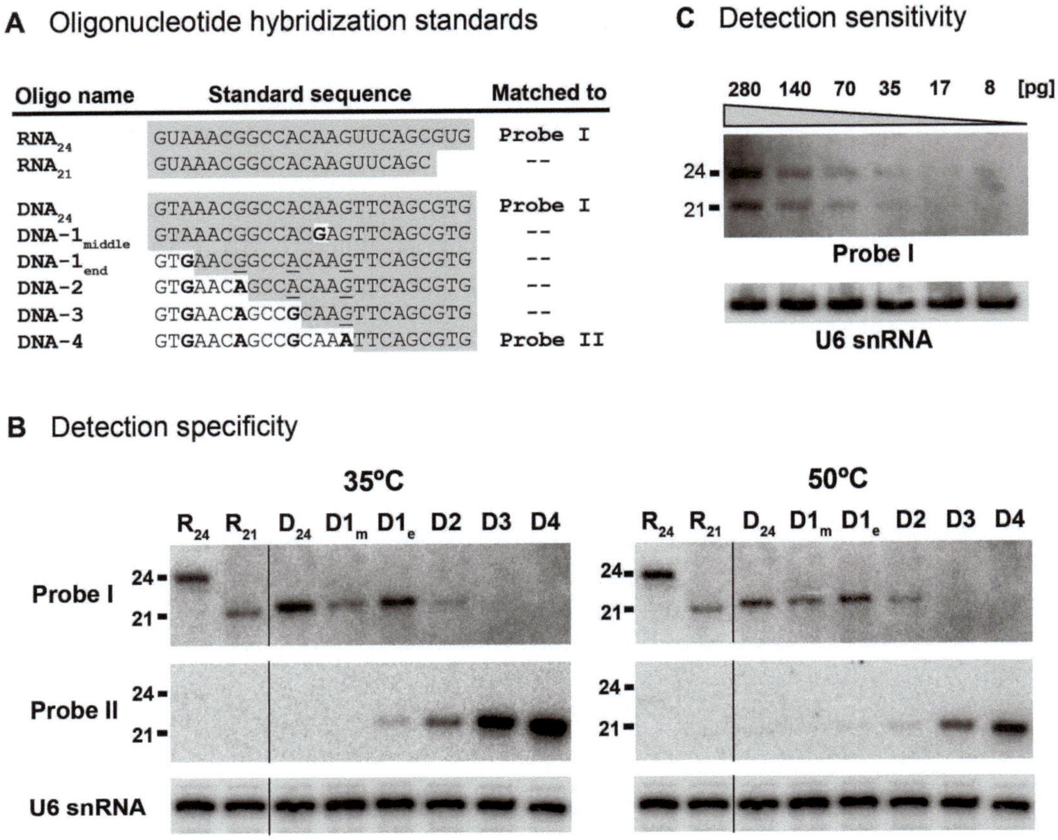

Fig. 2 Analysis of oligonucleotide hybridization specificity and sensitivity. (**a**) A series of DNA oligonucleotide (oligo) standards were designed such that each successive oligo diverged at an additional nucleotide (nt) position from the prototype, DNA_{24}. Two RNA oligos were included (RNA_{21} and RNA_{24}) with sequences derived from DNA_{24}. (**b**) Oligos were spiked into 4 μg aliquots of RNA from wild-type *Arabidopsis*. Samples were separated by PAGE and blotted to a nylon membrane. Hybridization at 35 °C with Probe I (designed for RNA_{24}, RNA_{21}, and DNA_{24}) detected standards that diverged from DNA_{24} at up to two different base positions; similar results were obtained at 50 °C, although an overall signal was reduced. In contrast, Probe II (designed for DNA-4) showed a higher specificity at 50 °C, and minimal to no signal was detected from standards that differed from DNA-4 at two or more positions. The best specificity resulted when a probe possessed internally mismatched bases with respect to undesired hybrids: i.e., Probe I detected variant DNA-1_{end} more strongly than DNA-1_{middle}. The DNA oligo standards (each 24 nt) migrated more rapidly than an RNA oligo of a similar sequence (RNA_{24}), as documented previously [63]. U6 small nuclear RNA (snRNA) signals detected in the *Arabidopsis* carrier RNA serve as a loading control; this result was intentionally duplicated in the left-hand and right-hand halves of panel **b**. (**c**) To estimate the sensitivity of probe hybridization, equal amounts of RNA_{21} and RNA_{24} standards were mixed and then spiked as a dilution series into *Arabidopsis* RNA for PAGE separation. The concentration of the standard decreases by a factor of 2 from left to right (280, 140, 70, 35, 17, and 8 pg). A clear detection required ~35 pg of RNA

10. A Hybond-N+ positively charged nylon membrane (Amersham/ GE Healthcare) or a similar positively charged nylon membrane [51].
11. A polyacrylamide gel-transfer apparatus: e.g., a Trans-Blot semidry transfer cell (Bio-Rad, Hercules, CA, USA).

12. An ultraviolet cross-linking device: e.g., Stratalinker apparatus (Stratagene, La Jolla, CA, USA) or GS Gene Linker apparatus (Bio-Rad).

2.4 Radiolabeled Probe Preparation and Hybridization

1. PerfectHyb Plus buffer (Sigma-Aldrich) or, alternatively, MicroHyb buffer (Invitrogen). Store at RT.
2. A hybridization oven and tubes with a 20–30 mL buffer/wash capacity: the tubes need to accommodate ~12 cm membranes.
3. Oligonucleotide (oligo) probes, 25 nmol scale, resuspended in bidistilled water to 100 μM. Oligo probes used in this chapter are listed in Table 1 and were synthesized by Integrated DNA Technologies (Coralville, IA, USA).
4. T4 Polynucleotide Kinase (PNK) and the supplied 10× Kinase Buffer (Promega). Store at −20 °C.
5. [Gamma-^{32}P] adenosine-5′-triphosphate ([γ-^{32}P]ATP) with a specific activity of 3000–6000 Ci/mmol and the radioactive concentration of 10 mCi/mL (PerkinElmer, Waltham, MA, USA). *CAUTION: Radiation protection measures must be taken for handling γ-32P and all derived materials.* Store in a shielded container in a dedicated freezer at −20 °C.
6. Performa DTR Gel Filtration Cartridges (Edge Biosystems, Gaithersburg, MD, USA) or, alternatively, MicroSpin G-25 Columns (GE Healthcare). Store at 4 °C.

2.5 Washing, Detection, Stripping, and Reprobing

1. SSC stock solution (20×): 3 M Sodium chloride, 0.3 M sodium citrate, adjust pH to 7.0 with concentrated HCl [1]. Store at RT.
2. Sodium dodecyl sulfate (SDS) stock solution, 10% (w/v) in bidistilled water. Store at RT.
3. Washing solution (1×): 2× SSC, 0.5% SDS. Store at RT; incubate at 65 °C for 15–20 min to resuspend SDS if it precipitates during storage.
4. Heavy-duty plastic kitchen wrap.
5. Phosphorimager screen with a cassette and an appropriate scanner (e.g., Personal Molecular Imager, Bio-Rad; or Typhoon Variable Mode Imager, Molecular Dynamics/GE Healthcare).
6. Kodak BioMax MR Film (Kodak, Rochester, NY, USA), an enhancer screen, and an automated film developer (only required if hybridization signals are very weak).
7. SDS stripping solution (1%), boiling or at 85 °C.

2.6 Biological Material Used in Examples

1. *Arabidopsis thaliana* grown from seed and collected 4 weeks post-germination. Wild-type material is of Columbia-0 ecotype, and *dicer-like* mutant (*dcl*) lines originate from the SALK and GABI T-DNA insertion collections [52, 53].

Table 1
Oligonucleotide probes used for northern blot detection in this chapter

Name	Sequence	Size [nt]
RNA24	*GUAAACGGCCACAAGUUCAGCGUG*	*24*
RNA21	*GUAAACGGCCACAAGUUCAGC*	*21*
DNA24	*GTAAACGGCCACAAGTTCAGCGTG*	*24*
Standard Probe I	**CACGCTGAACTTGTGGCCGTTTAC**	**24**
DNA-4	*GTGAACAGCCGCAAATTCAGCGTG*	*24*
Standard Probe II	**CACGCTGAATTTGCGGCTGTTCAC**	**24**
At-miR160a	*UGCCUGGCUCCCUGUAUGCCA*	*21*
Os-miR160a	*UGCCUGGCUCCCUGUAUGCCA*	*21*
Zm-miR160a	*UGCCUGGCUCCCUGUAUGCCA*	*21*
Pp-miR160a	*UGCCUGGCUCCCUGUAUGCCA*	*21*
miR160a_probe	**TGGCATACAGGGAGCCAGGCA**	**21**
At-miR165a	*UCGGACCAGGCUUCAU C CCCC*	*21*
At-miR166a	*UCGGACCAGGCUUCAUUCCCC*	*21*
Os-miR166a	*UCGGACCAGGCUUCAUUCCCC*	*21*
Zm-miR166a	*UCGGACCAGGCUUCAUUCCCC*	*21*
Pp-miR166a	*UCGGACCAGGCUUCAUUCCCC*	*21*
miR165_probe	**GGGGGATGAAGCCTGGTCCGA**	**21**
At-miR393a	*UCCAAAGGGAUCGCAUUGAUCC*	*22*
Bn-miR393	*UCCAAAGGGAUCGCAUUGAUC*	*21*
Os-miR393	*UCCAAAGGGAUCGCAUUGAUC*	*21*
Zm-miR393	*UCCAAAGGGAUCGCAUUGAUCU*	*22*
miR393a_probe	**GGAUCAAUGCGAUCCCUUUGGA**	**22**
At-miR824	*UAGACCAUUUGUGAGAAGGGA*	*21*
Bn-miR824	*UAGACCAUUUGUGAGAAGGGA*	*21*
miR824_probe	**TCCCTTCTCACAAATGGTCTA**	**21**
At-siR1003	*AGACCGUGAGGCCAAACUUGGCAU*	*24*
siR1003_probe	**ATGCCAAGTTTGGCCTCACGGTCT**	**24**
Ce-let-7	*UGAGGUAGUAGGUUGUAUAGUU*	*22*
Dm-let-7	*UGAGGUAGUAGGUUGUAUAGU*	*21*
Mm-let-7a	*UGAGGUAGUAGGUUGUAUAGUU*	*22*
Hs-let-7a	*UGAGGUAGUAGGUUGUAUAGUU*	*22*
let-7_probe	**AACTATACAACCTACTACCTCA**	**22**
Mm-miR-122a	*UGGAGUGUGACAAUGGUGUUUGU*	*23*
miR-122a_probe	**ACAAACACCATTGTCACACTCCA**	**23**

(continued)

Table 1
(continued)

Name	Sequence	Size [nt]
Dm-miR-124	*UAAGGCACGCGGUGAAUGCCAAG*	*23*
miR-124_probe	**CTTGGCATTCACCGCGTGCCTTA**	**23**
At-miR173	*UUCGCUUGCAGAGAGAAAUCAC*	*22*
miR173_probe	**GTGATTTCTCTCTGCAAGCGAA**	**22**
Probes for cabbage leaf curl virus (CaLCuV) small RNAs:		
Viral reg. 1 sense	**AATATGGTTGATCTTCCTTT GGGTGCAACAG**	**Detects 21–24 nt species**
Viral reg. 1 a/sense	**CTGTTGCACCCAAA GGAAGATCAACCATATT**	**Detects 21–24 nt species**
Viral reg. 2 sense	**TGGTGATGTAATTCTT GACGGCATTGGTGTCT**	**Detects 21–24 nt species**
Viral reg. 2 a/sense	**AGACACCAATGCCGTCAA GAATTACATCACCA**	**Detects 21–24 nt species**
U6 snRNAsequences (A.t., D.m. and M.m. U6 snRNA, respectively):		
5′------GUCCCUUCGGGGA CAUCCGA UAAAAUU GGAACGAUACAGAGAA GAUU - AGCAUGGCCCCUG CGCAAGGAUGACACGCA *UAAAUCGAGAAAU GGUCCAAAUUUU-3′*		
5′-GUUCUUGCUUCGGCAGA ACAUAUAC UAAAAUUGG AACGAUACAGAGAAGAUU - AGCAUGGCCCCUGC GCAAGGAUGACACGCA *AAAUCGUGAAGCGUU CCACAUUUU--3′*		
5′-GUGCUCGCUUCGGCAGC ACAUAUAC UAAAAUUG GAACGAUACAGAGAA GAUU *U* AGCAUGGCC CCUGCGCAAGGAUGA CACGCA *AAUUCGUGAA GCGUUCCAUAUUUUU-3′*		
Cons_U6-probeI	**AATCTTCTCTGTATCGTTCC AATTTTA**	**Detects 102–108 nt species**
Cons_U6-probeII	**TGCGTGTCATCCTTGCGCA GGGGCCATGCT**	**Detects 102–108 nt species**

Italic sequences are RNA targets, while bold sequences are probes used to detect them in data summarized by Figs. 2, 3, and 4. U6 small nuclear RNA (snRNA) sequences are displayed for the species *Arabidopsis thaliana* (*A.t.*), *Drosophila melanogaster* (*D.m.*), and *Mus musculus* (*M.m.*). Bold U6 snRNA regions are highly conserved across plants and animals, and were thus used to design conserved probes for the purpose of RNA loading controls

2. Cabbage Leaf Curl Virus, CaLCuV; viral constructs provided by Dominique Robertson and infected *Arabidopsis* plants provided by Mikhail Pooggin and Thomas Hohn [54].
3. *Physcomitrella patens* subspecies patens (WT06) streaked onto plates and cultured for 7 days. This material is primarily protonemal tissue (provided by Pierre-François Perroud).
4. Other green plants grown from seed in a greenhouse and collected 9 days post-germination. The material includes both cotyledons and true leaves (grown by Mike Dyer).
5. Wild-type *Drosophila melanogaster* embryos, 12–17 h (provided by Kathryn Huisinga).
6. Healthy mouse liver tissue (provided by Tatiana Simon and Luciano Marpegan).

3 Methods

3.1 RNA Isolation and Size Fractionation

1. Total RNA is isolated from plant or other biological tissue using TRI Reagent, a phenol-based reagent ideal for large-scale RNA extraction [55, 56]. Column or glass fiber-based purification is not appropriate at this stage in the protocol because they exclude low-molecular-weight (LMW) RNA. About 1 g tissue is ground to a fine powder in liquid nitrogen using a mortar and pestle (*see* **Note 2**). All implements and tubes are precooled in liquid nitrogen before contact with frozen powder. Each ~1.5 mL aliquot of powder is transferred to a 13 mL centrifuge tube. 10 mL TRI Reagent is added to each tube, and these capped tubes are vortexed until the powder melts and is evenly suspended. *CAUTION: Phenol is toxic and corrosive. Wear gloves while handling TRI Reagent and during subsequent steps. Work in a fume hood for pipetting TRI Reagent and further sample manipulations until RNA pellets are obtained.* These mixtures are incubated for 5 min at room temperature (RT). Then 2 mL cold chloroform is added, and the tubes are vortexed for 20 s and incubated for 3 min at RT. *CAUTION: Chloroform is toxic and a suspected carcinogen.*
2. The samples are centrifuged for 15 min (at 8000 × *g* and 4 °C). *WARNING: Excessive centrifugation speeds (>10,000 × g) may rupture polypropylene tubes.* The aqueous phase (~6 mL) is transferred to a new centrifuge tube, carefully avoiding contamination by the protein-rich interphase, and 6 mL cold isopropanol is added. The capped tube is gently inverted to mix and incubated for 15–30 min on ice. RNA is sedimented by centrifugation for 30 min (at 8000 × *g* and 4 °C). After decanting isopropanol into a waste container, the pellets are washed

with 75 % ethanol (prepared from DEPC-treated water). RNA pellets can be stored in 75 % ethanol overnight at −20 °C or for several days if necessary.

3. 75 % Ethanol is discarded, and the tubes are air-dried for 15 min at RT. Then, 60 μL DEPC-treated water preheated to 65 °C is pipetted into each tube and agitated across the entire interior surface. The tubes are centrifuged at low speed to collect the resuspended RNA. This is transferred to 1.5 mL microfuge tubes and kept on ice, while an additional 60 μL of DEPC-treated water is added to the larger tubes, repeating agitation and centrifugation steps. The second aliquots are combined with those already in the microfuge tubes. Nucleic acid concentrations are estimated using absorbance at the 260 nm wavelength in a spectrophotometer (*see* **Note 3**). The RNA samples can be stored at −20 °C for 1 month or at −80 °C for at least 6 months.
4. Total RNA is size-fractionated by means of RNeasy Mini Spin columns (optional—*see* **Note 4**) roughly following the manufacturer's RNA cleanup protocol. This improves the sensitivity and resolution during polyacrylamide gel electrophoresis by removing higher molecular weight RNA, thus concentrating small RNA within the loaded samples (*see* **Note 5**). 80–100 μg total RNA is brought to a volume of 100 μL with DEPC-treated water, and mixed with 350 μL RLT buffer (provided with columns), and then with 250 μL absolute ethanol. The mixtures are pipetted onto RNeasy Spin columns and centrifuged for 30 s (at 10,000 × *g* and RT). Flow-through fractions contain LMW RNA, as do the two subsequent washes with RPE buffer (supplied with columns).
5. Concentrated LMW RNA is recovered by combining the column flow-through and washes (~1.5 mL) and mixing this with an equal volume of isopropanol. This step is easily accomplished by dividing each combined sample into two microfuge tubes and adding 700 μL cold isopropanol to both. The tubes are inverted to mix and incubated for 2 h on ice (or to maximize recovery, overnight at −20 °C). LMW RNA is sedimented by centrifugation for 30 min (at 16,000 × *g* and 4 °C). The isopropanol is discarded, and the pellets are washed twice with 75 % ethanol in DEPC-treated water. The ethanol is discarded, and the pellets are air-dried for 15 min. Residual liquid is evaporated by placing open microfuge tubes for 10 min on a 65 °C block. LMW RNA can be resuspended in 30–60 μL of DEPC-treated water, with the final volume depending on the pellet's size. Store LMW RNA at −20 °C if not used immediately; it can be stored there at least for 1 month, or at −80 °C for at least 6 months.

3.2 Polyacrylamide Gel Preparation

1. Stock solution for 18 % polyacrylamide urea gels is prepared as follows: 42 g urea, 60 mL AccuGel 30 % acrylamide (19:1) solution, and 10 mL of 10× TBE are combined in a 250 mL flask for a final volume of ~100 mL. The solution is stirred until the urea dissolves completely, which is accelerated by placing it over low heat. It is then vacuum-filtered into a Millipore Express container and stored at RT. *CAUTION: Acrylamide monomer is a neurotoxin and a potential carcinogen. Wear gloves while handling acrylamide and clean any spillage thoroughly.*
2. Gel plates are prepared by cleaning the inner surfaces with ethanol and wiping away excess moisture. Plastic spacers are sandwiched between the left and right edges of these plates, and the whole assembly is clamped together by binder clips. With the assembly standing upright in a plastic receptacle (e.g., a pipette tip box cover), 1 % agarose is poured into the receptacle until it seeps 5–6 mm into the assembly from below. Then, 700 μL agarose is pipetted along both spacer edges. This agarose-sealed assembly is allowed to cool for 15 min before proceeding to pour the polyacrylamide gel.
3. For a single gel, 25 mL of acrylamide/urea stock solution is transferred into a small beaker. Then, 25 μL TEMED and 250 μL 10 % APS are added in quick succession (a fresh APS aliquot is used for each gel). A 30 mL syringe (without a needle) is used to mix the solution, drawing it up and back into the beaker and then up again. Without hesitation, the liquid is steadily injected into the assembly from above until 3–4 mm space remains. The comb is delicately inserted, taking care to avoid trapping bubbles around the teeth. The comb can be removed and reinserted two to three times to exclude such bubbles. Numerous attempts should be avoided, however, because they cause distortions as polymerization advances. The gel typically solidifies within 15–30 min.
4. The gel plate assembly is locked into its vertical gel rig using screws or binder clips, depending on the apparatus. Both upper and lower reservoirs are filled with 1× TBE buffer. Once buffer submerges the gel top, the comb is removed by slowly pulling up to avoid distorting or tearing wells. Connect the gel rig to the power supply and set it to ~15 W, which requires around 450–550 V. Look for small bubbles emerging from the electrodes. Pre-run the gel in this manner for 30 min.

3.3 RNA Sample Preparation, Electrophoresis, and Electroblotting

1. Samples to be loaded are thawed for 2 min at 65 °C to fully resuspend RNA and then placed on ice. Nucleic acid concentrations are estimated via the absorbance at the 260 nm wavelength in a spectrophotometer. Equal amounts of RNA must be loaded in the next step: if any sample contains less than 8 μg

total, then the smallest total mass will define the amount to be aliquoted from each sample.

2. Volumes for 8 μg aliquots of LMW RNA are calculated based on sample concentrations and each transferred to new microfuge tubes—alternatively, total RNA can be used (*see* **Note 4**). In addition, 1 μL each of 21 nt and 24 nt RNA oligos (100 μM stocks) are mixed in a microfuge tube to serve as size standards. All samples and markers are completely dried using the Speed-Vac (medium heating) and resuspended in 8 μL of RNA loading buffer. The samples are incubated for 3 min at 95 °C to minimize secondary structure folding, and placed on ice until loading.
3. After the 30-min gel pre-run, the power supply is disconnected, and gel slots are thoroughly rinsed with 1× TBE using a needled syringe: urea that has leached out into the wells during the pre-run can lead to uneven migration and therefore must be completely washed into the buffer reservoir. RNA is loaded using microcapillary pipette tips, slowly layering each sample at the bottom of its well, avoiding the generation of air bubbles (*see* **Note 4**).
4. The power supply is reconnected, and gel electrophoresis is performed for 1–2 h at 450–550 V, so as to maintain 15 W; this keeps the gel hot and enhances size resolution. The bromophenol blue marker will run off into the lower buffer reservoir; electrophoresis may be complete when the xylene cyanol FF marker has migrated past the midpoint of the gel. The optimal end point must be empirically determined for each gel rig, power supply and desired RNA size range. The power supply is disconnected, and then the plate assembly is removed and placed on a square of paper towel.
5. A thin metal spatula is used to wedge the plates apart. Whichever plate the gel adheres to is used as a support. Together, the gel and plates are transferred to a Pyrex dish containing 300 mL ethidium bromide stain (30 μL of 10 mg/mL ethidium bromide in 300 mL 1× TBE). *CAUTION: Ethidium bromide is an irritant, a mutagen, and a suspected carcinogen. Wear nitrile gloves for handling this chemical because latex gloves are too porous.* Staining is conducted on an orbital shaker for 15–20 min. The gel and plates are removed, draining excess stain into the dish, and RNA migration is documented under UV transillumination. Strong 5S rRNA and tRNA bands should be visible in test sample lanes (*see* Fig. 3a). A non-distinct smear would indicate degradation of the RNA sample. The migration of 21 and 24 nt standards in the size marker lane is important for subsequent comparison to blot hybridization results.

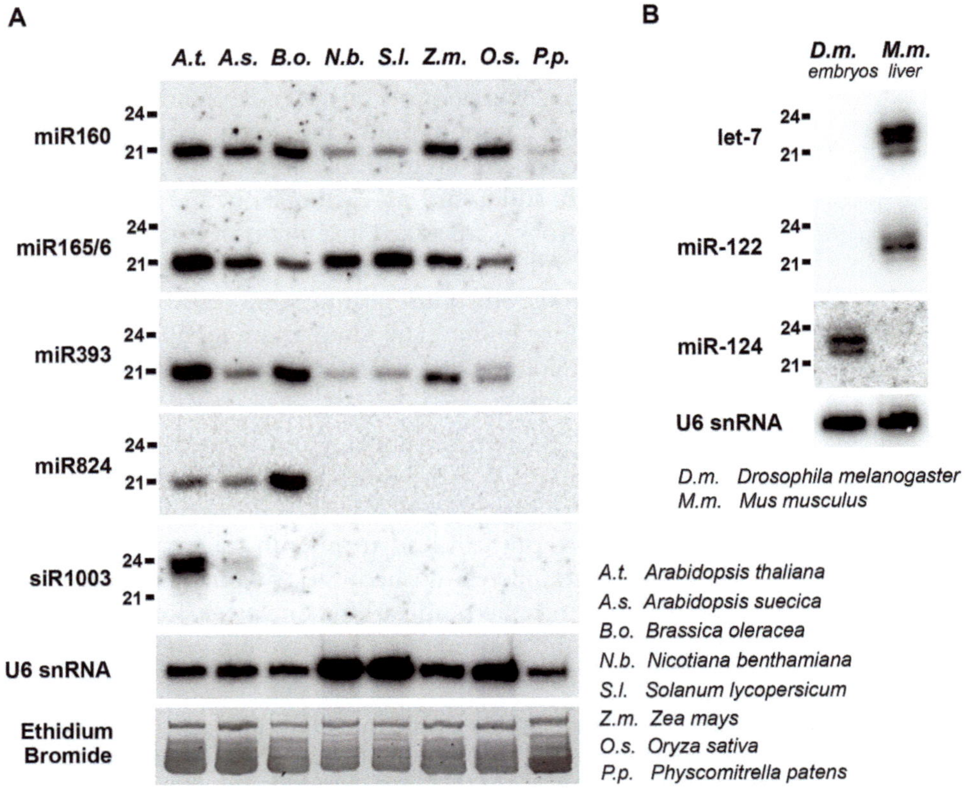

Fig. 3 Blot hybridization analysis of small RNA isolated from different model organisms. (**a**) Low-molecular-weight (LMW) RNA was isolated from a panel of plant species: included were three members of the Brassicaceae family (*Arabidopsis thaliana*, *Arabidopsis suecica*, and *Brassica oleracea*), two Solanaceous plants (*Nicotiana benthamiana* and *Solanum lycopersicum*), two monocots (*Zea mays* and *Oryza sativa*), and the moss *Physcomitrella patens*. miR160 is expressed from an evolutionarily ancient *MIR* gene family conserved from Brassicaceae to moss [36]. In contrast, miR824 is Brassicaceae specific [38] and the heterochromatic siR1003 is only detectable in *Arabidopsis* species. (**b**) LMW RNA was isolated from fruit fly (*Drosophila melanogaster*, 12–17-h embryos) and adult mouse (*Mus musculus*, liver). *let-7* is a prototypical animal miRNA, evolutionarily conserved across much of that kingdom. It is not yet expressed at the embryo stage in *Drosophila*, while being abundant in mouse liver [32]. miR-122 is not conserved in *Drosophila* but is expressed in vertebrate liver. Finally, miR-124 expression peaks at Drosophila embryo stage 12–17 h, but is not encoded in mammalian genomes [64]. U6 snRNA detection and ethidium bromide staining serve as loading controls

6. Remaining polyacrylamide well dividers and the agarose gel bottom are cut away using a razor blade. One square of a nylon membrane and two identically sized pieces of 3 MW paper (~12 × 14 cm) are cut to fit the gel. The membrane's upper edge is labeled in pencil to indicate experiment name and sample order and identify the side to which RNA will be blotted. All blot components are briefly soaked in 1× TBE. Removing the top electrode of the semidry transfer cell (here, the Bio-Rad Trans-Blot system), one 3 MW square is laid on the bottom electrode and doused with 1× TBE. Air bubbles are smoothed out by rolling a serological pipette (broken to fit)

across the 3 MW paper surface. The membrane is carefully laid atop the paper followed by the gel and a second square of 3 MW paper, smoothing out air bubbles between steps. Finally, the top electrode is locked in place and electroblotting is carried out for 3 h at 10 V. *WARNING: Orientation of the membrane-gel stack depends on the specific apparatus. Consult the manufacturer's instructions to avoid RNA loss from transfer in the wrong direction.*

7. While still damp, the membrane is UV cross-linked with an energy of 140 mJ (*see* **Note 6**). Over-cross-linking should be avoided because this leads to a decreased hybridization efficiency [51, 57]. The membrane can now be stored in a plastic sleeve at RT until use (or between hybridizations).

3.4 Probe Preparation and Hybridization

1. The membrane is slid into a hybridization tube with the RNA-side facing to the interior, and 10 mL of PerfectHyb Plus buffer is added. Prehybridization is carried out for 2–6 h at 35 °C.
2. A DNA oligo (a reverse complement of the small RNA to be detected) is resuspended in bidistilled water to a stock concentration of 100 μM. Table 1 lists the sequences of probes used to detect small RNA species in Figs. 2, 3, and 4 (*see* **Note 7**). The oligo end-labeling reaction is assembled in a microfuge tube as follows:
   - 12 μL Bidistilled water
   - 2 μL Kinase buffer (10×, provided with PNK)
   - 0.2 μL Oligo (i.e., 20 pmol)
   - 1 μL Polynucleotide kinase (PNK, 10 U/μL)
   - This partial reaction is mixed by tapping the tube gently.
3. *CAUTION: Radiation protection measures must be taken for probe preparation, hybridization, detection, and stripping steps. Wear gloves and a lab coat and monitor the work area regularly for contamination.* In a properly shielded radioisotope work area, 5 μL of [γ-^{32}P]ATP (3000 or 6000 Ci/mmol; 10 mCi/mL) is added to the partial reaction (*see* **Note 8**), pipetting up and down to mix. The complete mixture is incubated for 30 min at 37 °C.
4. A Performa DTR gel filtration cartridge (or MicroSpin G-25 Column) is placed in a microfuge tube and centrifuged for 2 min at 850 × *g* (~3000 rpm in a microcentrifuge). The cartridge is transferred to a new microfuge tube, and the entire end-labeling reaction mixture is pipetted onto the packed matrix. The cartridge and tube are centrifuged for 2 min at 850 × *g*. Unincorporated ^{32}P is retained with adenosine-5′-triphosphate in the matrix, while both labeled and unlabeled oligos pass into the eluate. A quick verification of

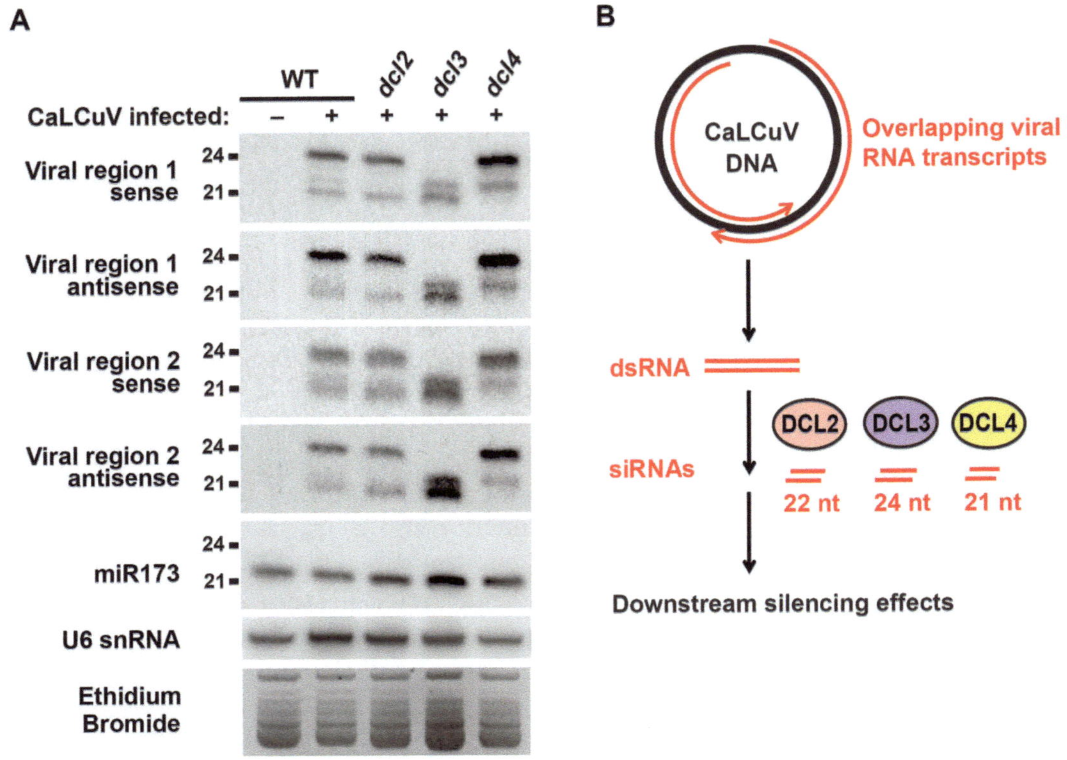

Fig. 4 Function of *Arabidopsis* Dicer-like (DCL) proteins in viral siRNA biogenesis. (**a**) Wild-type (WT), *dcl2*, *dcl3*, and *dcl4* mutant plants were inoculated (+) with the DNA virus, *cabbage leaf curl virus* (CaLCuV). An uninfected pool of WT plants (–) was used as a negative control. RNA from these samples was analyzed by northern blot hybridization using four DNA oligonucleotide probes to detect viral siRNAs from two regions of the viral genome. Three size classes of viral species (21, 22, and 24 nt in length) accumulated in the infected WT sample, as detected in both sense and antisense polarities. By contrast, each individual *dcl* mutant showed a deficiency in the accumulation of a specific size class of viral siRNA: *dcl2* was deficient in 22 nt siRNAs, *dcl3* in 24 nt siRNAs, and *dcl4* in 21 nt siRNAs. Biogenesis of miR173 is known to require DCL1, but not the three DCLs mutated in these lines, and was thus included as a positive control. U6 snRNA detection and ethidium bromide staining serve as loading controls (reproduced from Blevins et al. 2006 with permission from Oxford University Press). (**b**) This data supports a model wherein three *Arabidopsis* DCL proteins process viral dsRNA into a distinct size class of siRNA. Double-stranded RNA substrates for DCL processing appear to be overlapping viral transcripts in this particular system

^{32}P-incorporation can be made using a Geiger-Müller counter: the counts per minute reading with a detector pointed at the cartridge alone should be less than or equal to the reading when pointed at the eluate (with distances held constant). The cartridge is then disposed of in a solid radioactive waste container and the eluate (i.e., the probe) is retained.

5. An eluted end-labeling reaction (20 μL) contains sufficient probe for hybridization with one to four membranes. To facili-

tate its transfer to hybridization tubes, 20 μL of bidistilled water is added to the eluted probe for each additional membrane. Then, 20 μL of diluted probe is added to each hybridization tube; probe droplets should land in the pre-hybridization buffer rather than directly on the membrane. Hybridization is performed at 35 °C for 10–18 h (or at 50 °C for a higher specificity; *see* Fig. 2 and **Note 9**).

3.5 Washing, Detection, Stripping, and Reprobing

1. The membrane is washed in the hybridization tube three times with 2× SSC, 0.5 % SDS for 30 min at 35 °C (or 50 °C for a higher stringency; *see* **Note 9**). Each time, the contents of the hybridization tube are carefully poured off into liquid radioactive waste, and 15–20 mL of wash buffer is added. Then the membrane is removed from the tube using forceps, allowing excess wash buffer to drip back into the tube.
2. The membrane is placed onto a rectangle of plastic wrap just over twice its size. The excess plastic is folded over, and wrinkles are smoothed out. The plastic-sealed membrane is taped into a cassette with an erased phosphorimager screen. The screen is removed after 1–3 h (or up to 5 days later) and scanned. Exposure duration must be optimized for each particular small RNA target and probe. Weak signals may also be detected by exposure to Kodak MR film for 2–7 days at −80 °C (*see* **Note 10**). Figure 3 documents hybridization results for a panel of plant and animal species using different miRNA probes; variation in signal for individual miRNAs reflects their species and tissue-specific expression patterns.
3. Before hybridization with a new probe, the membrane is removed from the plastic wrap, placed in a Pyrex dish on an orbital shaker, and stripped with 0.1 % SDS previously heated to 85 °C. The stripping step is complete once the solution returns to RT or after ~30 min. Residual radioactivity on the membrane should be checked using a Geiger-Müller counter or by film exposure overnight at −80 °C. If a significant signal is detectable in the size range of 20–30 nt, then a second stripping with 0.1 % SDS needs to be performed. *CAUTION: The used stripping solution contains probe and must be disposed of in radioactive waste.*
4. After stripping, the membrane is rinsed for 5 min with 2× SSC at RT to remove excess SDS, and transferred to a hybridization tube, and Subheadings 3.4 and 3.5 are repeated. In addition to probes for endogenous miRNAs or viral small RNAs, probes for the highly conserved U6 small nuclear RNA (snRNA) are generally included. Because these species (102–108 nt) are produced independently of RNA silencing pathways, they serve as an RNA loading control.

4 Notes

1. *RNase-free water*: Dimethylpyrocarbonate (DMPC) is a suitable replacement for DEPC and is thought to be less carcinogenic; use the same procedure as for DEPC.
2. *Tissue homogenization*: Different tissues may require alternative homogenization techniques before or during TRI Reagent extraction. Although plant leaf, Drosophila embryo, and mouse liver samples were sufficiently homogenized by grinding in liquid nitrogen, some tissue types benefit from passage through a 15 mL Dounce homogenizer after suspension in TRI Reagent but before chloroform addition. This procedure is performed on ice to reduce RNA degradation prior to TRI Reagent penetration of tissue fragments.
3. *Spectrophotometric measurements*: Absorbance at the 260 nm wavelength is used to estimate RNA concentration as follows: c [μg/μL] = $(OD_{260} \times d \times 40)/1000$, where d is a fold dilution with respect to the original RNA sample. Nucleic acid purity can be roughly assessed using the OD_{260}/OD_{280} ratio, which is 1.8–2.0 for good RNA preps. Ratios below 1.7 indicate a poor sample quality (contamination by protein and/or other impurities). Such samples often require an additional phenol:chloroform extraction and isopropanol precipitation before proceeding to the northern blot.
4. *Loading RNA*: Total RNA can be loaded directly onto 18% polyacrylamide gels avoiding the need for size fractionation, although gel resolution may suffer as a result. About 5–10 μg is adequate for the detection of high-titer small RNAs (e.g., many miRNAs). To detect low-titer small RNAs, load 20–30 μg total RNA. Such large amounts of RNA or low-purity samples become viscous when resuspended in loading buffer. Cutting 3–5 mm off the microcapillary pipette tip with a razor will facilitate loading these samples.
5. *An alternative size fractionation method*: Polyethylene glycol (PEG) precipitation—described by Hamilton and Baulcombe [7] and modified in Vazquez et al. [28]—is more scalable than the column-based method and produces similar results. Prepare a solution of 20% PEG8000 (Promega, Madison, WI, USA) and 3 M sodium chloride in bidistilled water treating with DEPC. ~200 μL total RNA and 200 μL 20% PEG/3 M NaCl are mixed gently, incubated on ice for 20–30 min, and then centrifuged for 10 min (at 12,000 × *g* and 4 °C). This selectively sediments high-molecular-weight RNA. Transfer the supernatant (containing enriched small RNA) to new microfuge tubes, add 3 volumes cold ethanol; incubate for 1–2 h at −80 °C, and centrifuge for 20 min (at 14,000 × *g* and 4 °C). Wash

the pellet twice with 75 % ethanol, air-dry the pellet, and resuspend in DEPC-treated water.

6. *An improved RNA cross-linking method*: Pall and Hamilton (2008) found that carbodiimide-mediated, chemical cross-linking enhances small RNA detection by up to 50-fold over the standard UV cross-linking method [57]. To use this alternative procedure, PAGE should be performed using MOPS–NaOH (pH 7) buffer rather than TBE.

7. *Alternative radioactive labeling methods*: (a) *mirVana Probe Construction Kit* (Ambion) uses T7 polymerase transcription of oligo templates in the presence of [α-^{32}P]CTP—e.g., Onodera et al. [29] and Pontes et al. [58]. This method will aid the detection of low-titer small RNAs by incorporating multiple radiolabeled phosphates into each probe molecule. (b) *Single-stranded RNA probes* generated by *in vitro* transcription [1] of linearized plasmid templates in the presence of [α-^{32}P]UTP or [α-^{32}P]CTP. These transcripts are hydrolyzed to an average of 50 nt before use, following Hamilton and Baulcombe [7]. This method is best suited for detecting small RNAs from 100 to 1000 bp regions, as opposed to individual small RNAs already characterized by cloning and sequencing.

8. *Specific activity*: For end-labeled probes, γ-^{32}P with 6000 Ci/mmol is preferable. Each labeled oligo incorporates only a single radioactive atom, so a higher specific activity γ-^{32}P maximizes per molecule activity. Additionally, probes for low-titer small RNAs should be synthesized and used immediately after γ-^{32}P arrives from the supplier.

9. *Hybridization time and temperature*: If fast turnover times are required for successive probings, hybridization can be shortened to 6 h for conventional DNA oligo probes, or 2 h for locked nucleic acid probes [59] (*see* **Note 11**). Using the protocol described in this chapter, hybridization and washing at 35 °C yielded the strongest signal from oligo standards but did not distinguish 2–3 nt variants thereof, whereas hybridization and washing at 50 °C improved probe specificity but somewhat reduced signal strength (*see* Fig. 2).

10. *Phosphorimaging versus film*: All data shown in this chapter were collected using phosphorimager screen exposures of 1–48 h. However, very low-titer small RNAs may require several-day phosphorimager screen or film exposures using conventional DNA oligo probes. Locked nucleic acid probes improve hybridization sensitivity and reduce the time necessary for the overall protocol [59], which could eliminate the need for lengthy exposures (*see* **Note 11**).

11. *Locked nucleic acid (LNA) oligo probes*: LNA oligo probes contain high-affinity RNA analogues (e.g., at every third nucleo-

tide position) that possess modified ribose moieties. Hybridization with LNA-modified oligos enhances detection sensitivity and specificity. Some applications were demonstrated for northern blot detection of miRNAs [59] and heterochromatic siRNAs [60, 61], amongst others [62].

Acknowledgements

Many thanks to Frederick Meins, Jr. and Craig Pikaard for providing support and facilities for experiments shown in this chapter. Azeddine Si-Ammour and Hanspeter Schöb refined the northern blot techniques described here. Mikhail Pooggin and Thomas Hohn provided materials for the viral experiments. Mike Dyer cared for leafy plants, Pierre-François Perroud provided moss tissue, and Kathryn Huisinga supplied Drosophila embryos. Tatiana Simon and Luciano Marpegan provided mouse liver. Franck Vazquez, Mikhail Pooggin, and Andrzej Wierzbicki provided critical comments on the first edition of this book chapter. This work was supported by a Friedrich Miescher Institute student fellowship, and postdoctoral fellowships from the Swiss National Foundation and Novartis Foundation.

References

1. Sambrook J, Russell DW (2001) Molecular cloning. A laboratory manual, 3rd edn. Cold Spring Harbor Laboratory Press, Cold Spring Harbor
2. Southern EM (1975) Detection of specific sequences among DNA fragments separated by gel electrophoresis. J Mol Biol 98(3):503–517
3. Southern E (2006) Southern blotting. Nat Protoc 1(2):518–525
4. Alwine JC, Kemp DJ, Stark GR (1977) Method for detection of specific RNAs in agarose gels by transfer to diazobenzyloxymethyl-paper and hybridization with DNA probes. Proc Natl Acad Sci U S A 74(12): 5350–5354
5. Thomas PS (1980) Hybridization of denatured RNA and small DNA fragments transferred to nitrocellulose. Proc Natl Acad Sci U S A 77(9):5201–5205
6. Brown T, Mackey K, Du T (2004) Analysis of RNA by northern and slot blot hybridization. Curr Protoc Mol Biol 4:49
7. Hamilton AJ, Baulcombe DC (1999) A species of small antisense RNA in posttranscriptional gene silencing in plants. Science 286(5441):950–952
8. Hutvagner G, Mlynarova L, Nap JP (2000) Detailed characterization of the posttranscriptional gene-silencing-related small RNA in a GUS gene-silenced tobacco. RNA 6(10): 1445–1454
9. Reinhart BJ, Weinstein EG, Rhoades MW, Bartel B, Bartel DP (2002) MicroRNAs in plants. Genes Dev 16(13):1616–1626
10. Lau NC, Lim LP, Weinstein EG, Bartel DP (2001) An abundant class of tiny RNAs with probable regulatory roles in Caenorhabditis elegans. Science 294(5543):858–862
11. Llave C, Kasschau KD, Rector MA, Carrington JC (2002) Endogenous and silencing-associated small RNAs in plants. Plant Cell 14(7): 1605–1619
12. Lagos-Quintana M, Rauhut R, Lendeckel W, Tuschl T (2001) Identification of novel genes coding for small expressed RNAs. Science 294(5543):853–858
13. Pfeffer S, Zavolan M, Grasser FA, Chien M, Russo JJ, Ju J, John B, Enright AJ, Marks D, Sander C, Tuschl T (2004) Identification of virus-encoded microRNAs. Science 304(5671):734–736
14. Gu W, Shirayama M, Conte D Jr, Vasale J, Batista PJ et al (2009) Distinct argonaute-mediated 22G-RNA pathways direct genome surveillance in the C. elegans germline. Mol Cell 36(2):231–244

15. Chalker DL, Fuller P, Yao MC (2005) Communication between parental and developing genomes during tetrahymena nuclear differentiation is likely mediated by homologous RNAs. Genetics 169(1):149–160
16. Girard A, Sachidanandam R, Hannon GJ, Carmell MA (2006) A germline-specific class of small RNAs binds mammalian Piwi proteins. Nature 442(7099):199–202
17. Pane A, Wehr K, Schupbach T (2007) zucchini and squash encode two putative nucleases required for rasiRNA production in the Drosophila germline. Dev Cell 12(6):851–862
18. Meins F Jr, Si-Ammour A, Blevins T (2005) RNA silencing systems and their relevance to plant development. Annu Rev Cell Dev Biol 21:297–318
19. Chapman EJ, Carrington JC (2007) Specialization and evolution of endogenous small RNA pathways. Nat Rev Genet 8(11):884–896
20. Pikaard CS, Haag JR, Pontes OM, Blevins T, Cocklin R (2012) A transcription fork model for Pol IV and Pol V-dependent RNA-directed DNA methylation. Cold Spring Harb Symp Quant Biol 77:205–212
21. Xie Z, Johansen LK, Gustafson AM, Kasschau KD, Lellis AD, Zilberman D, Jacobsen SE, Carrington JC (2004) Genetic and functional diversification of small RNA pathways in plants. PLoS Biol 2(5), E104
22. Gasciolli V, Mallory AC, Bartel DP, Vaucheret H (2005) Partially redundant functions of Arabidopsis DICER-like enzymes and a role for DCL4 in producing trans-acting siRNAs. Curr Biol 15(16):1494–1500
23. Dunoyer P, Himber C, Voinnet O (2005) DICER-LIKE 4 is required for RNA interference and produces the 21-nucleotide small interfering RNA component of the plant cell-to-cell silencing signal. Nat Genet 37(12):1356–1360
24. Zilberman D, Cao X, Jacobsen SE (2003) ARGONAUTE4 control of locus-specific siRNA accumulation and DNA and histone methylation. Science 299(5607):716–719
25. Mette MF, Aufsatz W, van der Winden J, Matzke MA, Matzke AJ (2000) Transcriptional silencing and promoter methylation triggered by double-stranded RNA. EMBO J 19(19): 5194–5201
26. Vazquez F, Vaucheret H, Rajagopalan R, Lepers C, Gasciolli V, Mallory AC, Hilbert JL, Bartel DP, Crete P (2004) Endogenous trans-acting siRNAs regulate the accumulation of Arabidopsis mRNAs. Mol Cell 16(1):69–79
27. Yu B, Yang Z, Li J, Minakhina S, Yang M, Padgett RW, Steward R, Chen X (2005) Methylation as a crucial step in plant microRNA biogenesis. Science 307(5711):932–935
28. Vazquez F, Gasciolli V, Crete P, Vaucheret H (2004) The nuclear dsRNA binding protein HYL1 is required for microRNA accumulation and plant development, but not posttranscriptional transgene silencing. Curr Biol 14(4):346–351
29. Onodera Y, Haag JR, Ream T, Nunes PC, Pontes O, Pikaard CS (2005) Plant nuclear RNA polymerase IV mediates siRNA and DNA methylation-dependent heterochromatin formation. Cell 120(5):613–622
30. Hutvagner G, McLachlan J, Pasquinelli AE, Balint E, Tuschl T, Zamore PD (2001) A cellular function for the RNA-interference enzyme Dicer in the maturation of the let-7 small temporal RNA. Science 293(5531):834–838
31. Reinhart BJ, Slack FJ, Basson M, Pasquinelli AE, Bettinger JC, Rougvie AE, Horvitz HR, Ruvkun G (2000) The 21-nucleotide let-7 RNA regulates developmental timing in Caenorhabditis elegans. Nature 403(6772):901–906
32. Pasquinelli AE, Reinhart BJ, Slack F, Martindale MQ, Kuroda MI et al (2000) Conservation of the sequence and temporal expression of let-7 heterochronic regulatory RNA. Nature 408(6808):86–89
33. Grishok A, Pasquinelli AE, Conte D, Li N, Parrish S, Ha I et al (2001) Genes and mechanisms related to RNA interference regulate expression of the small temporal RNAs that control C. elegans developmental timing. Cell 106(1):23–34
34. Ketting RF, Fischer SE, Bernstein E, Sijen T, Hannon GJ, Plasterk RH (2001) Dicer functions in RNA interference and in synthesis of small RNA involved in developmental timing in C. elegans. Genes Dev 15(20):2654–2659
35. Bartel DP (2004) MicroRNAs: genomics, biogenesis, mechanism, and function. Cell 116(2):281–297
36. Axtell MJ (2008) Evolution of microRNAs and their targets: Are all microRNAs biologically relevant? Biochim Biophys Acta 1779(11):725–734
37. Jones-Rhoades MW, Bartel DP, Bartel B (2006) MicroRNAS and their regulatory roles in plants. Annu Rev Plant Biol 57:19–53
38. Kutter C, Schob H, Stadler M, Meins F Jr, Si-Ammour A (2007) MicroRNA-mediated regulation of stomatal development in Arabidopsis. Plant Cell 19(8):2417–2429
39. Rajagopalan R, Vaucheret H, Trejo J, Bartel DP (2006) A diverse and evolutionarily fluid set of microRNAs in Arabidopsis thaliana. Genes Dev 20(24):3407–3425
40. Bologna NG, Voinnet O (2014) The diversity, biogenesis, and activities of endogenous silencing small RNAs in Arabidopsis. Annu Rev Plant Biol 65:473–503

41. Holoch D, Moazed D (2015) RNA-mediated epigenetic regulation of gene expression. Nat Rev Genet 16(2):71–84
42. Pikaard CS, Mittelsten Scheid O (2014) Epigenetic regulation in plants. Cold Spring Harb Perspect Biol 6(12):a019315
43. Blevins T, Pontvianne F, Cocklin R, Podicheti R, Chandrasekhara C, Yerneni S et al (2014) A two-step process for epigenetic inheritance in Arabidopsis. Mol Cell 54(1):30–42
44. Henderson IR, Jacobsen SE (2007) Epigenetic inheritance in plants. Nature 447(7143):418–424
45. Head SR, Komori HK, LaMere SA, Whisenant T, Van Nieuwerburgh F, Salomon DR, Ordoukhanian P (2014) Library construction for next-generation sequencing: overviews and challenges. Biotechniques 56(2):61–64, 66, 68, passim
46. Coruh C, Shahid S, Axtell MJ (2014) Seeing the forest for the trees: annotating small RNA producing genes in plants. Curr Opin Plant Biol 18:87–95
47. Carbonell A, Fahlgren N, Garcia-Ruiz H, Gilbert KB, Montgomery TA, Nguyen T, Cuperus JT, Carrington JC (2012) Functional analysis of three Arabidopsis ARGONAUTES using slicer-defective mutants. Plant Cell 24(9):3613–3629
48. Blevins T, Rajeswaran R, Aregger M, Borah BK, Schepetilnikov M, Baerlocher L et al (2011) Massive production of small RNAs from a non-coding region of Cauliflower mosaic virus in plant defense and viral counter-defense. Nucleic Acids Res 39(12):5003–5014
49. Mari-Ordonez A, Marchais A, Etcheverry M, Martin A, Colot V, Voinnet O (2013) Reconstructing de novo silencing of an active plant retrotransposon. Nat Genet 45(9): 1029–1039
50. Reimao-Pinto MM, Ignatova V, Burkard TR, Hung JH, Manzenreither RA, Sowemimo I, Herzog VA, Reichholf B, Farina-Lopez S, Ameres SL (2015) Uridylation of RNA hairpins by tailor confines the emergence of MicroRNAs in drosophila. Mol Cell 59(2):203–216
51. Reed KC, Mann DA (1985) Rapid transfer of DNA from agarose gels to nylon membranes. Nucleic Acids Res 13(20):7207–7221
52. Alonso JM, Stepanova AN, Leisse TJ, Kim CJ, Chen H, Shinn P et al (2003) Genome-wide insertional mutagenesis of Arabidopsis thaliana. Science 301(5633):653–657
53. Sessions A, Burke E, Presting G, Aux G, McElver J, Patton D et al (2002) A high-throughput Arabidopsis reverse genetics system. Plant Cell 14(12):2985–2994
54. Blevins T, Rajeswaran R, Shivaprasad PV, Beknazariants D, Si-Ammour A, Park HS et al (2006) Four plant dicers mediate viral small RNA biogenesis and DNA virus induced silencing. Nucleic Acids Res 34(21): 6233–6246
55. Chomczynski P, Sacchi N (1987) Single-step method of RNA isolation by acid guanidinium thiocyanate-phenol-chloroform extraction. Anal Biochem 162(1):156–159
56. Chomczynski P, Sacchi N (2006) The single-step method of RNA isolation by acid guanidinium thiocyanate-phenol-chloroform extraction: twenty-something years on. Nat Protoc 1(2):581–585
57. Pall GS, Hamilton AJ (2008) Improved northern blot method for enhanced detection of small RNA. Nat Protoc 3(6):1077–1084
58. Pontes O, Li CF, Nunes PC, Haag J, Ream T, Vitins A, Jacobsen SE, Pikaard CS (2006) The Arabidopsis chromatin-modifying nuclear siRNA pathway involves a nucleolar RNA processing center. Cell 126(1):79–92
59. Varallyay E, Burgyan J, Havelda Z (2008) MicroRNA detection by northern blotting using locked nucleic acid probes. Nat Protoc 3(2):190–196
60. Henderson IR, Jacobsen SE (2008) Tandem repeats upstream of the Arabidopsis endogene SDC recruit non-CG DNA methylation and initiate siRNA spreading. Genes Dev 22(12):1597–1606
61. Zheng B, Wang Z, Li S, Yu B, Liu JY, Chen X (2009) Intergenic transcription by RNA polymerase II coordinates Pol IV and Pol V in siRNA-directed transcriptional gene silencing in Arabidopsis. Genes Dev 23(24): 2850–2860
62. Castoldi M, Schmidt S, Benes V, Noerholm M, Kulozik AE, Hentze MW, Muckenthaler MU (2006) A sensitive array for microRNA expression profiling (miChip) based on locked nucleic acids (LNA). RNA 12(5):913–920
63. Bonifacio GF, Brown T, Conn GL, Lane AN (1997) Comparison of the electrophoretic and hydrodynamic properties of DNA and RNA oligonucleotide duplexes. Biophys J 73(3):1532–1538
64. Aravin AA, Lagos-Quintana M, Yalcin A, Zavolan M, Marks D, Snyder B, Gaasterland T, Meyer J, Tuschl T (2003) The small RNA profile during Drosophila melanogaster development. Dev Cell 5(2):337–350

Chapter 13

Stem-Loop qRT-PCR for the Detection of Plant microRNAs

Erika Varkonyi-Gasic

Abstract

Plant microRNAs (miRNAs) play important roles in the posttranscriptional regulation of protein-coding genes, and they are essential for a normal development and survival. Mature miRNAs are cleaved from larger precursor RNAs and are typically 21–22 nt long.

The small size, the lack of a common feature like a poly(A) tail, 3′ end-modifications, and presence of a precursor—all these factors affect the detection and hinder the quantification of miRNAs. The stem-loop qRT-PCR method described here is designed to detect and quantify mature miRNAs in a fast, specific, accurate, and reliable manner. Firstly, a miRNA-specific stem-loop RT primer is hybridized to miRNA and then reverse transcribed. Next, the RT product is amplified and monitored in real time using a miRNA-specific forward primer and a universal reverse primer. This method enables miRNA expression profiling from as little as 10 pg of total RNA, and it is suitable for a relatively high-throughput analysis of miRNA expression.

Key words miRNA, Stem-loop RT, qPCR, SYBR Green, UPL

1 Introduction

MicroRNAs (miRNAs), a class of 20- to 24-nucleotide endogenous small RNAs, are the evolutionarily conserved key regulators of gene expression in eukaryotes [1]. They arise from larger precursor RNAs with a characteristic hairpin structure and act as repressors of gene activity by targeting complementary sequences in the target RNA. Animal miRNAs interact with their RNA targets through imprecise base-pairing, resulting in an arrest of translation [2]. In plants, most miRNAs interact with their targets through near-perfect base-pairing [3], resulting in target degradation [3, 4], translational repression [5–7], or DNA methylation [8, 9].

Plant miRNAs control the expression of genes encoding transcription factors [3], stress response proteins [10], and other proteins essential for the normal plant growth, development, and physiology. Their biogenesis and regulatory action are complex and highly regulated [11], and the diversification of paralogs within

Igor Kovalchuck (ed.), *Plant Epigenetics: Methods and Protocols*, Methods in Molecular Biology, vol. 1456,
DOI 10.1007/978-1-4899-7708-3_13, © Springer Science+Business Media New York 2017

miRNA families may result in further differences in target gene regulation [12]. The presence of a miRNA may result in a complete removal of the corresponding target resulting in call-fate changes [13, 14], the reduction in target mRNA levels and the co-expression of both the miRNA and the target mRNA in the same tissue [15, 16], or the miRNA itself can also serve as a systemic signal [17, 18].

Because of an important role that miRNAs play in the regulation of gene expression, it is of great interest to reliably detect individual miRNAs and quantitatively determine their expression level. Each of the methods commonly used for the detection of plant miRNA has advantages and limitations (Table 1), but

Table 1
Comparison of plant miRNA profiling methods

Method	Purpose	Throughput	Sensitivity	Specificity	Advantages	Limitations
Cloning	Discovery and confirmation	Low	Low	High	Discovery of new miRNAs	High complexity, time, and cost
Northern blot	Confirmation	Low	Low	High	"Golden standard" in early miRNA studies	High complexity, time, and RNA input
In situ hybridization	Confirmation	Low	Very low	Low	Precise spatial distribution	High complexity, potential background problems
Microarray	Confirmation	High	Low	Semilow	High throughput	Relatively high RNA input, low specificity for related miRNA
qRT-PCR	Confirmation	Semi-high	Very high	High	Fast, low RNA requirement, high sensitivity, wide dynamic range, quantitative	Potential contamination
NGS	Discovery and confirmation	High	High	High	High throughput, quantitative, discovery of new miRNAs	High complexity, time, RNA input, and cost

quantitative reverse transcription PCR (qRT-PCR) combines high speed, throughput, sensitivity, specificity, and affordability, thus remaining a method of choice for most researchers. Mature miRNAs are too short to serve as templates for standard qRT-PCR and different methods have been developed to extend the miRNA template and incorporate additional nucleotides during the reverse transcription (RT) step. Two most commonly used methods are (1) in vitro polyadenylation followed by oligodT-primed RT and (2) the stem-loop RT-PCR method that requires the usage of a hairpin primer that is complementary to the 3′ end of miRNA. The efficiency of polyadenylation is affected by the methylation of the 3′ termini of plant miRNAs [19]. Also the stem-loop reverse transcription primers provide better specificity and sensitivity than linear primers [20]. The sensitivity of miRNA detection can be further increased by a pulsed RT reaction, using short cycles of incubation at gradually increasing annealing temperatures to ensure correct pairing [21]. These features are combined within the two-step miRNA detection method (Fig. 1). First, the stem-loop RT primer is hybridized to a specific miRNA molecule and then reverse transcribed in a pulsed RT reaction. Next, the RT product is amplified using the miRNA-specific forward primer and the universal reverse primer. The product can be visualized by gel electrophoresis upon a set number of PCR cycles or monitored in real time using quantitative PCR (qPCR). The quantification of miRNA expression can be performed using the SYBR Green detection technology or the hydrolysis probe technology for an increased specificity such as the Universal ProbeLibrary (UPL; Roche) or TaqMan® (Life Technologies).

In addition to expression analysis of endogenous plant miRNAs, this method is amenable for the detection and quantification of animal miRNAs and other small RNAs, including artificial miRNAs and synthetic siRNAs.

2 Materials

2.1 Plant Material and RNA

1. Plant tissue frozen in liquid nitrogen and handled according to standard practices to prevent RNA degradation.
2. Total RNA extracted by standard methods and handled according to standard laboratory practices to avoid RNase contamination. Avoid RNA purification by using RNA-binding glass-fiber filters (*see* **Note 1**). Prior to reverse transcription, RNA should be quantified and evaluated (*see* **Note 2**).

2.2 Stem-Loop Pulsed RT

1. Stem-loop RT primers: Prepare 100 μM stocks for long-term storage and 1 μM dilutions for an immediate use.

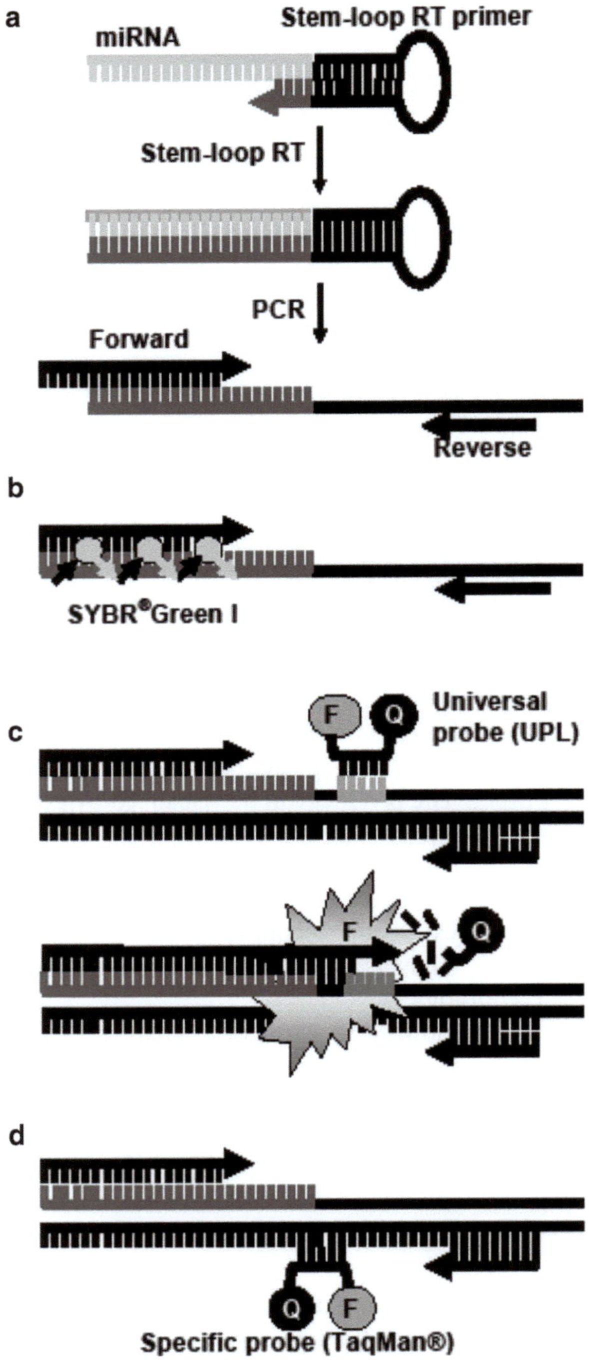

Fig. 1 Schematic showing the stem-loop RT-PCR for the detection of microRNAs. (**a**) A stem-loop RT primer binds to the 3′ portion of the miRNA, initiating reverse transcription. The RT product is amplified using a miRNA-specific forward primer and a universal reverse primer. (**b**) The quantification is achieved through SYBR Green I incorporation during the amplification. (**c**) The quantification is achieved by the fluorescence generated upon cleavage of an UPL probe, a hydrolysis probe of eight nucleotides including one locked nucleic acid (LNA) that increases a binding specificity. This probe is designed to hybridize to a region within the amplicon and is dual-labeled with a reporter dye and a quenching dye. The quenching dye is suppressing the fluorescence of the reporter dye while in close proximity. Once the probe is degraded by the exonuclease activity of Taq polymerase, the fluorescence of the reporter increases at a rate proportional to the amplification level. (**d**) A miRNA-specific TaqMan probe can be used in the same manner as the universal UPL probe

2. 10 mM dNTP mix: Prepare it by mixing dATP, dCTP, dGTP, and dCTP stock solutions, aliquot out, and store at −20 °C.
3. Reverse transcriptase, e.g., SuperScript® III RT, 200 U/μL; supplied with 5× first-strand buffer for cDNA synthesis and 0.1 M DTT (Life Technologies).
4. RNase inhibitor, e.g., RNaseOUT™, 40 U/μL (Life Technologies).
5. Nuclease-free water.

2.3 miRNA SYBR Green qPCR Assay

1. The SYBR Green master mix: It is prepared according to the manufacturer's instructions, e.g., 5× LightCycler® FastStart SYBR Green I Master (Roche) for carousel-based instruments or 2× LightCycler® 480 SYBR Green I Master (Roche) for the LightCycler® 480 instruments.
2. The universal reverse primer: Prepare 100 μM stock for long-term storage and 10 μM dilution for an immediate use.
3. The forward miRNA-specific primer: Prepare 100 μM stock for long-term storage and 10 μM dilution for an immediate use.
4. The 10 mM dNTP mix, prepared as described above.
5. Nuclease-free water.

2.4 miRNA Hydrolysis Probe qPCR Assay

1. A TaqMan® master mix is prepared according to the manufacturer's instructions, e.g., LightCycler® TaqMan® Master (Roche) for carousel-based instruments and LightCycler® 480 Probes Master (Roche) or TaqMan® Fast Advanced Master Mix (Life technologies) for the LightCycler® 480 instruments.
2. UPL probe #21 is prepared in the same way as 10 μM stock (Roche) (*see* **Note 3**).
3. The universal reverse primer: Prepare 100 μM stock for long-term storage and 10 μM dilution for an immediate use.
4. The forward miRNA-specific oligonucleotide: Prepare 100 μM stock for long-term storage and 10 μM dilution for an immediate use.
5. The 10 mM dNTP mix, prepared as described above.
6. Nuclease-free water.

2.5 Equipment

1. Standard laboratory equipment for the isolation of RNA (a fume hood, a centrifuge, tubes, pipettes, and tips).
2. A spectrophotometer for RNA quantification, e.g., a NanoDrop Spectrophotometer (Thermo Scientific) (*see* **Note 2**).
3. Standard gel electrophoresis equipment (casting trays, gel tanks, power supply, a UV transilluminator).
4. A thermal cycler for pulsed reverse transcription.
5. A real-time thermal cycler, e.g., LightCycler® (Roche).

3 Methods

3.1 Primer Design

The primers are designed according to Chen et al. [20] with some modifications [22] (Fig. 2). The stem-loop RT primers have a universal backbone and a specific extension. The universal backbone sequence is designed to form a stem-loop structure because of the complementarity between nucleotides at the 5′ and 3′ ends. It includes a reverse complement of UPL probe #21 and the universal reverse primer site in the loop region. The specificity of a stem-loop RT primer to an individual miRNA is conferred by a six-nucleotide extension at the 3′ end. This extension is a reverse complement of the last six nucleotides at the 3′ end of the miRNA. The forward primers are specific to the miRNA sequence but exclude the last six nucleotides at the 3′ end of the miRNA. A 5′ extension of 3–7 nucleotides is added to each forward primer to

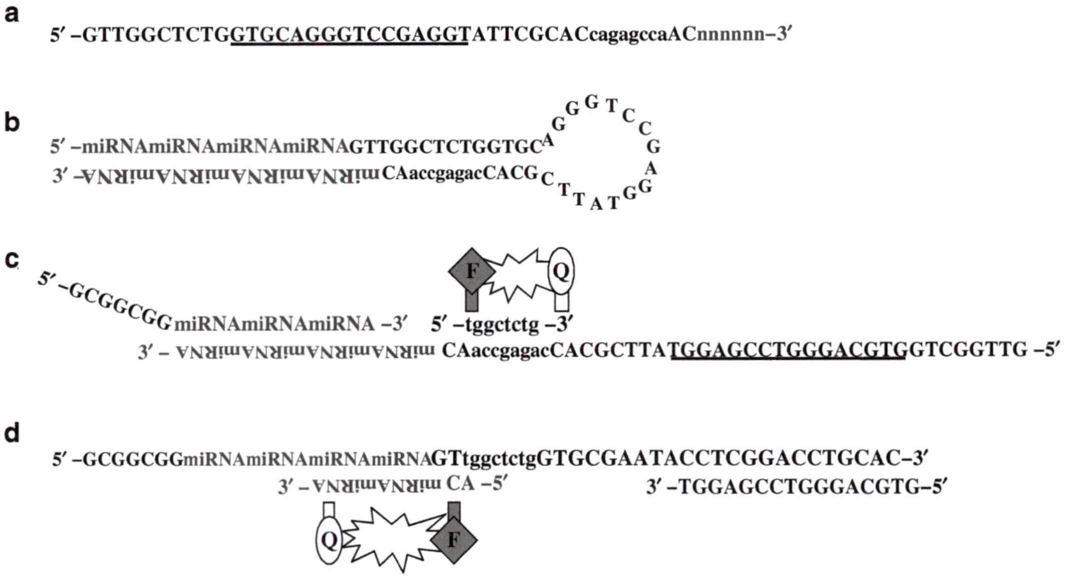

Fig. 2 Schematic showing primer design for stem-loop RT-PCR for the detection of microRNAs. (**a**) The stem-loop RT primers have a universal backbone and a miRNA-specific extension. The specificity of the stem-loop RT primer to an individual miRNA is conferred by a six-nucleotide extension at the 3′ end (nnnnnn) which is a reverse complement of the last six nucleotides at the 3′ end of the miRNA. The backbone includes the reverse complement of the UPL probe #21 (in a *lower case*) and the universal reverse primer site (*underlined*). (**b**) The backbone sequence can form a stem-loop structure because of the complementarity between nucleotides at the 5′ and 3′ ends. Reverse transcription is initiated upon annealing to the six nucleotides at the 3′ end of the miRNA. (**c**) Forward primers are specific to the miRNA sequence but exclude the six nucleotides at the 3′ end of the miRNA. A random 5′ extension of 3–7 nucleotides is added to each forward primer to increase the length and melting temperature. The 5′ extension sequences are usually relatively GC rich, bringing the GC content of the forward primer to 50–60 %. The UPL probe #21 sequence can hybridize to the DNA but is removed by the exonuclease activity of Taq polymerase, resulting in the detectable fluorescence of the reporter dye. (**d**) A miRNA-specific TaqMan probe can be designed to distinguish between highly homologous targets

optimize the length and melting temperature, thus enhancing the assay specificity. The above guidelines have been recently used to develop a primer design software available online [23].

3.2 Stem-Loop Pulsed RT

The protocol is designed to quantify either the expression of a specific miRNA in multiple biological samples or the expression of multiple miRNAs in one sample. The most reproducible results are obtained with 1–10 ng of total RNA per reaction, but moderately and highly abundant miRNAs can be detected from as little as 10 pg of total RNA. When handling large number of samples, keep the reactions on ice or work in a cold room.

To quantify the expression of one miRNA in multiple samples:

1. Denature the appropriate 1 μM stem-loop RT primer by heating to 65 °C for 5 min.
2. Incubate on ice for 2 min.
3. Centrifuge briefly to bring the solution to the bottom of the tube. Incubate on ice.
4. Prepare the required amount of "no-RNA" master mix by scaling the volumes for a single "no-RNA" RT reaction mix (listed below) to the desired number of RT reactions. At least three replicates per RT reaction are recommended. It is a good practise to prepare additional 10% of master mix to cover pipetting errors. The single "no-RNA" RT reaction mix is prepared by adding the following components to a nuclease-free microcentrifuge tube:

   0.5 μL 10 mM dNTP mix,

   11.15 μL Nuclease-free water,

   1 μL of denatured stem-loop RT primer,

   4 μL 5× First-strand buffer,

   2 μL 0.1 M DTT,

   0.1 μL RNaseOUT (40 U/μL) and

   0.25 μL SuperScript III RT (200 U/μL).
5. Mix gently and centrifuge to bring the solution to the bottom of the tube.
6. Assemble the RT reactions by aliquoting 19 μL of the "no-RNA" master mix and adding 1 μL RNA template (*see* **Note 4**).
7. Mix gently and centrifuge to bring the solution to the bottom of the tube.
8. Prepare the "minus RT" controls by omitting the reverse transcriptase from the reactions and "no-template" controls by adding nuclease-free water in place of RNA.

9. Load a thermal cycler and incubate for 30 min at 16 °C followed by pulsed RT of 60 cycles at 30 °C for 30 s, 42 °C for 30 s, and 50 °C for 1 s.
10. Incubate at 85 °C for 5 min to inactivate the reverse transcriptase.

To quantify the expression of multiple miRNAs in one sample, follow the same steps, but prepare a "no-RT" master mix (**step 4**) by adding an appropriate RNA template in place of the denatured stem-loop RT primer. To assemble the RT reaction (**step 6**), aliquot 19 μL of the "no-RT" master mix and add 1 μL of an appropriate denatured stem-loop RT primer.

3.3 miRNA SYBR Green qPCR Assay

The SYBR Green I assay provides a good specificity if the number of PCR cycles is limited to 35 for minimizing nonspecific amplification. This number of cycles is sufficient for the quantification of moderately abundant miRNAs.

1. Prepare a PCR master mix by scaling the volumes listed below to a desired number of amplification reactions. Prepare additional 10% of master mix to cover pipetting errors. For a single reaction, add the following components to a nuclease-free microcentrifuge tube:

   3 μL Nuclease-free water.

   5 μL 2× SYBR Green I master mix.

   0.5 μL of the forward (miRNA-specific) primer (10 μM).

   0.5 μL of the reverse (universal) primer (10 μM).
2. Mix gently and centrifuge to bring the solution to the bottom of the tube.
3. Store in a cooling block or on ice.
4. Pipette 9 μL master mix into a well plate or a capillary.
5. Add 1 μL of the RT product.
6. Seal the multi-well plate (or the capillary) and place it into the LightCycler® instrument.
7. Incubate the samples at 95 °C for 5 min followed by 35 cycles of 95 °C for 5 s and 60 °C for 10 s.
8. For the melting curve analysis, denature the samples at 95 °C, and then cool them to 65 °C at 20 °C/s. Collect fluorescence signals at 530 nm wavelength continuously from 65 to 95 °C at 0.2 °C/s.

3.4 miRNA Hydrolysis Probe qPCR Assay

For miRNA sequences that are expressed at low levels or when a particular set of primers produces the background amplification, a hydrolysis probe assay provides a higher specificity. The protocol for the UPL probe assay is presented here that utilizes a universal single-hydrolysis probe, UPL probe #21 (Roche), to distinguish between specific amplicons and primer-dimers. MicroRNA assays for the detection of individual miRNAs using individual miRNA-specific TaqMan® probes are available commercially (*see* **Note 3**).

1. Prepare a PCR master mix by scaling the volumes listed below to a desired number of amplification reactions. Prepare additional 10 % of master mix to cover pipetting errors. For a single reaction, add the following components to a nuclease-free microcentrifuge tube:

   2.9 μL Nuclease-free water.

   5 μL 2× TaqMan® master mix.

   0.5 μL of the forward (miRNA specific) primer (10 μM).

   0.5 μL of the reverse (universal) primer (10 μM).

   0.1 μL UPL probe #21 (10 μM).
2. Mix gently and centrifuge to bring the solution to the bottom of the tube.
3. Store in a cooling block or on ice.
4. Pipette 9 μL master mix into a well plate or a capillary.
5. Add 1 μL of the RT product.
6. Seal the multi-well plate (or the capillary) and place it into the LightCycler® instrument.
7. Incubate the samples at 95 °C for 5 min followed by 35–45 cycles of 95 °C for 5 s and 60 °C for 10 s.

3.5 Data Analysis

Data analysis consists of data processing, quality assessment, normalization, and calculation. In this section, some general considerations for some aspects of data analysis are provided. For detailed instructions, refer to the appropriate instrument and software user's manuals.

1. Commercial qPCR instruments are equipped with software for raw data analysis and visualization. A range of open-access tools are also available and may suit specific applications [24].
2. Quality assessment includes the analysis of internal controls and replicate performance.
3. If the miRNA SYBR Green qPCR assay is used, perform the melting curve analysis. This analysis is done to determine that each of the primer pairs amplifies a single predominant product with a distinct melting temperature (Tm). Follow the instructions from the instrument user's manual for the melting curve analysis and Tm calling. If a single melting peak is observed for a particular primer pair, it is likely that a single product with a distinct Tm was amplified. Evaluate by gel electrophoresis (*see* **Note 5**).

4. At present, there is no consensus normalization approach for any of the miRNA profiling methods, and several normalization techniques have been used, but all of them have limitations [25, 26]. The advantages and limitations of potential normalization methods are listed in Table 2, and they need to be carefully considered for data analysis and interpretation. The most common approach for the analysis of qRT-PCR data is its normalization to one or more endogenous reference genes with the expression that is constant across tissues and cell types. A suitable control for the normalization of miRNA expression should be similar to the analyzed miRNAs in terms of size and stability, and it should be amenable to the miRNA assay design (*see* **Note 6**). The chosen endogenous controls

Table 2
Comparison of normalization methods

Method	Purpose/description	Examples of reference genes	Advantages	Limitations
Endogenous control RNA	Normalization to endogenous miRNA	miR156	Amenable to the miRNA assay design	Require validation for each experimental design
	Normalization to endogenous small RNA	*Arabidopsis* snoR101, snoR41Y, snoRNA65, snoR66, snoR85, U6 snRNA, 5S RNA	Amenable to the miRNA assay design, although often significantly longer	Disproportionally high levels, validation for each experimental design required
	Normalization to one or more mRNA	Actin, tubulin, ubiquitin, elongation factor 1α, GAPDH	Indicator of general RNA quality, quantification, and technical variations	Not amenable to miRNA assay design
Mean expression value	Global measure of the miRNA expression data used as the normalizer		Accurate for large-scale miRNA profiling	Not accurate for small-medium-scale miRNA profiling
Synthetic control RNA (spiked-in)	Synthetic RNA introduced into the RNA sample		Accurate and rigorous normalization over a range of concentrations	Costly, time consuming, no control over some technical variations such as RNA quality or quantification

may include a specific small nuclear RNA (snRNA), a small nucleolar RNA (snoRNA), or a miRNA that demonstrate the least variability across tissues and in the experimental conditions (*see* **Note 7**).

5. Stem loop qRT-PCR can be used for an absolute quantification of miRNA expression. In that case, a standard curve is generated from serial dilutions of an external standard (e.g., synthetic RNA oligonucleotides of known concentration) to determine the concentration of miRNA.

4 Notes

1. Whenever possible, we used the TRIzol® Reagent (Life Technologies) for the isolation of RNA because of its convenience, good RNA quality, and speed. Some plant tissues may not be amenable to RNA isolation by this method. Other methods for RNA isolation may be used; however, avoid RNA purification methods that use RNA-binding glass-fiber filters that do not recover small RNA species quantitatively (e.g., Qiagen RNeasy® mini and midi kits). If you are unfamiliar with the method for RNA isolation, it is recommended to use the subsequent isolation, quantification, and polyacrylamide gel electrophoresis of the low-molecular-weight RNA fraction to evaluate its quantity and quality.
2. Spectrophotometry followed by gel electrophoresis is still the most widely used method for assessing RNA yield, purity, and quality. Microfluidic systems (e.g., Agilent 2100 Bioanalyzer) can be used to determine miRNA yield and quality. The accuracy of estimation of miRNA abundance is strongly influenced by total RNA integrity; RNA degradation results in a large overestimation of miRNA amount on the chip.
3. The presented miRNA hydrolysis probe assay utilizes UPL probe #21 (Roche), a universal hydrolysis probe of eight nucleotides including one locked nucleic acid (LNA), that is specific to the stem-loop RT primer backbone. Individual miRNA-specific TaqMan® probes may provide an increased specificity for the quantification of highly similar miRNAs. MicroRNA Assay kits for the detection of individual miRNAs using individual miRNA-specific TaqMan® probes are available commercially (e.g., Life Technologies). They are designed for the quantification of published miRNA sequences available in miRBase (http://www.mirbase.org/) or with custom sequences.
4. It has been suggested that the denaturation of RNA may reduce the yield of cDNA for some miRNAs. In our hands, both non-denatured RNA and RNA denatured by incubation at 65 °C for 5 min produced similar results.

5. The melting curve analysis needs to be combined with gel electrophoresis. Due to a small size of fragments, a primer-dimer product generated from the minus RT and no-template controls often has a very similar Tm to that of an appropriate miRNA amplification fragment. This becomes an issue with lowly abundant miRNAs that require a large number of PCR amplification cycles. In that case, the UPL assay is recommended.
6. In standard RT-PCR, endogenous controls are amplified from the same RT template as the genes of interest, thus taking into account a possible variation in the efficiency of reverse transcription between samples. In stem-loop RT-PCR, each RT reaction is performed using a specific stem-loop RT primer and therefore may not accurately reflect a possible variation in the efficiency of reverse transcription between samples.
7. Stable expression patterns have been demonstrated for *Arabidopsis* snoR41Y, snoR65, snoR66, and snoR85 [27], but a large-scale analysis of plant snoRNA and snRNA to evaluate their suitability for plant miRNA quantification is still outstanding. Therefore, using endogenous miRNA genes, structural RNAs (e.g., U6 snRNA, 5S RNA), and standard housekeeping genes in addition to snoRNA reference is recommended. For a more accurate evaluation, spike-in synthetic RNA may be chosen.

References

1. Bartel DP (2004) MicroRNAs: genomics, biogenesis, mechanism, and function. Cell 116:281–297
2. Fabian MR, Sonenberg N, Filipowicz W (2010) Regulation of mRNA translation and stability by microRNAs. Annu Rev Biochem 79:351–379
3. Rhoades MW, Reinhart BJ, Lim LP, Burge CB, Bartel B, Bartel DP (2002) Prediction of plant microRNA targets. Cell 110:513–520
4. Llave C, Xie Z, Kasschau KD, Carrington JC (2002) Cleavage of *Scarecrow-like* mRNA targets directed by a class of *Arabidopsis* miRNA. Science 297:2053–2056
5. Aukerman MJ, Sakai H (2003) Regulation of flowering time and floral organ identity by a MicroRNA and its *APETALA2*-like target genes. Plant Cell 15:2730–2741
6. Brodersen P, Sakvarelidze-Achard L, Bruun-Rasmussen M, Dunoyer P, Yamamoto YY, Sieburth L, Voinnet O (2008) Widespread translational inhibition by plant miRNAs and siRNAs. Science 320:1185–1190
7. Li S, Liu L, Zhuang X, Yu Y, Liu X, Cui X, Ji L, Pan Z, Cao X, Mo B, Zhang F, Raikhel N, Jiang L, Chen X (2013) MicroRNAs inhibit the translation of target mRNAs on the endoplasmic reticulum in *Arabidopsis*. Cell 153:562–574
8. Wu L, Zhou H, Zhang Q, Zhang J, Ni F, Liu C, Qi Y (2010) DNA methylation mediated by a microRNA pathway. Mol Cell 38:465–475
9. Bao N, Lye KW, Barton MK (2004) MicroRNA binding sites in *Arabidopsis* class III HD-ZIP mRNAs are required for methylation of the template chromosome. Dev Cell 7:653–662
10. Sunkar R, Zhu JK (2004) Novel and stress-regulated microRNAs and other small RNAs from *Arabidopsis*. Plant Cell 16:2001–2019
11. Rogers K, Chen X (2013) Biogenesis, turnover, and mode of action of plant microRNAs. Plant Cell 25:2383–2399
12. Li AL, Mao L (2007) Evolution of plant microRNA gene families. Cell Res 17:212–218
13. Juarez MT, Kui JS, Thomas J, Heller BA, Timmermans MC (2004) MicroRNA-mediated repression of *rolled leaf1* specifies maize leaf polarity. Nature 428:84–88
14. Kidner CA, Martienssen RA (2004) Spatially restricted microRNA directs leaf polarity through ARGONAUTE1. Nature 428:81–84
15. Tang G, Reinhart BJ, Bartel DP, Zamore PD (2003) A biochemical framework for RNA silencing in plants. Genes Dev 17:49–63
16. Mallory AC, Reinhart BJ, Jones-Rhoades MW, Tang G, Zamore PD, Barton MK, Bartel DP

(2004) MicroRNA control of *PHABULOSA* in leaf development: importance of pairing to the microRNA 5' region. EMBO J 23:3356–3364
17. Pant BD, Buhtz A, Kehr J, Scheible WR (2008) MicroRNA399 is a long-distance signal for the regulation of plant phosphate homeostasis. Plant J 53:731–738
18. Varkonyi-Gasic E, Gould N, Sandanayaka M, Sutherland P, MacDiarmid RM (2010) Characterisation of microRNAs from apple (*Malusdomestica* 'Royal Gala') vascular tissue and phloem sap. BMC Plant Biol 10:159
19. Adhikari S, Turner M, Subramanian S (2013) Hairpin priming is better suited than *in vitro* polyadenylation to generate cDNA for plant miRNA qPCR. Mol Plant 6:229–231
20. Chen C, Ridzon DA, Broomer AJ, Zhou Z, Lee DH, Nguyen JT, Barbisin M, Xu NL, Mahuvakar VR, Andersen MR, Lao KQ, Livak KJ, Guegler KJ (2005) Real-time quantification of microRNAs by stem-loop RT-PCR. Nucleic Acids Res 33:e179
21. Tang F, Hajkova P, Barton SC, Lao K, Surani MA (2006) MicroRNA expression profiling of single whole embryonic stem cells. Nucleic Acids Res 34:e9
22. Varkonyi-Gasic E, Wu R, Wood M, Walton EF, Hellens RP (2007) Protocol: a highly sensitive RT-PCR method for detection and quantification of microRNAs. Plant Methods 3:12
23. Czimmerer Z, Hulvely J, Simandi Z, Varallyay E, Havelda Z, Szabo E, Varga A, Dezso B, Balogh M, Horvath A, Domokos B, Torok Z, Nagy L, Balint BL (2013) A versatile method to design stem-loop primer-based quantitative PCR assays for detecting small regulatory RNA molecules. PLoS One 8:e55168
24. Pabinger S, Rödiger S, Kriegner A, Vierlinger K, Weinhäusel A (2014) A survey of tools for the analysis of quantitative PCR (qPCR) data. BDQ 1:23–33
25. Meyer S, Pfaffl M, Ulbrich S (2010) Normalization strategies for microRNA profiling experiments: a 'normal' way to a hidden layer of complexity? Biotechnol Lett 32:1777–1788
26. Pritchard CC, Cheng HH, Tewari M (2012) MicroRNA profiling: approaches and considerations. Nat Rev Genet 13:358–369
27. Kahl G, Meksem K (eds) (2008) The handbook of plant functional genomics, "Real-Time Quantitation of MicroRNAs by TaqMan® MicroRNA Assays,". Wiley-VCH, Weinheim

Chapter 14

Profiling New Small RNA Sequences

Masayuki Tsuzuki and Yuichiro Watanabe

Abstract

Small RNAs are key molecules in RNA silencing pathways that exert the sequence-specific regulation of gene expression and chromatin modifications in many eukaryotes. In plants, endogenous small RNAs, including microRNAs (miRNAs), *trans*-acting short interfering RNAs (tasiRNAs), and heterochromatic siRNAs (hc-siRNAs), play an important role in switching or orchestrating biological processes during the development and at the onset of stress responses. These endogenous and exogenous small RNAs are mainly 20–24 nucleotides in length. In addition, viral genome-derived siRNAs of similar lengths are produced during viral infection, and they exhibit anti-viral defense activity in RNA silencing pathway.

Here, we introduce a method to isolate and characterize small RNA molecules possibly applicable to a wide range of plant resources and tissues. After purification from total RNAs, small RNAs were subjected to Illumina sequencing analysis using compatible reagents kits. Following the sample preparation protocol, small RNAs are ligated first at the 3′- and then at the 5′-end to the respective RNA adapters followed by reverse transcription with a set of primers to produce cDNAs with Index sequences at ends. After PCR amplification, cDNAs are subjected (after gel purification) to RNA-seq analysis. This method could be applied to isolate small RNAs from different sources and characterize small RNA profiles to compare different sets of samples, e.g., wild-type and mutant plants, plants under different stress environments, and virus-infected plants because the starting RNA material is free of contaminated starch or similar material which would block further analysis.

Key words Cloning, Small RNA, siRNA, miRNA, Virus-derived siRNA, Sequencing

1 Introduction

It has been established that many types of noncoding RNAs are expressed in different organisms and play many essential roles. The functions of the major types of small RNAs (20–24 nucleotides in length) have been analyzed. These are key molecules in RNA silencing pathways that regulate gene expression transcriptionally or posttranscriptionally in eukaryotes. Small regulatory RNAs are classified into several groups based on their biogenesis and functions. They include microRNAs (miRNAs), trans-acting short interfering RNAs (tasiRNAs), and heterochromatic siRNAs (hc-siRNAs) [1]. Target mRNAs of miRNAs include many transcription factors that are important for plant development in Arabidopsis

Igor Kovalchuck (ed.), *Plant Epigenetics: Methods and Protocols*, Methods in Molecular Biology, vol. 1456, DOI 10.1007/978-1-4899-7708-3_14, © Springer Science+Business Media New York 2017

[2]; therefore proper biogenesis and functions of miRNAs are essential for normal development and are tightly regulated in various organs [3]. In addition to the endogenous small RNAs, viral genome-derived siRNAs are produced during viral infection [4]. The viral genome is targeted by these siRNAs and silenced using the RNA silencing pathway. Thus, RNA silencing also functions as an anti-viral defense mechanism.

In this chapter, we describe a method for analyzing small RNA molecules from plant tissues such as leaves and flowers using the Illumina systems. Following this method, we obtained and characterized small RNA sequences from leaves of Arabidopsis and thalli of *Marchantia polymorpha* which is an emerging model liverwort recently established. Our protocol is modified from a previously published protocol [5]. This procedure does not require cloning of each miRNA as our earlier protocol but require preparing libraries for direct sequencing.

The procedure outline is shown in Fig. 1. Total RNA is isolated from plant tissue. RNA adapters are ligated to RNA at its 5′- and 3′-ends for the following reverse transcription and PCR amplification (*see* **Note 1**). Then the product is subjected to gel purification and pooled as a library product followed by cluster generation, sequencing, and bioinformatics analysis.

2 Materials

2.1 RNA Isolation

1. 1 g of Arabidopsis and *Marchantia polymorpha* tissue.
2. A mortar and pestle.
3. TRIzol® Reagent (Invitrogen, Carlsbad, CA) or RNA isoPlus (TAKARA, Otsu, Japan). *CAUTION: Phenol is toxic and corrosive. Store at 4°*C.
4. Liquid nitrogen.
5. Chloroform. *CAUTION: This is a probable carcinogen. Store at room temperature.*
6. Isopropanol. *CAUTION: This is flammable and harmful. Store at room temperature.*
7. 70 % (v/v) Ethanol.
8. RNase-free water (*see* **Note 2**).

2.2 Polyacrylamide Gel Electrophoresis (PAGE)

1. 5× TBE buffer: 445 mM Tris, 445 mM boric acid, 10 mM EDTA, pH 8.0. Store at room temperature.
2. 40 % (w/v) Acrylamide/bis solution (19:1). *CAUTION: This is a neurotoxin when unpolymerized.*
3. 20 % Ammonium persulfate (APS). *CAUTION: This is harmful. Store at −20 °C and use within 1 month.*

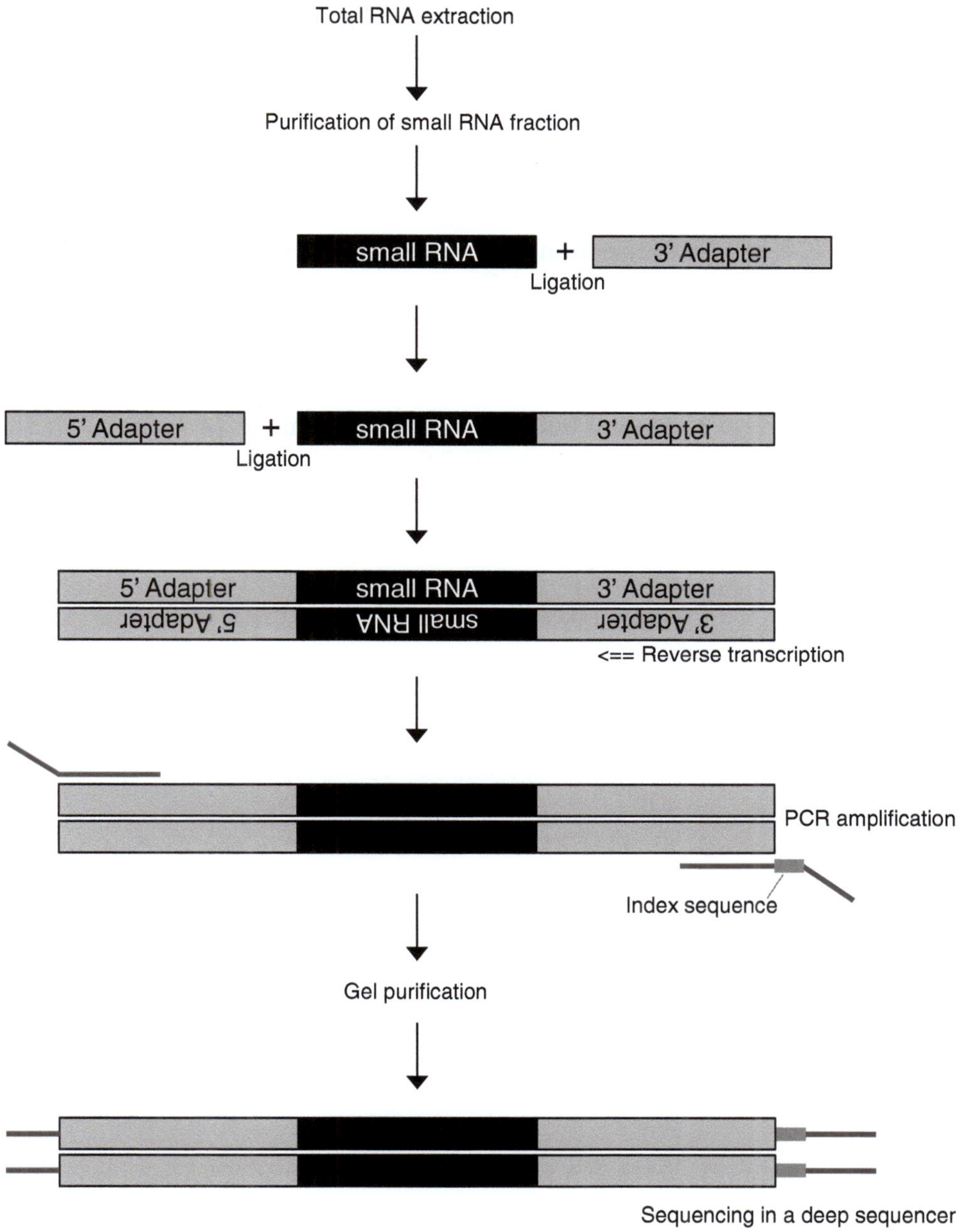

Fig. 1 A flow diagram of small RNA characterization

4. *N,N,N,N*-Tetramethylethylenediamine (TEMED). *CAUTION: This is corrosive.*
5. Urea.
6. Glass plates (11 × 10 cm), 0.1 cm spacers, and 14-well combs.
7. 4× Loading dye: 50% Glycerol, 0.03% bromophenol blue (BPB), 50 mM Tris, pH 7.7, 5 mM EDTA, pH 8.0. Store at room temperature.

8. Formamide. *CAUTION: This is a probable carcinogen. Store at −20 °C.*
9. 21 and 24 nt RNA size markers (*see* **Note 3**).
10. High-resolution ladder marker in the TruSeq Small RNA Sample Preparation Kit (Illumina, San Diego, CA).
11. Custom ladder marker in the TruSeq Small RNA Sample Preparation Kit.

2.3 Gel Purification and Precipitation of Small RNA or DNA Fraction

1. Ethidium bromide (EtBr) solution: Prepare 1 μg/mL solution. *CAUTION: This is a mutagen; handle with nitrile gloves at all times.*
2. Elution buffer for RNA or DNA: 0.5 M Ammonium acetate, 1 mM EDTA, pH 8.0.
3. Homogenization pestle for a 1.5 mL tube.
4. Ethanol, 3 M sodium acetate, and 20 μg/μL glycogen for ethanol precipitation.
5. TE buffer: 10 mM Tris, 1 mM EDTA, pH 8.0.
6. RNase-free water (*see* **Note 2**).

2.4 3′ Adapter Ligation

1. RNase-free water (*see* **Note 2**).
2. RNA 3′ Adapter in the TruSeq Small RNA Sample Preparation Kit.
3. 5× HM ligation buffer (HML) in the TruSeq Small RNA Sample Preparation Kit.
4. RNase Inhibitor in the TruSeq Small RNA Sample Preparation Kit.
5. T4 RNA ligase 2, truncated in the TruSeq Small RNA Sample Preparation Kit.
6. Stop solution in the TruSeq Small RNA Sample Preparation Kit.

2.5 5′ Adapter Ligation

1. RNA 5′ Adapter, in the TruSeq Small RNA Sample Preparation Kit.
2. 10 mM ATP, in the TruSeq Small RNA Sample Preparation Kit.
3. T4 RNA ligase in the TruSeq Small RNA Sample Preparation Kit.

2.6 Reverse Transcription

1. dNTP mixture (25 mM each) in the Tru Seq small RNA Sample Preparation Kit.
2. RNase-free water (*see* **Note 2**).
3. RNA RT primer, in the TruSeq Small RNA Sample Preparation Kit.

4. 5× First-strand buffer, in the TruSeq Small RNA Sample Preparation Kit.
5. 100 mM DTT.
6. RNase inhibitor, in the TruSeq Small RNA Sample Preparation Kit.
7. SuperScript II Reverse Transcriptase, in the TruSeq Small RNA Sample Preparation Kit.

2.7 PCR Amplification

1. RNase-free water (*see* **Note 2**).
2. 5× Phusion HF buffer, in the TruSeq Small RNA Sample Preparation Kit.
3. RNA PCR primer, in the TruSeq Small RNA Sample Preparation Kit.
4. RNA PCR primer Index, in the TruSeq Small RNA Sample Preparation Kit.
5. 25 mM dNTP mixture, in the TruSeq Small RNA Sample Preparation Kit.
6. Phusion DNA polymerase, in the TruSeq Small RNA Sample Preparation Kit.

3 Methods

3.1 RNA Isolation

To prepare RNA samples for cloning, crude total RNA samples are extracted from plant tissues using the TRIzol reagent or various equivalent RNA isolation kits (*see* **Note 4**). After this fractionation, RNAs of less than ~50 nt in length will be obtained. The procedures listed in this section should be performed using a chemical fume hood when handling TRIzol and chloroform.

3.1.1 Total RNA Extraction from Both Arabidopsis and Marchantia

1. Grind 1 g of *Arabidopsis* or *Marchantia polymorpha* tissues in liquid nitrogen to fine powder with a clean mortar and pestle.
2. Add 10 mL (for *Arabidopsis*) or 20 mL (for *Marchantia polymorpha*) of TRIzol reagent to the frozen powder and mix thoroughly (*see* **Note 5**). After it melted, transfer the sample to a 15 mL tube and incubate for 5 min at room temperature.
3. Add 2 mL of chloroform, vortex the tube, and incubate for 3 min at room temperature.
4. Centrifuge the sample at 4500 ×*g* for 20 min at 4 °C. Transfer the upper aqueous phase including RNA into a new 15 mL tube (*see* **Note 6**).
5. Add an equal volume (~5 mL) of isopropanol, vortex briefly, and keep at −20 °C for 30 min. Centrifuge the sample at 4500 ×*g* for 30 min at 4 °C.

6. Discard the supernatant and add 3 mL of 70% ethanol to the pellet. Centrifuge at 4500 × *g* for 10 min at 4 °C.
7. Discard the supernatant and air-dry the pellet. Do not vacuum-dry the RNA pellet because RNA becomes hard to dissolve if it is too dry.
8. Completely dissolve the pellet in 60 μL of RNase-free water.
9. Measure the RNA concentration using spectrometry instruments for small volumes like NanoDrop (Thermo Scientific).

3.2 Small RNA Purification and Adapter Ligation

In this section, small RNAs from total RNA isolated from plant tissues are gel-purified. To know where to excise in the gel, RNA size markers (21 and 24 nt in length) are used as the markers to place the proper molecular ranges of sample RNAs.

3.2.1 Gel Purification of Small RNA Fraction from Total RNA

1. Add 60 μL of formamide and 40 μL of 4× loading dye to total RNA sample (RNA:formamide:4 × loading dye = 1.5:1.5:1). Also add 4× loading dye to the 21 and 24 nt RNA size marker (2–3 nMs). Denature by incubating for 15 min at 65 °C. Place the tubes on ice. Load samples onto a denaturing 15% polyacrylamide gel (7 M urea, *see* **Note 7**).
2. Stain the gel in EtBr solution (*see* **Note 8**). Excise the gel containing 15–30 nt small RNAs (Fig. 2a, it is common that distinct bands are not visible) referring to the mobility of 21 and 24 nt RNA size markers. Transfer each gel slice to pre-weighed 1.5 mL tubes separately and determine the gel slice weight.
3. Add 2 volumes (v/w) of elution buffer to each gel slice and then crush with a pestle. Incubate the tubes overnight (~16 h) at 4 °C with rotation.
4. Transfer the eluate to new 1.5 mL tubes without sucking out any small gel pieces. Centrifuge at 15,000 × *g* for 5 min at 4 °C to remove any small gel remnants. Add 1 μL glycogen and 0.1 volume of 3 M sodium acetate, and mix well. Add 2.5 volumes of ethanol and mix by vortexing. Incubate the tubes at −30 °C for more than 1 h.
5. Precipitate RNAs by centrifugation at 15,000 × *g* for 30 min at 4 °C. Remove the supernatant completely. Do not wash the pellet with 70% EtOH.
6. Dissolve each RNA pellet in 10 μL of RNase-free water (*see* **Note 9**).

3.2.2 3′ Adapter Ligation

1. Transfer 5 μL of a small RNA sample to a nuclease-free 200 μL PCR tube and add 1 μL of 3′ RNA adapter. Denature RNA samples by incubating at 70 °C for 2 min and immediately place on ice.
2. Add 2 μL of 5× HM ligation buffer, 1 μL of RNase inhibitor, and 1 μL of T4 RNA Ligase 2, truncated. Mix well.

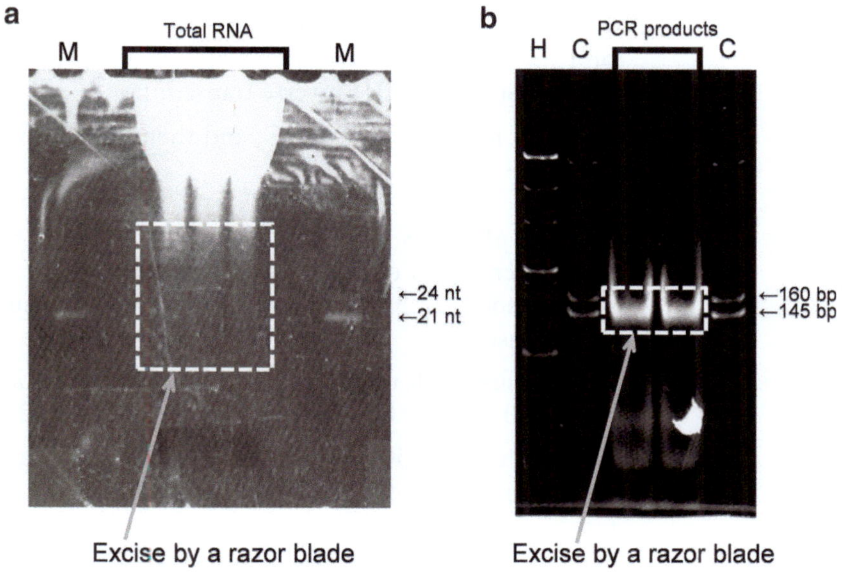

Fig. 2 Examples of band patterns in gel purification. *White dash line boxes* show areas excised by razor blades. (**a**) A 15 % acrylamide gel pattern after EtBr staining for small RNA purification. M: 21 and 24 nt small RNA markers. (**b**) A 6 % acrylamide gel pattern after EtBr staining for the purification of PCR products of adapter-ligated cDNA. *H*: High-resolution ladder marker, *C*: Custom ladder marker (both included in the TruSeq Small RNA Sample Preparation Kit)

3. Incubate the tube on the thermal cycler at 28 °C for 1 h.
4. Add 1 μL of stop solution and mix well. Continue to incubate the reaction tube on the thermal cycler at 28 °C for 15 min. And then place the tube on ice.

3.2.3 5′ Adapter Ligation

1. Transfer 1 μL of RNA 5′ adapter into a separate 200 μL PCR tube. Denature the adapter by incubating at 70 °C for 2 min and immediately place on ice.
2. Add 1 μL of 10 mM ATP and 1 μL of T4 RNA ligase to the separate 5′ RNA adapter tube and mix well.
3. Transfer 3 μL of the mix from the separate 5′ RNA adapter tube to the reaction from **step 4**, Subheading 3.2.2, of 3′ RNA adapter.
4. Incubate the tube on the thermal cycler at 28 °C for 1 h. And then place the tube on ice.

3.3 Reverse Transcription and PCR Amplification of Adapter-Ligated RNAs

In this section, 5′ and 3′ adapter-ligated small RNAs are reverse-transcribed to cDNAs. The cDNA is amplified in PCR rounds with primers. After the PCR, the bands derived from the direct ligation products of 5′ and 3′ adapters (no small RNAs in between) are removed by gel purification.

3.3.1 Reverse Transcription of the Ligation Product

1. Add 6 μL of 5′ and 3′ adapter-ligated RNA (Subheading 3.2.3) and 1 μL RT primer into a new 200 μL PCR tube. Denature RNA samples by incubating at 70 °C for 2 min and immediately place on ice.

2. Add 1 μL of 0.1 M DTT, 4 μL of 5× first-strand buffer, 0.5 μL of 12.5 mM dNTP mixture, 1 μL of RNase inhibitor, and 1 μL of SuperScriptIII reverse transcriptase.
3. Incubate for 30 min at 50 °C, and then for 15 min at 70 °C, and place on ice.

3.3.2 PCR Amplification of cDNA and Gel Purification

1. Add 0.5 μL of 25 mM dNTP mixture, 10 μL of 5× Phusion HF buffer, 2 μL of RNA PCR primer, 2 μL of RNA PCR primer index, 22.5 μL of water, and 0.5 μL of Phusion DNA polymerase.
2. Run the first PCR using the following cycle conditions: 11–15 cycles; 10 s at 98 °C, 30 s at 60 °C, and 15 s at 72 °C.
3. Run the total PCR product with the custom ladder and the high-resolution ladder markers on a 6% polyacrylamide gel (*see* **Note** 7, without urea).
4. Stain the gel in EtBr solution (*see* **Note 8**). The 145–150 bp bands are derived from 5′ and 3′ adapter-ligated 20–24 nt small RNAs (Fig. 2b). Excise the gel between 145 and 160 bp bands of the custom ladder marker.
5. Transfer each gel slice to pre-weighed 1.5 mL tubes separately and determine the gel slice weight.
6. Add 2 volumes (v/w) of elution buffer to each gel slice and then crush with a pestle. Incubate the tubes for 1 h at 4 °C with rotation.
7. Transfer the eluate to new 1.5 mL tubes without sucking out any small gel pieces. Centrifuge at 15,000 × *g* for 5 min at 4 °C to remove any small gel remnants. Add 1 μL glycogen and 0.1 volume of 3 M sodium acetate and mix well. Add 2.5 volumes of ethanol and mix by vortexing. Incubate the tubes at −30 °C for more than 1 h.
8. Precipitate cDNAs by centrifugation at 15,000 × *g* for 30 min at 4 °C. Remove the supernatant completely. Do not wash the pellet with 70% EtOH.
9. Dissolve the pellet in 10 μL TE.

3.4 Deep Sequencing of Small RNAs and Analysis of Results

3.4.1 Deep Sequencing by the Next-Generation Sequencer

For small RNA sequencing, we used Illumina sequencers (HiSeq or MiSeq, depending on the availability or the number of samples) because of its abundant outputs and the short-read sequencing system. It is recommended to refer to the manufacturer's manual for detailed information on the following process.

3.4.2 Analysis of Results

Because a deep sequencer provides abundant read outputs (usually >100 MBytes), a manual analysis is almost impossible. So, bioinformatic approaches are usually used for data analysis. Bioinformatic tools that we have used for the analysis of *Marchantia* miRNAs are listed below with brief comments. A normal personal computer (2 GB of memory is sufficient for the following analysis) with the Linux/UNIX system and R software installed is needed for the following processing.

Adapter Trimming

The sequenced small RNA reads usually contain an additional 3′ adapter sequence which was added to make the length of small RNA sequences longer than reading capacity of the next-generation sequencer machines. For removing such extra sequences from read sequences or trimming of adapter sequences, some R/Bioconductor packages are available and released for free (R: http://www.r--project.org) (Bioconductor: http://www.bioconductor.org):

- girafe (http://bioconductor.org/packages/release/bioc/html/girafe.html) [6]
- ShortRead (http://bioconductor.org/packages/release/bioc/html/ShortRead.html) [7]

Mapping Reads

Sequences obtained by the next-generation sequencer would be mapped to or aligned with reference sequences like genome or transcript data to guarantee their sequence origins. The mapping process would exclude contaminated sequences and provide various information for the following analyses.

- Bowtie: A fast mapping program on UNIX commands. Bowtie maps sequenced reads to reference sequences and output in SAM/BAM formats (http://bowtie-bio.sourceforge.net/index.shtml) [8]. (*Bowtie2 is another version of Bowtie but it is not appropriate to map reads shorter than 50 bases.)

miRNA Prediction

miRNA is one type of small RNAs which are processed from a primary precursor transcript having a characteristic hairpin structure [9]. Some programs perform prediction of *MIRNA* loci by scores of free energy of hairpin folding, the expression of miRNA* (*see* **Note 10**), the length of mature products, etc.

- ShortStack: A stand-alone Perl program performing a series of small RNA characterization like sequence mapping, the discovery of small RNA clusters, and prediction of *MIRNA* and phasing siRNA loci (http://axtell-lab-psu.weebly.com/shortstack.html) [9].
- miRDeep*: An integrated stand-alone application performing sequence alignment, pre-miRNA secondary structure calculation. Graphical displays are java-coded (http://www.australianprostatecentre.org/research/software/mirdeep-star) [10].

Other Software

Database for references and mapping for respective organisms

- miRBase: Database containing mature miRNA sequences and its precursor of all organisms (http://www.mirbase.org) [11].
- TAIR: Arabidopsis genome, gene, and protein sequence data are usable here (http://www.arabidopsis.org).
- Marchantia polymorpha: Phytozome 11 (https://phytozome.jgi.doe.gov/pz/portal.html#!info?alias=Org_Mpolymorpha_er).

Sequence browser

- Integrative Genomics Viewer (IGV): Visualizing mapping data with both sequence reads and reference sequences (https://www.broadinstitute.org/igv/) [12].

Converting file formats

- Samtools: A program working on UNIX commands which converts SAM/BAM file formats (http://samtools.sourceforge.net) [13].

4 Notes

1. In the case of Arabidopsis, DCL1 is involved in the cleavage of most miRNAs from their primary transcripts and precursors. Due to their coherent properties, most of the ever-characterized mature miRNAs have a 5′-phosphate and a 3′-hydroxyl group. In contrast, small RNAs degraded spontaneously by the inherent RNase attack have a 5′-hydroxyl and a 3′-phosphate group; thus such contaminating molecules would be neglected by the following ligation procedures and the adapters in the kit.
2. When handling RNA, all reagents should be prevented from being contaminated by RNase. All solutions should be made using commercial RNase-free water or DEPC-treated water until RNA is reverse-transcribed to cDNA. Bench tops and pipettes should be treated with RNase AWAY (Thermo Fisher Scientific, Waltham, MA).
3. 5′ Phosphorylated RNA oligos (21 and 24 nt in length): Any sequence can be used unless it includes biased nucleotide compositions or sequences. For example, we have used synthetic RNAoligos,21ntGFP:p-UGUGGCCGAGGAUGUUUCCGU, or miR164a: p-UGGAGAAGCAGGGCACGUGCA and 24 nt GFP: p-UUGUGGCCGAGGAUGUUUCCGUCC, or miR163: pUUGAAGAGGACUUGGAACUUCGAU (24 nt).
4. Common RNA purification kits using disposable spin columns are not recommended since most kits cannot trap small RNAs otherwise stated.

5. When we used 1 g *Marchantia* tissues, we could obtain 4–5 μg/μL solution, thus giving 250–300 μg RNA in total.
6. If the upper phase is not transparent or the white layer at the interface is contaminated, transfer the upper aqueous phase in a new 15 mL tube and add an equal volume of a phenol/chloroform mixture (1:1). Vortex briefly and centrifuge at 4500 ×*g* for 10 min at 4 °C. Then, transfer the upper phase to a new 15 mL tube.
7. A 15 % polyacrylamide gel (7 M or without urea): Mix 1 mL of 5× TBE buffer, 7.5 mL of 40 % acrylamide/bis solution, and 4.2 g or without urea. Mix up to 10 mL with RNase-free water. Incubate at 50 °C until urea dissolves, and then place on ice. Add 50 μL of 20 % APS and 5 μL of TEMED. Mix well by inversion, pour into glass plates, and wait for 30 min until it solidifies.
8. The gel is stained in 1 μg/mL EtBr solution for 20 min on a shaker. Handle with extreme care wearing disposable nitrile gloves. EtBr intercalates between bases of DNA and RNA. DNA or RNA bands are visualized by exposing to a UV light transilluminator.
9. Not always required, but the measurement of RNA concentration might work as a progress checkpoint. When we started from 1 g *Marchantia* tissues, we could obtain 20–25 μg/μL solution, thus giving about 200 ng small RNA in total.
10. miRNA* means a complementary side of miRNA mature sequences in a hairpin precursor. Both miRNA and miRNA* are cleaved by DCL1, and only miRNA* strand would be discarded when forming the RNA-induced silencing complex (RISC) [14–16].

Acknowledgments

We thank Drs. Minami Matsui and Yukio Kurihara at RIKEN CSRS for kind advice on sequencing. We also thank Drs. Takayuki Kohchi at Kyoto University and John Bowman at Monash University for sharing genome data information about *M. polymorpha*.

References

1. Axtell MJ (2013) Classification and comparison of small RNAs from plants. Ann Rev Plant Biol 64:137–159
2. Willmann MR, Poethig RS (2007) Conservation and evolution of miRNA regulatory programs in plant development. Curr Opin Plant Biol 10:503–511
3. Schauer SE, Jacobsen SE, Meinke DW, Ray A (2002) DICER-LIKE1: blind men and elephants in Arabidopsis development. Trends Plant Sci 7:487–491
4. Peláez P, Sanchez F (2013) Small RNAs in plant defense responses during viral and bacterial interactions: similarities and differences. Front Plant Sci 4:343

5. Tagami Y, Inaba N, Watanabe Y (2010) Cloning new small RNA sequences. In: Kovalchuk I, Zemp FJ (eds) Plant epigenetics, Methods in molecular biology. Springer, New York, NY, pp 123–138
6. Toedling J, Ciaudo C, Voinnet O, Heard E, Barillot E (2010) girafe - an R/Bioconductor package for functional exploration of aligned next-generation sequencing reads. Bioinformatics 26:2902–2903
7. Morgan M, Anders S, Lawrence M, Aboyoun P, Pagès H, Gentleman R (2009) ShortRead: a Bioconductor package for input, quality assessment and exploration of high-throughput sequence data. Bioinformatics 25: 2607–2608
8. Langmead B, Trapnell C, Pop M, Salzberg SL (2009) Ultrafast and memory-efficient alignment of short DNA sequences to the human genome. Genome Biol 10:R25
9. Axtell MJ (2013) ShortStack: comprehensive annotation and quantification of small RNA genes. RNA 19:740–751
10. An J, Lai J, Lehman ML, Nelson CC (2013) miRDeep*: an integrated application tool for miRNA identification from RNA sequencing data. Nucleic Acids Res 41:727–737
11. Kozomara A, Griffiths-Jones S (2014) miRBase: annotating high confidence microRNAs using deep sequencing data. Nucleic Acids Res 42:D68–D73
12. Robinson JT, Thorvaldsdóttir H, Winckler W, Guttman M, Lander ES, Getz G, Mesirov JP (2011) Integrative genomics viewer. Nat Biotechnol 29:24–26
13. Li H, Handsaker B, Wysoker A, Fennell T, Ruan J, Homer N, Marth G, Abecasis G, Durbin R, 1000 Genome Project Data Processing Subgroup (2009) The sequence alignment/map (SAM) format and SAMtools. Bioinformatics 25:2078–2079
14. Song L, Axtell MJ, Fedoroff NV (2010) RNA secondary structural determinants of miRNA precursor processing in Arabidopsis. Curr Biol 20:37–41
15. Iki T, Yoshikawa M, Nishikiori M, Jaudal MC, Matsumoto-Yokoyama E, Mitsuhara I, Meshi T, Ishikawa M (2010) In vitro assembly of plant RNA-induced silencing complexes facilitated by molecular chaperone HSP90. Mol Cell 39:282–291
16. Tsuzuki et al. (2016) The result of analysis of *Marchantia* miRNA appeared as a paper, Plant Cell Physiol 57:359–372

Chapter 15

Small RNA Library Preparation and Illumina Sequencing in Plants

Andriy Bilichak, Andrey Golubov, and Igor Kovalchuk

Abstract

The discovery of small RNAs in plants and animals almost two decades ago attracted a significant interest towards epigenetic regulation of gene expression and the practical implementation of the gained knowledge in applied studies. New and sometimes unexpected functions have been ascribed to sRNAs almost every couple of years since their discovery, hence indicating that the complete role of sRNAs in plant and animal physiology is still barely understood. Next-generation sequencing technologies allow to generate high-resolution profiles of sRNAs for the consequent analysis and possibly to discover novel functions of sRNAs. In this chapter, we provide brief guidelines for sRNA library preparation in plants and a practical approach that can be implemented to overcome possible difficulties with sequencing library generation.

Key words Epigenetics, Small RNAs, Sequencing library preparation, Next-generation sequencing, Illumina sequencing

1 Introduction

Small regulatory RNAs comprise an additional class of noncoding RNAs that complements the epigenetic mechanism of sequence-specific regulation of gene expression. The small RNAs (sRNAs) range in size from 20 to 27 nt. Initially, they are generated as double-stranded fragments dissected from the helical regions of the larger RNA precursors by the endonuclease activities of Dicer-like (DCL) proteins [1]. A number of regulatory functions are ascribed to sRNAs in plants, including responses to abiotic stress, immune defence against pathogens, guiding the DNA double-strand break repair machinery and the global maintenance of genome integrity both in somatic tissues and during periconceptional development that covers gametogenesis, fertilization, and early zygotic development [2–8].

Predominantly, sRNAs act by the sequence-specific suppression of gene expression through either transcriptional (DNA methylation of promoter regions) or posttranscriptional (mRNA

Igor Kovalchuck (ed.), *Plant Epigenetics: Methods and Protocols*, Methods in Molecular Biology, vol. 1456, DOI 10.1007/978-1-4899-7708-3_15,

degradation) gene silencing of complementary DNA and RNA, respectively. In both pathways, sRNAs are used as guides that direct effector proteins to the target nucleic acid molecules through base-pairing interactions [9].

Surprisingly, there is a flip side of sRNA action that has been recently discovered in animals, and the emerging evidence indicates that the similar pathway may also exist in plants. RNA activation is a mechanism of up-regulation of gene expression prompted by sRNAs termed "small activating RNAs" (saRNAs) [10–12]. In plants, saRNAs have been reported to act either through targeting DNA methylation to a putative negative cis-element or DNA demethylation of the coding sequence of the gene [13, 14]. The field of RNA activation is still in its infancy, but it holds enormous potential for applied studies if the complete mechanism of saRNA action is revealed.

A massive amount of data produced by the next-generation sequencing technologies allowed to discover a variety of sRNAs expressed from either endogenous loci or invading viral genomes. The most recent attempt to classify sRNAs in plants has divided them into two primary classes based on their origin: small interfering RNAs (siRNAs) and hairpin RNAs (hpRNAs) [1]. Whereas the former ones derive from the processing of dsRNA, the latter ones are dissected from the helical region of a self-complementary single-stranded RNA. hpRNAs are further subdivided into microRNAs (miRNAs) and others. siRNAs diverge into heterochromatic siRNAs, secondary siRNAs, and natural antisense transcript siRNAs.

The NGS technology offers the possibility to capture a snapshot of the sRNA abundance in cells or tissues under specific physiological conditions. This information can be eventually related to the transcript level of a given gene [15], DNA methylation and histone modifications [16], or any other physiological condition of interest. The library preparation for sRNA sequencing includes six major steps: total RNA isolation, RNA 3′-adapter ligation, RNA 5′-adapter ligation, reverse-transcription PCR for the first-strand synthesis, PCR amplification with the barcoded primers, and gel purification of the pooled library. The specific ligation of the adapters to sRNAs that bear the 5′-phosphate and 3′-hydroxyl groups as a result of the activity of DCL proteins is used to selectively enrich sRNA fraction during library preparation. The Illumuna Company offers an exhaustive protocol for sRNA library preparation for NGS platforms like Genome Analyzer and others (TruSeq® Small RNA Sample Preparation Guide, Illumina, Inc.). Therefore in this chapter, we provide a brief overview of this protocol with the emphasis on sRNA library preparation for plants and the potential pitfalls and solutions that one can implement for the efficient and reproducible construction of sRNA libraries.

2 Materials

2.1 Adapter Ligation

1. TruSeq Small RNA Sample Preparation Kit (Illumina, USA).
2. 5 μl (200 ng/μl) of total RNA.
3. T4 RNA Ligase 2 (Deletion Mutant) (Epicentre, USA).
4. Nuclease-free water.

2.2 cDNA Synthesis and Amplification

1. SuperScript II Reverse Transcriptase (Invitrogen, USA).
2. Nuclease-free water.

2.3 PCR Fragment Purification and Library Validation

1. 6% Polyacrylamide gel (for two gels: 4 ml of 40% acrylamide/Bis solution 19:1, 15.8 ml of the 1× TBE buffer, 0.2 ml of 10% ammonium persulfate, 12 μl of TEMED).
2. 1× TBE: 90 mM Tris, pH 8.0, 90 mM boric acid, 2 mM EDTA.
3. 6× DNA gel loading buffer.
4. Nuclease-free water.
5. Razor blade.
6. Gel Breaker tube (IST Engineering Inc., USA).
7. 5 μm Filter (IST Engineering Inc., USA).
8. 100% Ethanol.
9. 70% Ethanol.
10. 10 mM Tris–HCl, pH 8.5.
11. Agilent High Sensitivity DNA Kit (Agilent Technologies, USA).

3 Methods

3.1 Adapter Ligation

1. Ligate the 3′-adapter (*see* **Notes 1–3**).
   Mix thoroughly 1 μl of RNA 3′ adapter (RA3) and 5 μl (200 ng/μl) of total RNA. Incubate it at 70 °C for 2 min in the preheated thermal cycler, and then immediately place it on ice. In a separate tube, prepare a master mix consisting of 2 μl of the ligation buffer (HML), 2 μl of RNase inhibitor, and 1 μl of T4 RNA ligase 2 (deletion mutant). Prepare a ligation mixture: mix by pipetting the master mix and the RA3/total RNA mix; place it in the preheated thermal cycler, and incubate it at 28 °C for 1 h. Add 1 μl of stop solution (STP) to the ligation mixture (the ligation mixture should remain in the thermo cycler), mix them thoroughly by pipetting, and continue to incubate at 28 °C for 15 min. Place it immediately on ice after incubation.
2. Ligate the 5′-adapter.

Prepare the 5′-adapter master mix:

1 μl of the RNA 5′-adapter (RA5) per sample should be heat-denatured in a thermal cycler at 70 °C for 2 min. Place it immediately on ice after incubation. Add 1 μl of 10 mM ATP and 1 μl of T4 RNA ligase (per sample) to the denatured RNA 5′-adapter and mix them by pipetting.

Ligation:

Add 3 μl of the 5′-adapter master mix to each 3′-adapter ligation mixture, mix them by pipetting, and incubate it at 28 °C for 1 h in a preheated thermal cycler. Place it immediately on ice after incubation.

3.2 cDNA Synthesis and Amplification

1. cDNA preparation

   Mix thoroughly by pipetting 6 μl of 5′/3′-adapter-ligated RNA and 1 μl of RNA RT primer (RTP). Incubate the mixture at 70 °C for 2 min in a preheated thermal cycler and then immediately place the tube on ice.

   Prepare cDNA synthesis master mix:

   Mix together on ice: 2 μl of 5× first-strand buffer, 0.5 μl of 12.5 mM dNTP mix, 1 μl of 100 mM DTT, 1 μl of RNase inhibitor, and 1 μl of SuperScript II Reverse Transcriptase.

   Add 5.5 μl of the master mix to the reaction tube with 5′/3′-adapter-ligated RNA and RNA RT Primer (RTP), mix them by pipetting, and incubate it at 50 °C for 1 h in a preheated thermal cycler. Place it immediately on ice after incubation.

2. PCR amplification

   Prepare PCR master mix:

   Combine together 25 μl of PCR mix (PML), 2 μl of RNA PCR primer (RP1), 8.5 μl of ultrapure water.

   Add 2 μl of a unique RNA PCR Primer Index (RPI) and 35.5 μl of PCR master mix into a tube with cDNA, and mix them by pipetting. Run PCR in the thermal cycler using the following PCR cycling conditions: the initial denaturation at 98 °C for 30 s, 11 cycles of 98 °C for 10 s, 60 °C for 30 s, and 72 °C for 15 s followed by a final extension at 72 °C for 10 min; samples can be held at 4 °C in the cycler.

3.3 PCR Fragment Purification and Library Validation

1. Purify PCR fragments

   Mix all the PCR fragments with 6× DNA loading dye (the total volume should not exceed 50 μl), 2 μl of custom ladder with 2 μl of DNA loading dye, and 2 μl of high-resolution ladder with 2 μl of DNA loading dye. Prepare 6% polyacrylamide gel. Load wells 1 and 6 with 2 μl of high-resolution ladder/loading dye mix, wells 2 and 5 with 2 μl of custom ladder/loading dye mix, and wells 3 and 4 with a mix of PCR fragments with loading dye (maximum 25 μl per well). Run PAGE in 1× TBE buffer for 60 min at 145 V or until the blue front dye exits the gel.

Stain the gel with ethidium bromide (0.5 μg/ml in water) for 2–3 min, rinse it with distilled water, and visualize DNA on a UV transilluminator. There are three dsDNA fragments of 145 bp, 160 bp, and 500 bp in the custom ladder.

Excise the gel fragments (lanes 3 and 4) between the 160 and 145 bp bands of the custom ladder using a razor blade and place them into the 0.5 ml Gel Breaker tube (IST Engineering Inc., USA) inserted into a 2 ml tube. Centrifuge the tubes at 20,000 ×*g* for 2 min at room temperature. To elute DNA, add 300 μl of ultrapure water to the gel debris in the 2 ml tube and incubate them with shaking for at least 2 h (preferably overnight) at room temperature. Transfer the eluate/gel debris mix to the top of a 5 μm filter (IST Engineering Inc., USA). Centrifuge them for 10 s at 600 ×*g*, and then discard the filter. Precipitate DNA by adding to the eluate 2 μl of glycogen, 30 μl of 3 M NaOAc, and 975 μl of 100% ethanol. Mix them by vortexing and spin them down at 20,000 ×*g* for 20 min at 4 °C. Discard the supernatant and wash the DNA pellet once in 500 μl of 70% ethanol. Air-dry the pellet at room temperature for 20–30 min, and dissolve it in 10 μl of 10 mM Tris–HCl, pH 8.5 (*see* **Note 4**).

2. Validate library
   Validate library on an Agilent Technologies 2100 Bioanalyzer using High Sensitivity DNA chip. There might be more than one peak. In this case, a summary of all peak molarities should be used to calculate the total molarity (*see* Figs. 1, 2, and 3 and **Note 5**). DNA should be diluted to 2 nM with Tris–HCl 10 mM, pH 8.5.

4 Notes

1. You have to double-check that you have two consumables that are not supplied with the Illumina's TruSeq Small RNA Sample Preparation kit: T4 RNA Ligase 2 (Deletion Mutant) (Epicentre, USA) and SuperScript II Reverse Transcriptase (Invitrogen, USA).
2. The kit is optimized for 1 μg of total RNA with RIN higher than 8, dissolved in 5 μl of nuclease-free water.
3. Add extra 10% of each reagent in the case of master mix preparation: all recipes are given for one sample.

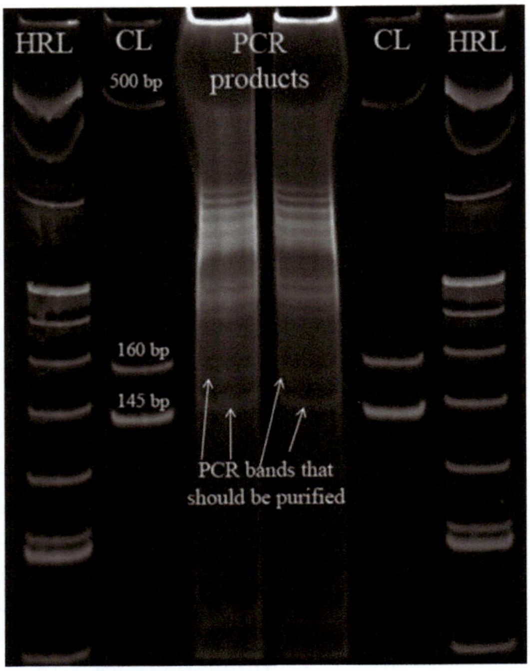

Fig. 1 6 % PAGE before cutting. *HRL* high-resolution ladder, *CL* custom ladder, PCR products—two lanes loaded with PCR reaction products, PCR bands that should be purified—the *white arrows* point to four faint bands that should be purified

4. It is possible to use EB buffer (elution buffer) from either QIAGEN PCR Purification Kit or QIAGEN DNA Miniprep Kit.
5. Four white arrows point on the PCR fragments that should be purified (Fig. 1). The gel can be cut between 160 and 145 bp bands ("CL" lane, Fig. 2). Figure 3 shows the electropherogram of a typical High Sensitivity DNA chip with the purified PCR fragments (144 and 154 bp peaks, the purified PCR bands). The Bioanalyzer software for data analysis will also show concentrations for each peak, for example, 12 nm for 144 bp peak and 8 nm for 154 bp peak. In this case, molarities of both peaks should be summarized (12+8=20 nM). The sample then should be diluted to 2 nM with Tris–HCl 10 mM, pH 8.5.

Acknowledgements

We acknowledge the financial support of Alberta Innovates Biosolutions and Natural Sciences and Engineering Research Council of Canada grants to Igor Kovalchuk and Alberta Innovates Technology Futures for scholarship to Andriy Bilichak. We thank Valentina Titova for proofreading the manuscript.

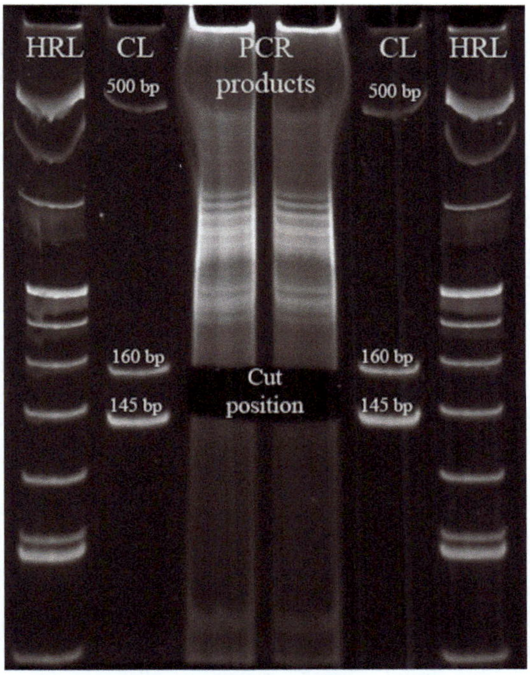

Fig. 2 6 % PAGE after cutting. *HRL* high-resolution ladder, *CL* custom ladder, PCR products—two lanes loaded with PCR reaction products, cut position—the gel slice containing PCR fragments that were cut out to undergo purification

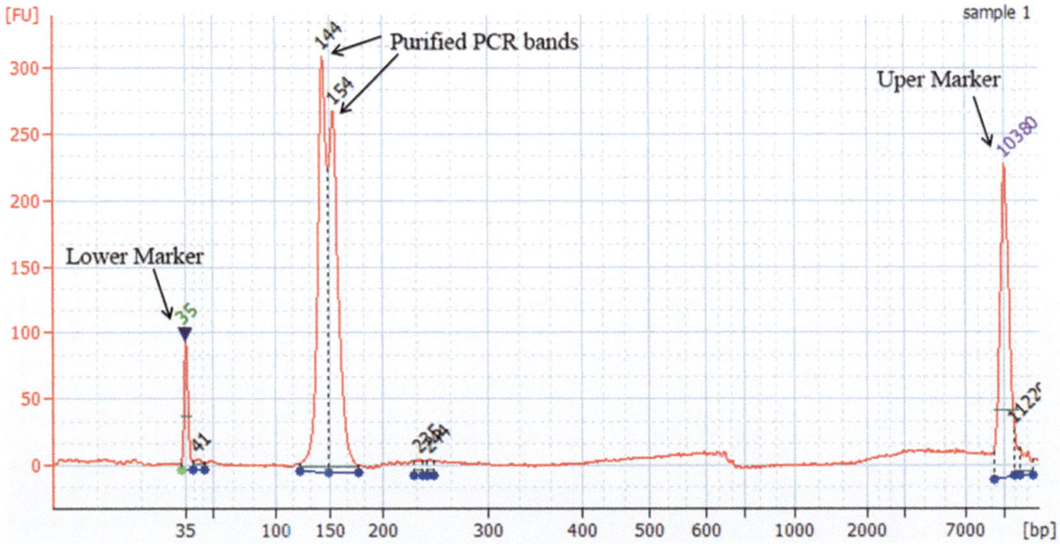

Fig. 3 Electropherograms of high-sensitivity DNA chips. *FU* fluorescent units, *bp* base pairs, 144—a 144 bp PCR fragment, 154—a 154 bp PCR fragment

References

1. Axtell MJ (2013) Classification and comparison of small RNAs from plants. Annu Rev Plant Biol 64:137–159
2. Slotkin RK et al (2009) Epigenetic reprogramming and small RNA silencing of transposable elements in pollen. Cell 136(3):461–472
3. Bourc'his D, Voinnet O (2010) A small-RNA perspective on gametogenesis, fertilization, and early zygotic development. Science 330(6004):617–622
4. Sunkar R, Chinnusamy V, Zhu J, Zhu JK (2007) Small RNAs as big players in plant abiotic stress responses and nutrient deprivation. Trends Plant Sci 12(7):301–309
5. Seo JK et al (2013) Contribution of small RNA pathway components in plant immunity. Mol Plant Microbe Interact 26(6):617–625
6. Jin H (2008) Endogenous small RNAs and antibacterial immunity in plants. FEBS Lett 582(18):2679–2684
7. Wei W et al (2012) A role for small RNAs in DNA double-strand break repair. Cell 149(1):101–112
8. Gao M et al (2014) Ago2 facilitates Rad51 recruitment and DNA double-strand break repair by homologous recombination. Cell Res 24(5):532–541
9. Carthew RW, Sontheimer EJ (2009) Origins and mechanisms of miRNAs and siRNAs. Cell 136(4):642–655
10. Li LC et al (2006) Small dsRNAs induce transcriptional activation in human cells. Proc Natl Acad Sci U S A 103(46):17337–17342
11. Janowski BA et al (2007) Activating gene expression in mammalian cells with promoter-targeted duplex RNAs. Nat Chem Biol 3(3):166–173
12. Huang V et al (2010) RNAa is conserved in mammalian cells. PLoS One 5(1):e8848
13. Shibuya K, Fukushima S, Takatsuji H (2009) RNA-directed DNA methylation induces transcriptional activation in plants. Proc Natl Acad Sci U S A 106(5):1660–1665
14. Wojtasik W et al (2014) Oligonucleotide treatment causes flax beta-glucanase up-regulation via changes in gene-body methylation. BMC Plant Biol 14:261
15. Bilichak A et al (2015) The elucidation of stress memory inheritance in Brassica rapa plants. Front Plant Sci 6:5
16. Saze H et al (2012) DNA methylation in plants: relationship to small RNAs and histone modifications, and functions in transposon inactivation. Plant Cell Physiol 53(5):766–784

Chapter 16

Bioinformatics Analysis of Small RNA Transcriptomes: The Detailed Workflow

Slava Ilnytskyy and Andriy Bilichak

Abstract

Next-generation sequencing became a method of choice for the investigation of small RNA transcriptomes in plants and animals. Although a technical side of sequencing itself is becoming routine, and experimental costs are affordable, data analysis still remains a challenge, especially for researchers with limited computational experience. Here, we present a detailed description of a computational workflow designed to take raw sequencing reads as input, to obtain small RNA predictions, and to detect the differentially expressed microRNAs as a result. The exact commands and pieces of code are provided and hopefully can be adapted and used by other researchers to facilitate the study of small RNA regulation.

Key words Bioinformatics, Small RNA, microRNA, DESeq, Differential expression, Next-generation sequencing, Illumina

1 Introduction

The development of next-generation sequencing (NGS) has opened a world of possibilities to study small RNA (sRNA) transcriptome in living cells. The previously existing microarray platforms have allowed to obtain simultaneous expression estimates for thousands of transcribed entities, but their application is limited only to the known annotated features, cross-hybridization may pose a problem when examining short transcripts (only 21–24 nucleotides (nt) in length), the reliance on fluorescence intensity measurements leads to a limited dynamic range, and finally the study of transcript variance and transcription structure using microarrays is very difficult. NGS, on the other hand, is largely devoid of such issues. Sequencing does not require a prior knowledge of the transcript sequence, thus making a discovery of novel transcripts or transcript classes as well as hypothesis-free investigations possible. The digital nature of NGS (the read is either present or not) increases a dynamic range of transcript expression estimation;

Igor Kovalchuck (ed.), *Plant Epigenetics: Methods and Protocols*, Methods in Molecular Biology, vol. 1456, DOI 10.1007/978-1-4899-7708-3_16,

therefore features with very high and very low expression levels can be detected and compared with a greater precision.

These features make NGS a method of choice in studying sRNA transcriptomes where transcripts under investigation are short in length (17–40 nt); they have diverse cellular functions and biogenesis; they originate from various coding and noncoding loci that are frequently with no annotation available; and, finally, they often undergo posttranscriptional processing and RNA editing that lead to the formation of functionally distinctive isoforms that can only be differentiated by a detailed sequence analysis.

Today, the preparation and sequencing of libraries from sRNA fragments have become a routine procedure accessible to any researcher with some experience in experimental molecular biology. A number of commercial companies offer reagent kits that make library preparations fast and reliable. As the technology matures, NGS experiments are also becoming much cheaper, and labeling fragment libraries with unique DNA barcodes (multiplexing) allows for the simultaneous sequencing of multiple samples, thus making the use of "sequencing real estate" even more efficient.

The main issue with the use of NGS in biological experiments arises from its power—a single sequencing run can generate hundreds of millions of sequencing reads. The analysis of this mass of data requires a researcher to develop and use much more advanced computational and statistical skills to answer biological questions, discover novel phenomena, and make decisions to reject or accept hypotheses. In this text, we would like to introduce researchers who are interested in the analysis of small RNA sequencing to the computational workflow that starts with raw data taking a collection of sequencing reads and becomes a tool that provides quality control for sequencing libraries by trimming, filtering, and mapping reads to the genome, predicting novel sRNAs, and detecting deferentially expressed miRNAs (Fig. 1). The workflow description contains exact commands and pieces of code that can be copied and run directly after generic file names are substituted by the actual file names used by a researcher.

This workflow is applicable to single-end sRNA sequencing libraries obtained with the use of Illumina sequencing platform. Bioinformatics software used to complete various steps of the workflow is freely available, published, and tested in real-life research (*see* Subheading 2.3 for the list of software used). Most of academic bioinformatics software is written for Linux environment and used as command-line tools; commands are typed in Linux terminal rather than executed with a click of the mouse. Statistical approaches used for the detection of differentially expressed entities are frequently implemented as function packages for R language; therefore some familiarity with Linux command line and R is absolutely required to complete NGS data set analysis. We invite readers to check some of the material freely available on the Web to introduce themselves to Linux shell and R syntax and the overall questions that arise in the process of working with NGS (Table 1).

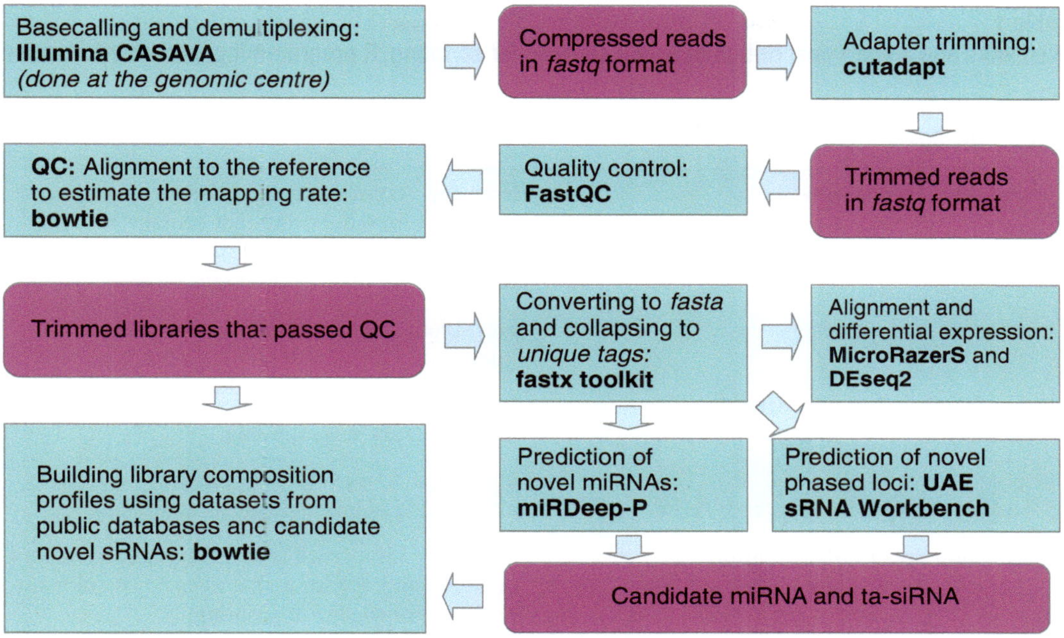

Fig. 1 Schematic outline of steps of a bioinformatics workflow designed to perform a comprehensive analysis of single-end Illumina small RNA libraries

Commands entered in Linux shell start with ">", which served as command prompt, and it may appear differently on your system. The names of files or folders that the operations are performed on are shown like this: <file_name>, and they have to be substituted by the actual file or folder names in order for commands to work. Whenever we are using a software application as a part of this workflow example, we are trying to discuss some of its uses and limitations and to explain options invoked to modify its behavior. More information can be always gained from manuals available through software distributions.

2 Data, Workstation, and Software Requirements

2.1 Sequencing Libraries and Raw Data

Small RNA sequencing libraries used in this workflow were assembled with the TruSeq small RNA sample preparation kit; samples were multiplexed and sequenced on Illumina GAIIx genome analyzer. The sequencing was single end with the total length of sequencing reads at 36 bases out of which 7 bases belonged to the barcode, leaving 29 base pairs (bps) of usable read length.

Base-calling and demultiplexing were performed using Illumina CASAVA-1.8.2 pipeline with default settings, which resulted in the generation of raw sequencing data in fastq format. Base quality score encoding in CASAVA 1.8 versions or higher uses the standard offset

Table 1
Web links and a short description of useful sources for scripting, R programming, and bioinformatics

Web page link	Description
http://seqanswers.com/	A large and very active community forum of people working with sequencing that contains a wealth of answered questions, discussions, and example workflows for many applications of NGS.
https://www.biostars.org/	A great bioinformatics board to ask and answer questions.
http://rosalind.info/problems/locations/	A free platform for learning bioinformatics and bioinformatics programming.
http://korflab.ucdavis.edu/unix_and_Perl/	A free course to introduce people with no prior experience to Unix and perl programming.
http://tldp.org/LDP/abs/html/	An extensive Linux shell (bash) tutorial suitable for reference and self-study that does not require any previous knowledge of scripting.
http://stackoverflow.com/	A great question and answer board for scripting and programming.
http://tryr.codeschool.com/	An interactive introduction to R for complete beginners.
http://swirlstats.com/	Swirl is a neat package that provides an interactive and self-paced learning environment to learn R (requires RStudio).
http://www.ubuntu.com/	A great Linux distribution that is very easy to use and install, an active community of users who share the information.
http://cran.r-project.org/	Home page of R, with the newest R distributions, instruction manuals, tutorials, and packages.
http://www.rstudio.com/	A very convenient development environment for R.
http://bioconductor.org/	A huge collection of R packages used in bioinformatics, biostatistics, and life science.

value of 33 rather than 64 in previous versions. In order to increase the efficiency of memory usage, CASAVA 1.8 splits the output of reads into several files and compresses it in GNU zip format (*see* **Note 1**).

Short sequencing reads generated by CASAVA 1.8 serve as a primary input into the computational workflow described in this chapter, and they can be considered raw data for our purpose.

2.2 Workstation

The bioinformatics workflow described here was completed on HP Pavilion Elite workstation with Intel(R) Core(TM) i7 CPU 930 @

2.80 Ghz and 8 Gb of RAM running an Ubuntu 11.04 (Natty Narwahl) 64-bit operation system.

2.3 A List of Software Used

cutadapt-0.9.5—https://code.google.com/p/cutadapt/

fastqc-0.10.0—http://www.bioinformatics.babraham.ac.uk/projects/fastqc/

bowtie version 0.12.7—http://bowtie-bio.sourceforge.net/index.shtml

miRDP1.3—http://faculty.virginia.edu/lilab/miRDP/

ViennaRNA-1.8.5—http://www.tbi.univie.ac.at/RNA/

srna-workbench 2.3.2—http://srna-workbench.cmp.uea.ac.uk/

samtools-0.1.18—http://samtools.sourceforge.net/

bedtools v2.16.2—http://bedtools.readthedocs.org/en/latest/

MicroRazerS version 0.1—https://www.seqan.de/projects/microrazers/

FASTX Toolkit 0.0.13—http://hannonlab.cshl.edu/fastx_toolkit/

3 Small RNA Analysis Workflow

3.1 Adapter Trimming

By definition, small RNAs fall in the range of approximately 15–40 base pairs, while the usable short read length generated by sequencing is 29 bps; therefore in many instances, we would expect the sequencing to run past the endogenous DNA fragment and into Illumina adapter. The presence of adapter as a part of a sequencing read drastically reduces alignment efficiency; hence adapters have to be removed prior to the genomic alignment stage of the workflow.

The adapters were trimmed using cutadapt software [1] which can be downloaded at this link: https://code.google.com/p/cutadapt/. Cutadapt is used as a command-line utility in linux shell:

```
> cutadapt -b <adapter_fasta> -m 17 -q 20 <input_fastq> >
  <trimmed_fastq>
```

Here, cutadapt was instructed to search for adapter sequence (in fasta format) anywhere within the read (-b flag, multiple -b flags can be used), to retain sequences with at least 17 bps (-m) and perform quality trimming with a quality cutoff of 20 on the Sanger scale (-q). The trimmed sequences are saved in fastq format, and all reads shorter than 17 bps are discarded in the process.

3.2 Initial Quality Control with FastQC Software

A number of useful metrics of short read libraries are visualized with FastQC software developed by Simon Andrews and available at http://www.bioinformatics.babraham.ac.uk/projects/fastqc/.

FastQC takes sequences in various formats (fastqc, fastqc.gz, SAM, BAM) and produces a comprehensive sequence quality report that includes the information on basic parameters of the library such as the total number of reads and file type, encoding a range of sequence lengths. FastQC is platform independent and can be run as both graphical user interface (GUI) and command-line utility.

Detailed reports are provided in both text and html format and include the information on base quality score distributions, per-sequence base content, GC content, N content, sequence length distribution, and sequence duplication levels. In addition, FastQC also lists sequences that occupy over 0.1 % of the total library size and matches them against the database of adapters and primers used in library construction for various NGS platforms.

Finally, it examines k-mer distribution profiles where k-mers are short oligomers (5 bps in length in the case of FastQC). FastQC counts the number of occurrences for each possible k-mer and calculates the observed-to-expected ratio of occurrence at every nucleotide position. It reports a list of all k-mer sequences with the associated expected-to-observed ratios and nucleotide positions where the encountered observed-to-expected ratios were the highest. The relative enrichment for the top six most over-represented k-mers is displayed on a line graph to allow a convenient examination of k-mer enrichment biases. For instance, a dramatic increase of k-mer bias at the 5′ or 3′ ends of the sequence may indicate the presence of adapters or polyA tails.

When using FastQC with sRNA sequencing, we have to keep in mind that their characteristics are substantially different from genomic and mRNA short-read libraries. These differences arise primarily from a short length of sRNA fragments and their lower diversity as compared to mRNA transcriptome- or genome-derived fragment libraries; therefore the quality criteria applied to sRNA libraries are different from those used in other common NGS workflows. When applying FastQC to sRNA libraries, we pay attention to the total number of reads in sample libraries before and after adapter and quality trimming to estimate the loss of reads due to trimming and spot the libraries with an unusually high adapter contamination and/or a lower sequence quality. Next, we examine per-base and per-sequence quality plots to ensure that the majority of bases have low error rates (Fig. 2).

Base quality over 30 on the Sanger scale (*see* **Note 2**) is considered to be high for NGS projects. In our experience, sequence quality is usually not a problem with sRNA sequences because sRNA fragments are quite short, whereas quality values in the case of Illumina sequencing protocols tend to deteriorate with an increase in read lengths. Since sRNA libraries are dominated by a small number of very abundant sequences, their per-base sequence and GC content will be severely biased compared to mRNA-seq and genome sequencing libraries; therefore, base content imbalance should not raise alarms with sRNA-seq.

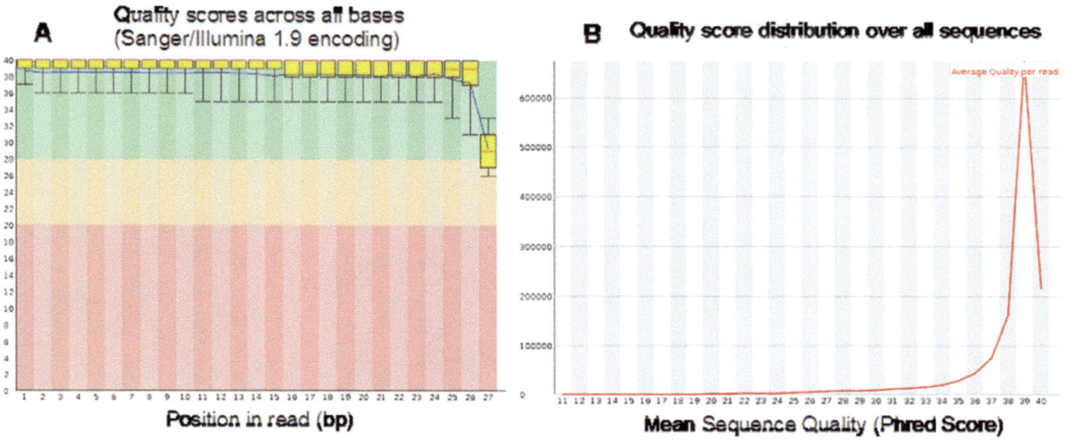

Fig. 2 Per-base (**a**) and per-sequence quality scores (**b**) calculated and charted by FastQC. This particular library was obtained from the apical meristem of Brassica rapa and sequenced using the Illumina GAIIx genome analyzer. The quality score distribution demonstrates high-quality, low-error-rate reads where the majority of the individual base qualities and mean sequence qualities are over 30 on the Sanger scale (1/1000 probability of a wrong base call). Axis titles were reprinted with the larger font to enhance presentation quality

It is always instructive to examine the read length distribution profile after trimming in order to gain an initial insight into sRNA library composition (Fig. 3). For instance, in plants, we will frequently observe two distinctive dominant peaks at 21 and 24 nt where the 21-nt-long fraction will most likely consist of lineage-specific miRNAs, while the 24-nt peak may be formed by long miRNAs and heterochromatic siRNAs [2]. The length distribution profile from small RNA libraries may vary significantly between different tissues as it reflects genuine differences in their biological makeup [3].

Duplicate reads arise as PCR artifacts during the library amplification step, or they can be sequenced from identical endogenous DNA or RNA fragments, and the removal of the former ones helps mitigate PCR biases introduced at the library amplification step. Although duplicate removal is performed routinely in a number of NGS workflows [4, 5], this step should not be included into sRNA sequencing analysis. Since sRNA libraries tend to be composed of the highly expressed short RNA molecules with identical or almost identical sequences, duplicate read filtering would completely remove a large number of highly expressed sRNAs from a sample's transcriptional profile, thus creating a completely skewed and unrealistic picture of sRNA expression.

Low diversity of sRNA sequencing libraries causes FastQC to detect a large number of over-represented sequences that occupy over 0.1% of the total number of sequences in the library. This should not be considered a quality problem unless some of those sequences match the adapters or primers used in library construction. In such cases, the second round of trimming should be performed by contaminating sequences to finalize library cleanup. We

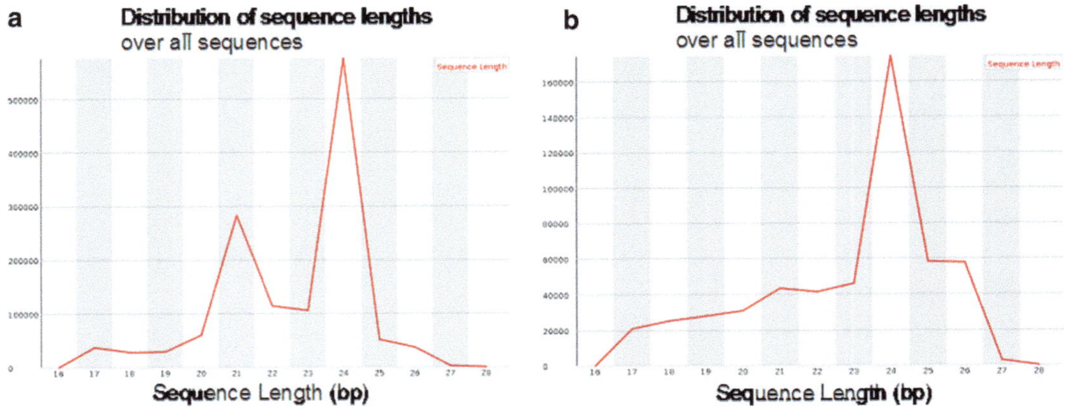

Fig. 3 Examples of length distribution profiles built using FastQC. The sequencing libraries were obtained from apical meristem (**a**) and ovules (**b**) of Brassica rapa. Notice the difference between tissues: two distinctive 21 nt and 24 nt peaks in apical meristem sample that probably correspond to miRNAs and heterochromatic siRNAs, and a single dominant 24 nt peak in ovules, pointing at dramatic differences in sRNA-directed regulation between cell types. Axis titles were reprinted with the larger font to enhance presentation quality

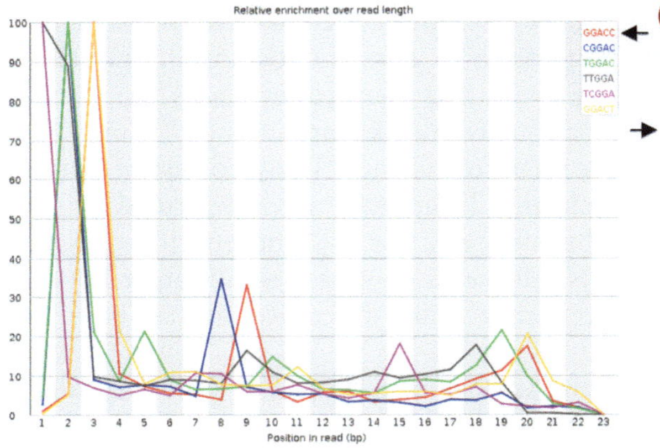

Overrepresented sequences

Sequence	Count	Percentage	Possible Source
TGTCGTCCAGCGGTTAGGATATCT	30389	2.2967630371477177	No Hit
TCGGACCAGGCTTCATTCCCC	29044	2.1951096005435624	No Hit
TTTCCAAATGTAGACAAAGCA	18450	1.3944281824138798	No Hit
TTTGGATTGAAGGGAGCTCTA	18216	1.3767427518076552	No Hit
TTGGACTGAAGGGAGCTCCCT	13455	1.016912259857927	No Hit
GCGAGAGCGGACCGCCT	9195	0.694946728308706	No Hit
AGTTACTAATTCATGATCTGGC	7909	0.597752438737744	No Hit
ACGACCTGAGAACGATGAAA	6617	0.5001046765871351	No Hit
TTGGACTGAAGGGAGCTCCC	5351	0.4044219622816624	No Hit
TTAGATTCACGCACAAACTCG	5106	0.3859051652794185	No Hit

Fig. 4 The relative enrichment of 5 nt k-mers along the length of the read sequence and top ten over-represented sequences detected by FastQC. The top enriched k-mers originate from the highly over-represented sequences pointed by arrows; for example, the enriched k-mer GGACC originates from position 3 in the second most enriched read. Such k-mer imbalance is unavoidable in the sRNA library due to their low diversity and the presence of highly expressed identical short sequences

still expect to observe a significant enrichment of certain k-mers even when contaminating sequences are fully removed; however, it is easy to confirm that they originate from the top over-represented endogenous sRNA reads (Fig. 4).

3.3 Aligning Reads to the Reference Genome

Although FastQC analysis is an extremely useful step in quality control in sample libraries, it cannot provide all the necessary information because it offers no data on the fraction of reads that can be aligned to the reference genome (*see* **Note 3**). In order to do this, a genomic alignment step has to be introduced. A genomic alignment of short reads is performed using a specialized software—a short read aligner. Numerous aligners have been developed to date; most of them are the freely available command-line tools designed for the Unix-like operating systems (for a useful review and performance comparison, see Lindner and Friedel [6]). Bowtie is a stable, easy-to-use, and well-documented aligner which combines extremely low runtime and memory requirements with a high precision and recall rate [6, 7]; it serves as our aligner of choice when matching sRNA sequences to the reference genome is required. Bowtie is also fully fastq and sam compatible (*see* **Note 4**).

Prior to use, bowtie requires building an index of reference sequence (in this case, the genome of interest), which can be done with the following command:

```
> bowtie-build <genome_fasta> <index_name>
```

In order to match sRNA to the genome, we use bowtie as follows:

```
> bowtie -v 2 --best -p 4 -m 1 --al <uniquely_aligned.fastq> --un
    <unaligned.fastq> --max <multiple.fastq> <index> <trimmed_
    reads.fastq>
```

Since short RNAs can undergo RNA editing and due to a relatively high error rate of NGS, we allowed up to two mismatches in the alignment (-v 2); --best option made sure that the reported alignments were always the best in terms of stratum and quality (*see* **Note 5**); -p specified the number of processor cores to use; only output unique alignments (-m 1) used in combination with --max allowed to write all multi-matching reads to a file; --al and --un options allowed to write the uniquely aligning and non-matching reads, respectively, to a file.

After all reads with unique and multiple alignments as well as those with no match at all are written to separate files, their numbers can be counted in order to obtain a basic idea about mapping efficiency. Libraries with the dramatically reduced mapping efficiencies may have technical problems that avoid the detection during the previous quality control steps.

The number of reads in the fastq file can be counted using the following shell command:

```
> wc -l <fastq_file> | cut -f1 -d ' ' | awk '{print $1/4}'
```

Note that we had to divide the number of lines in the file by 4—**awk '{print $1/4}'**; this was required because every sequence in the fastq format is represented by four lines of text: the name of the read, read sequence itself, a spacer, and the base quality string (Fig. 5).

Mapping efficiency can be calculated as a ratio of reads aligning to the genome to the total reads in a library:

$$E = (U + M) / (U + M + N)$$

```
@HWUSI-EAS1814:22:FC:2:120:12767:21186 1:Y:0:CGGAAT
GGTGGCTGTAGTTTAGTGGT
+
BB@@@@B@@B@BBB@@@B@@
@HWUSI-EAS1814:22:FC:2:120:16069:21186 1:Y:0:CGGAAT
GGTGGCTGTAGTTTAGTGGT
+
BB@@@AAAAAAAAAAAAAAA
```

Fig. 5 An example of two sequencing reads generated by the CASAVA Illumina base-calling and demultiplexing pipeline. Each read consists of four lines: a sequence identifier formed according to the Illumina conventions, a sequence itself, a spacer, and coded quality string

where

E—is mapping efficiency.

U—the number of reads with unique alignments.

M—the number of reads with multiple alignments.

N—the number of reads with no match to the genome.

4 Prediction of Novel miRNAs

Prediction of novel miRNAs is a common task in sRNA profiling which can provide a researcher with a more complete and clear picture of regulatory networks supporting a plant cell's function. Currently, a number of software tools are available to deal with this task; examples include miRDeep-P [8], miRDeepFinder [9], miR-Cat [10], and ShortStack Michael J Axtell [11].

Here, we describe the application of miRDeep-P as an efficient tool for the discovery of novel miRNA in plants [8]. MiRDeep-P is a modification of the popular miRDeep* pipeline used for the prediction of novel miRNAs in animals [12]. It takes into account specific characteristics of plant miRNAs such as longer pre-miRNA transcripts with more variable lengths and the widespread presence of paralogs in the plant genome which form the families of identical or nearly identical miRNAs [8].

MiRDeep-P distribution can be obtained from SourceForge at http://sourceforge.net/projects/mirdp/. The software does not require its installation because it is organized as a set of separate perl scripts that have to be executed in a specific sequence. MiRDeep-P dependencies that need to be installed separately include the bowtie and ViennaRNA package [13]; the latter can be obtained at http://www.tbi.univie.ac.at/RNA/. The miRDeep-P zip archive includes all the necessary perl scripts, the detailed manual, and sample data that can be used to test the software.

In order to increase a pool of potential miRNA candidates, we suggest combining all the reads mapped to the genome (valid unique and multiple matches) obtained during the genomic alignment step from all the sample libraries into a single fastq file. This can be done using the shell cat command:

```
> cat file1.fastq file2.fastq … > combined.fastq
```

The combined reads are then fed into the miRDeep-P pipeline, but prior to this, they have to be converted to a fasta format and collapsed to unique tags. This can be done with a perl one liner or a combination of shell commands; here is an example of a shell solution found at https://www.biostars.org/p/85929/:

```
> cat <reads.fastq> | paste - - - - | sed 's/^@/>/g' | cut -f1-2 | tr
    '\t' '\n' > <reads.fasta>
```

Next, the short reads contained in a fasta file have to be collapsed to unique tags, which entails counting how many times the given read is present in the file and writing its unique ID and sequence to a separate file in fasta format. Fasta id for a unique tag must contain its count number. One of the convenient tools for tag counting is fastx_collapser from fastx_toolkit that can be obtained at http://hannonlab.cshl.edu/fastx_toolkit/ along with a number of other useful scripts. To use fastx_collapser, enter the following command in Linux terminal:

```
> fastx_collapser -i <reads.fasta> -o <unique_tags.fasta>
```

Finally, to make unique tags usable by miRDeep-P, we have to introduce one last edition to fasta headers; specifically fasta identifiers have to be made in the form of ">1_x5", which requires a small substitution from their current format that looks like this - ">1-5" (Fig. 6). The file with unique tags can be modified with the sed command:

```
> sed 's/-/_x/' <unique_tags.fasta> > <modified_tags.fasta>
```

After sRNA reads are converted to fasta, and collapsed to unique tags, and their identifiers are modified to conform to miRDeep-P requirements, they are mapped to the genome using bowtie:

```
> bowtie -a -v 0 -f <genome_index> <modified_tags.fasta>
    <aligned_reads>
```

Here, bowtie is instructed to map reads in fasta format (-f) with no mismatches allowed (-v 0) and to output all valid alignments (-a) in the bowtie-formatted alignment file. Next, bowtie alignments are converted to blast format using the perl script supplied with the software; the script requires a file with bowtie alignments, unique tags, and the genomic sequence in fasta format:

```
> convert_bowtie_to_blast.pl <aligned_reads> <modified_tags.
    fasta> <genome.fasta> > <aligned_reads.bst>
```

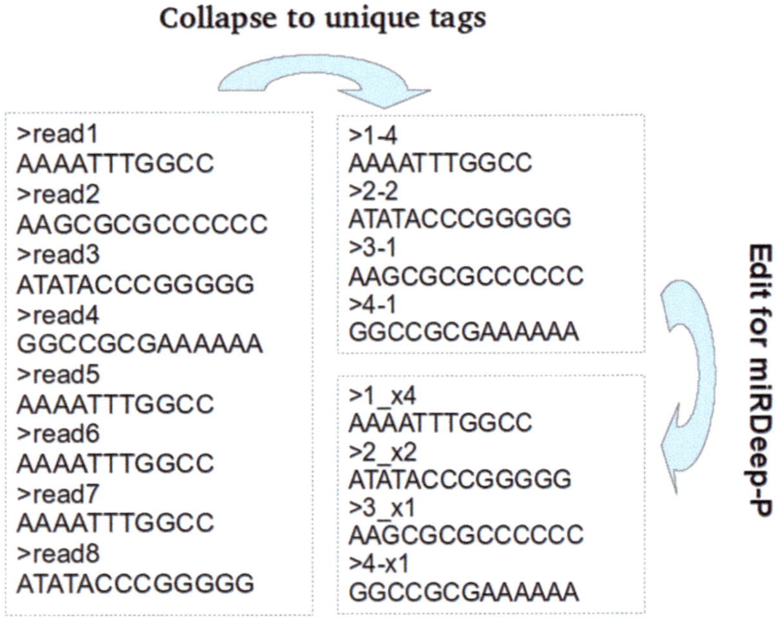

Fig. 6 Preparation of sequencing reads in fasta format for miRNA prediction using miRDeep-P. First, short reads are collapsed to unique tags, and the headers are modified to conform to the software format requirements

The next script filters the alignments and retains only those with 100 % identity, full-length alignment, and a number of matches that do not exceed a user-specified cutoff. The authors of the software recommend using the multi-match cutoff of 15 because the largest miRNA family in Arabidopsis, miR169, was shown to have 14 members [14]. This can be adjusted for other species based on the current knowledge of their miRNA composition. The script is executed using the following command:

```
> filter_alignments.pl <aligned_reads.bst> -c 15 > <filtered_reads.bst>
```

In order to increase prediction efficiency, it is useful to further filter the reads based on annotation, i.e., to discard all the alignments that overlap with the known exons (rRNAs, tRNAs, snoRNAs, etc.). The corresponding annotations can be obtained from the appropriate public databases depending on the species of interest; for example, annotations for Arabidopsis can be retrieved here: ftp://ftp.arabidopsis.org/home/tair/Genes/. Convenience scripts designed to perform annotation-based filtering are also provided with miRDeep-P:

```
> overlap.pl <filtered_reads.bst> <annotation.gff> -b > <overlap_ids>
```

This script will output all read identifiers that overlap with the known annotations provided in gff format. Note that the annotations have to belong to the same reference genome assembly used throughout the project. The next script discards all the reads that overlap with the known annotated features:

```
> alignedselected.pl <filtered_reads.bst> -g <overlap_ids> > <annotation_filtered.bst>
```

The same operation is performed on read sequences in fasta format:

```
> filter_alignments.pl <annotation_filtered.bst> -b <modified_tags.fasta> > <filtered_tags.fasta>
```

The remaining alignments are used to excise the potential miRNA precursor sequences from the genome (*see* **Note 6**); the length of precursors is set at 250 base pairs:

```
> excise_candidate.pl <genome.fasta> <annotation_filtered.bst> 250 > <precursors.fasta>
```

The secondary structure of potential precursors is predicted by the RNAfold software from the Vienna RNA package; -noPS option is specified to prevent the graphical output:

```
> cat <precursors.fasta> | RNAfold -noPS > <structures>
```

Now, the filtered reads have to be aligned to potential precursors in order to generate miRNA signatures and to be converted into blast format:

```
> bowtie-build <precursors.fasta> <precursors_index>
> bowtie -a -v 0 -f <precursors_index> <filtered_tags> > <tags_aligned_to_precursors.aln>
> convert_bowtie_to_blast.pl <tags_aligned_to_precursors.aln> <filtered_tags> <precursors.fasta> >
<tags_aligned_to_precursors.bst>
```

The blast-formatted file that contains alignments of precursors to unique tags has to be sorted; prior to the final step of miRNA prediction, the pipeline is executed, and this step incorporates RNA secondary structure information obtained by RNAfold:

```
> sort +3 -25 <tags_aligned_to_precursors> > <signatures>
> miRDeep.pl <signatures> <structures> > <miRNA_predictions>
```

5 Prediction of Phased siRNA Loci

A substantial fraction of plant siRNAs is produced from specific PolII transcripts (TAS genes) in a very distinctive phased manner [2]. In this case, phasing refers to the precise cleavage of long double-stranded RNA precursors in 21-nt increments with a predefined terminus. Phased cleavage was shown to be triggered by complementary binding of the TAS transcript to specific miRNAs; after the initial cleavage event, both the 5′ and 3′ products are converted to dsRNA with the help of RDR6 and SGS3 [15]. Double-stranded RNA is then processed by Dicer-like protein DCL4 into phased 21-nt fragments—trans-acting small interfering RNAs (ta-siRNAs); these fragments subsequently trigger the degradation of endogenous transcripts in a sequence-specific manner [15, 16].

The phased nature of siRNAs makes a computational prediction of TAS loci the relatively straightforward problem. One of such algorithms developed by Chen et al. [17] was implemented as a part of the UEA small RNA workbench—a set of tools for preprocessing of short sequencing reads and for an advanced analysis and visualization of small RNA data [18].

The SRNA Workbench is implemented in Java and is composed of a number of tools that can be launched independently from either a command-line interface or inside a graphical user interface (GUI). The SRNA workbench is a Java application that does not require installation; the distribution can be downloaded at http://srna-workbench.cmp.uea.ac.uk/. To see a list of tools available for sRNA analysis, enter the following command in linux shell:

```
> java -jar Workbench.jar -tool
```

We will select from the following analysis modules: adaptor, paresnip, filter, mircat, siloco, mirprof, tasi, seq_align. To predict TAS loci, tasi analysis module is used that can be launched using the following command:

```
> java -jar Workbench.jar -tool tasi -srna_file <srna_tags.fasta>
    -genome <genome.fasta> -out_file <prefix> -params <parameters.cfg>
```

The ta-siRNA prediction tool has to be provided with the following files:

Path to a file with small RNA tags in fasta format (-srna_file option).

Path to a multi-fasta file with the genome of interest (-genome).

Prefix that will be attached to all output files (-out_file).

An optional file with parameters, if not specified, "tasi" module will use a default parameter file located in data directory of sRNA workbench folder; by default, "tasi" module will use the *p*-value cutoff of 0.0001 and a minimal sRNA abundance of 2.

Fasta headers in the file with sRNA tags have to be modified to comply with the following format:

```
>1(153371)
TGTCGTCCAGCGGTTAGGATATCT
```

In this case, the fasta header consists of a unique identifier, and a number of times a particular sequence is present in the library (the number in brackets). The fasta file with unique tags used for miRNA prediction had to be prepared to comply with the following format:

```
>1_x153371
TGTCGTCCAGCGGTTAGGATATCT
```

To make this file usable by the sRNA workbench, its fasta headers have to be modified by the following combination of sed commands:

```
> cat <modified_tags.fasta> | sed 's/_x/(/' - | sed '/[0-9]/s/$/)/' -
  > <srna_tags.fasta>
```

The ta-siRNA prediction tool produces two output files: <prefix>_locuslist.csv and <prefix>_srnas.txt. The locus list files contain comma-separated entries for every candidate ta-siRNA-producing locus displaying their genomic coordinates, the number of reads overlapping the locus, the number of phased sequences, and a *p*-value assigned by the program. The second text file generated by the software contains a list of unique phased sRNA sequences for every predicted TAS gene.

After prediction of TAS loci is completed, it may be necessary to extract their sequences in fasta format based on their genomic coordinates. This task is easily achieved with bedtools—an excellent suite of utilities to perform various operations on sets of genomic coordinates [19]. The primary file format used by bedtools is a bed-tab-delimited text format designed to store genomic intervals. A bed format file in its simplest form requires only three columns—a chromosome name, and the start and end of the interval. Additional columns specify the name of the region, a numerical score, and a genomic strand; full specifications for bed format can be found at this link: http://genome.ucsc.edu/FAQ/FAQformat.html#format1.

Genomic coordinates of the predicted TAS loci have to be extracted from locuslist.csv file generated by the ta-siRNA prediction tool and saved in bed format. The following combination of shell commands and a perl one liner will achieve this by skipping the first three lines of the file with no interval data, replacing commas with tabs, and printing out genomic coordinates and unique names of the regions:

```
> tail -n+4 <locuslist.csv> | perl -nle '@parts=split(",");print "$parts[0]\
  t$parts[1]\t$parts[2]\t", join('::', @parts)' > <TAS_loci.bed>
```

The obtained bed file along with the genomic sequence is an input into the bedtools getfasta program to produce fasta sequences for the predicted TAS loci:

```
> bedtools getfasta -fi <genome.fasta> -bed <TAS_loci.bed> -fo
  <TAS_loci.fasta>
```

It is always instructive to visualize the aligned reads in order to validate predictions and in general to gain a better understanding of sRNA expression relative to other features (coding genes, other noncoding RNAs, repeats, the condensed heterochromatin, etc.). Read alignments, the genomic sequence, and annotations can be displayed in genome browsers. A large number of genome browsers have been developed to date; the well-known examples include Integrative Genomic Viewer (IGV, http://www.broadinstitute.org/igv/), Integrative Genomic Browser (IGB, http://bioviz.org/igb/index.html), GBrowse (http://gmod.org/wiki/GBrowse), Savant (http://genomesavant.com/p/home/index), and others. Here, we discuss the visualization of the predicted TAS loci along with the aligned reads using a popular IGV software [20].

In order to visualize raw read alignments, they have to be converted to bam format that is a more compact and faster binary variant of sam. Reads can be aligned to the genome using bowtie. Here, we allow up to two mismatches in the read length and instruct bowtie to find and output the best valid alignments in sam format:

```
> bowtie -v 2 --best -p 4 -S <index> <trimmed_reads.fastq> <aligned.sam>
```

Now, the generated sam files can be easily converted to bam format with samtools [21]; after that, bam files should be sorted and indexed to increase access speed:

```
> samtools view -bS <aligned.sam> > <aligned.bam>
> samtools sort <aligned.bam> <aligned_sorted.bam>
> samtools index <aligned_sorted.bam>
```

IGV can be launched directly from the software's website, or it can be downloaded as a binary distribution and launched from the command line with the following command:

```
> java -jar -Xmx2048m igv.jar
```

-Xmx option here specifies the size of the allocated random access memory (RAM). IGV offers a broad selection of reference genomes that can be uploaded from a remote server; if needed, a reference genome sequence can be imported from a local fasta file. IGV accepts a broad variety of file formats capable of storing sequence alignments, genomic intervals, genome coverage, and other information including bam and bed formats necessary to display sRNA expression relative to the predicted TAS loci (Fig. 7).

6 Building sRNA Library Composition Profiles

Interesting insights into biology and functional significance of sRNA regulation in plants may be gained through a detailed analysis of sequencing library composition profiles. In this case, building

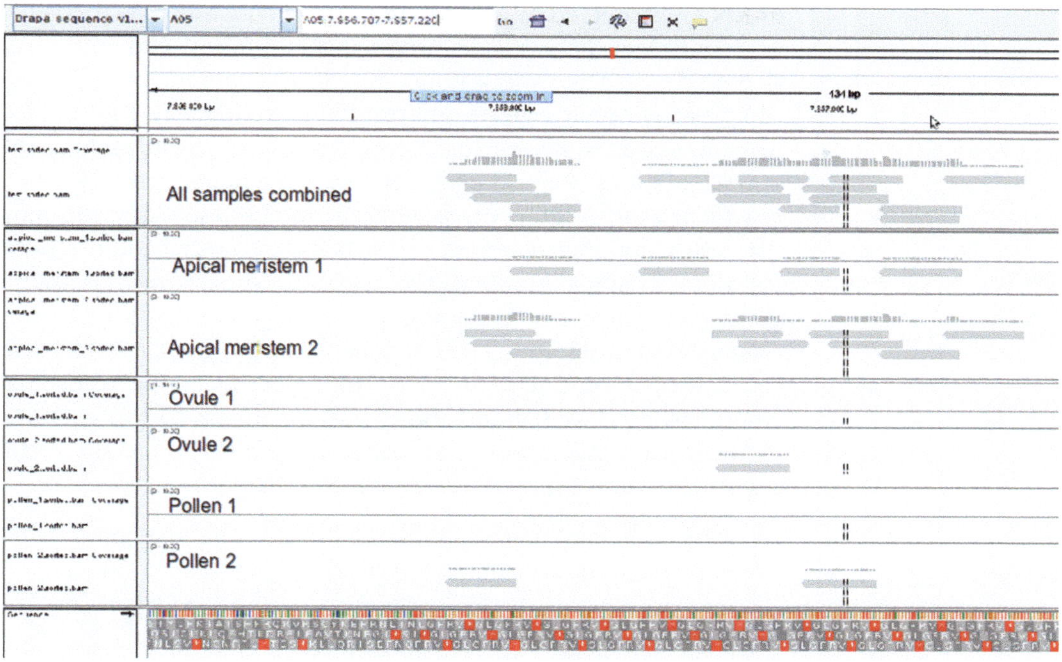

Fig. 7 A screenshot taken from the IGV genomic browser that shows a location of one of the predicted TAS loci along bam files containing read alignments. The prediction of TAS loci was completed using the UAE sRNA Workbench based on six short read libraries combined: three Brassica rapa tissues (the apical meristem, ovules, and pollen) with two biological replicates in each. For visualizing the reads, they were aligned to the Brassica genome using Bowtie, converted to bam files, indexed, and loaded into the browser

a sequence composition profile refers to the process of dividing a sequencing library into fractions based on the known classes of various genomic features. The selection of features depends on a current state of knowledge about the plant's genome and interests pursued by a researcher. Some examples of such features basically include components of protein-coding genes (exons, introns, promoters, etc.), various classes of noncoding RNAs, transposable elements, repeats, intergenic regions, and actually any biologically meaningful categories that genomic sequences could fall into.

DNA sequences corresponding to features of interest can be from various sources; for instance, many genome annotations are available through Ensembl for plants http://plants.ensembl.org/index.html, PlantGDB http://www.plantgdb.org/, genomic projects dedicated to specific species such as Rice Genome Annotation Project http://rice.plantbiology.msu.edu/, or Brassica database http://brassicadb.org/brad/. The detailed descriptions and comparison of the available plant annotation databases and computational tools were reviewed elsewhere [22, 23].

The sequencing library composition profiles displayed here were based on a stepwise alignment of sRNA reads obtained from three different tissues of Brassica rapa plants. Annotation data sets used in the classification included fasta sequences of the known miRNAs discovered by Yu et al. [24], candidate miRNAs predicted with miRDeep-P, ta-siRNA-producing loci found with the UAE sRNA Workbench, various noncoding RNAs obtained from RFAM [25], and genomic repeats and protein-coding genes downloaded from Brassica database [26]. Short RNA reads from each of the sample libraries were aligned to various sequence classes using bowtie aligner. Alignments were performed sequentially, meaning that reads aligning to certain data sets were counted and excluded from subsequent alignments; thus a pool of initial reads was gradually depleted with the remaining reads falling outside of the known classification categories. The last data set used in the alignment had to be the genome itself in order to differentiate the reads that could not be classified and those with no match to the genome. Stepwise alignments were performed with bowtie; the successful alignments were counted using shell commands:

```
> bowtie -v 2 --best -p 4 --al <aligned.fastq> --un <unaligned.fastq>
    <bowtie_index> <trimmed_reads.fastq> <aln>
> wc -l <aligned.fastq> | cut -f 1 -d ' ' | awk '{print $1/4}'
> bowtie -v 2 --best -p 4 --al <aligned.fastq> --un <unaligned.fastq>
    <bowtie_index> <unaligned.fastq> <aln>
...
```

In practice, the procedure of stepwise alignment has to be automated using, for example, bash or perl scripting; otherwise entering a large number of commands is error prone and time consuming. The library composition data obtained can be expressed either as a percentage of reads aligning to a certain category of genomic features from the total mapped reads or as reads per million (RPM) in the library and displayed as a bar graph (Fig. 8).

7 Identification of Differentially Expressed (DE)miRNAs

One of the typical tasks of small RNA sequencing is the detection of DE sRNAs between treatment groups. Bioinformatics workflow applied to achieve this goal can be divided into the following stages: (1) obtaining raw read counts; (2) scaling or normalizing the counts to the size of the sequencing library; and (3) the application of the appropriate statistical test to detect differentially expressed entities. Here, we illustrate this process in detail by comparing miRNA expression profiles between three different tissues of *Brassica rapa*: apical meristem, pollen, and ovules.

Cordero et al. [27] showed that in order to achieve an efficient alignment and counting of miRNA reads, it is best to use a specialized aligner designed to handle very short reads within the range of small RNA lengths, such as MicroRazers [28] or SHRiMP [29], and to align reads against the focused data set of miRNAs rather

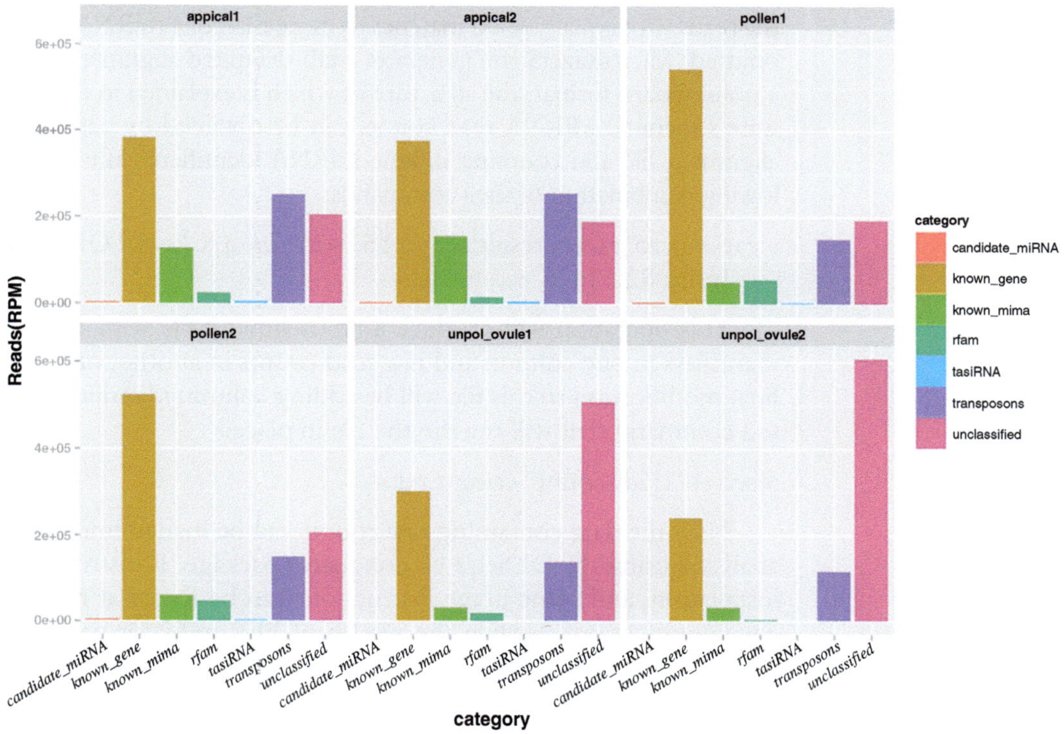

Fig. 8 Library composition profiles that were built using a stepwise alignment of sequencing reads to various functional categories of genomic sequences as described in Subheading 6. Short sequencing libraries were obtained from the apical meristem, pollen, and ovules of *Brassica rapa*; two biological replicates were available for each tissue type. The bar chart was drawn using the ggplot2 R package

than the whole genome. The same study identified Bayseq [30] and DESeq [31] as the best software solutions to perform the normalization between samples and statistical comparisons. To complete this workflow, we applied MicroRazerS to obtain raw miRNA counts and performed the normalization and statistical testing in R using DESeq2 Bioconductor package.

The reference data set can be downloaded from http://mirbase.org/ [32] which currently hosts 96 known mature miRNA sequences; candidate miRNAs predicted by miRDeep-P can be added to this list if desired. Prior to MicroRazerS alignment, the trimmed reads have to be converted to fasta format as described above. MicroRazerS can be obtained at https://www.seqan.de/projects/microrazers/: unzip the software in a desired directory and perform the installation following the instructions from README file. To align sRNA reads to mature miRNA sequences, enter the following command in Linux shell:

```
> micro_razers64 -m 1 -pa <mature-miRNA.fasta> <trimmed_
  reads.fasta>
```

With these options, MicroRazerS will search for alignments with no mismatches in the first 16 nt of the read; to avoid double-counting, the reads with equally good matches to more than one miRNA will be omitted. MicroRazerS run generates a tab-delimited alignment file in a nonstandard format, the structure of which is explained in the software's manual. MiRNA read counts can be obtained by parsing an alignment file and counting unique miRNA identifiers with the following combination of shell commands:

```
> cat <micro_razers.result> | cut -f5 | sort | uniq -c | awk ' OFS="\t"
    {print $2,$1}' > <count.txt>
```

This operation will produce a tab-delimited file with miRNA identifiers in one column and raw read counts in another one. The final modification to this file will be adding column identifiers with sed command that will rewrite the file in place:

```
> sed -i '1i id\tcount' <count.txt>
```

Now the data containing raw counts can be loaded into R and analyzed using the DESeq2 bioconductor package. R distribution, installation, and some usage instructions can be found at http://www.r-project.org/, while the wealth of Bioconductor packages, many of which are indispensable for the analysis of genomics data, can be found at http://www.bioconductor.org/. We recommend using RStudio as a freely available IDE http://www.rstudio.com/ to make a work in R environment more convenient.

In order to complete this workflow, we will need to download and install the following Bioconductor packages (all of the commands below are entered in R or R-Studio):

```
> library(DESeq2)
> library(edgeR)
> library(Biobase)
> library(arrayQualityMetrics)
> library(gplots)
> library(ggplot2)
```

DESeq2 requires a matrix of raw read counts to perform DE analysis; it is absolutely crucial that counts are not transformed or normalized at any way. Such matrix can be created from the separate count files using a convenient function implemented in the edgeR package [33]. First, we have to change into the working directory containing the count files and create vectors of file names and sample names:

```
> setwd("path/to/working directory")
> countFiles <- c("apical_meristem_1.count.txt", "apical_meristem_2.count.txt", "ovule_1.count.txt", "ovule_2.count.txt",
    "pollen_1.count.txt", "pollen_2.count.txt")
> labels <- c("apical_1", "apical_2", "ovule_1", "ovule_2", "pollen_1", "pollen_2")
```

A count matrix is prepared using the readDGE function from edgeR:

```
> countDGE <- readDGE(files=countFiles, columns=c(1,2), labels=labels)
> countData <- countDGE$counts
```

In order to make counts usable by DESeq2, apart from the count matrix (Table 2), we need to supply the software with sample data:

```
> colData <- data.frame(colnames(countMatrix), c("apical_meristem", "apical_meristem", "ovule", "ovule", "pollen", "pollen"))
> rownames(colData) <- colData[,1]
> names(colData) <- c("sample", "tissue")
```

Now, a DESeq2 data object can be created from a previously prepared count matrix and the data frame with sample information. In addition, we have to specify the experimental design formula; in this case, we are performing a one-way ANOVA-type experiment where we are comparing three types of plant tissues between each other (*see* **Note 7**):

```
> dds <- DESeqDataSetFromMatrix(countData = countData, colData = colData, design = ~ tissue)
```

Prior to a statistical comparison, it is necessary to perform some data exploration in order to get an impression about data quality and the relationship between samples and to spot possible outliers. To apply the standard data exploration methods, such as principal component analysis (PCA) or hierarchical clustering, the

Table 2
Top ten rows of a matrix of raw counts used as an input to detect differentially expressed miRNAs using the DESeq2 Bioconductor package

	Apical_1	Apical_2	Ovule_1	Ovule_2	Pollen_1	Pollen_2
bra-miR1140	5008	6517	227	137	45	29
bra-miR156a-3p	13	15	0	0	84	56
bra-miR156e-3p	38	36	1	1	1	0
bra-miR156f-3p	2	1	0	0	0	0
bra-miR157a	124	123	10	6	20	16
bra-miR158-3p	20738	31651	1273	981	23461	27208
bra-miR158-5p	301	435	99	71	53	62
bra-miR159a	19344	18016	735	429	142	139
bra-miR160a-3p	1	1	3	1	1	0

data should be homoscedastic, at least approximately. DESeq2 achieves this by applying the regularized logarithmic transformation with rlog function; the matrix of transformed values can be extracted with the assay accessor function:

```
> rld <- rlog(dds)
> norm <- assay(rld)
```

The arrayQualityMetrics package [34] was developed to perform a comprehensive exploratory analysis and detect outlying samples in gene expression microarray data, but with small tweaks, it can be efficiently applied to RNA-seq experiments. The first matrix containing rlog-transformed values has to be converted into ExpressionSet object; the sample information can be added to it as an object of AnnotatedDataFrame:

```
> phenoData <- as(colData, "AnnotatedDataFrame")
> normExp <- new("ExpressionSet", exprs = norm, phenoData =
    phenoData)
> arrayQualityMetrics(normExp, outdir="arrayQualityMetrics_
    results", intgroup="tissue")
```

ArrayQualityMetrics saves the results as an html report that can be opened in any web browser; all the related files and images are saved in the directory specified with the 'outdir=' option; the 'intgroup=' option specifies a variable defining colouring of data points (Fig. 9).

Another convenient way to investigate relationships between samples is to perform the unsupervised hierarchical clustering of

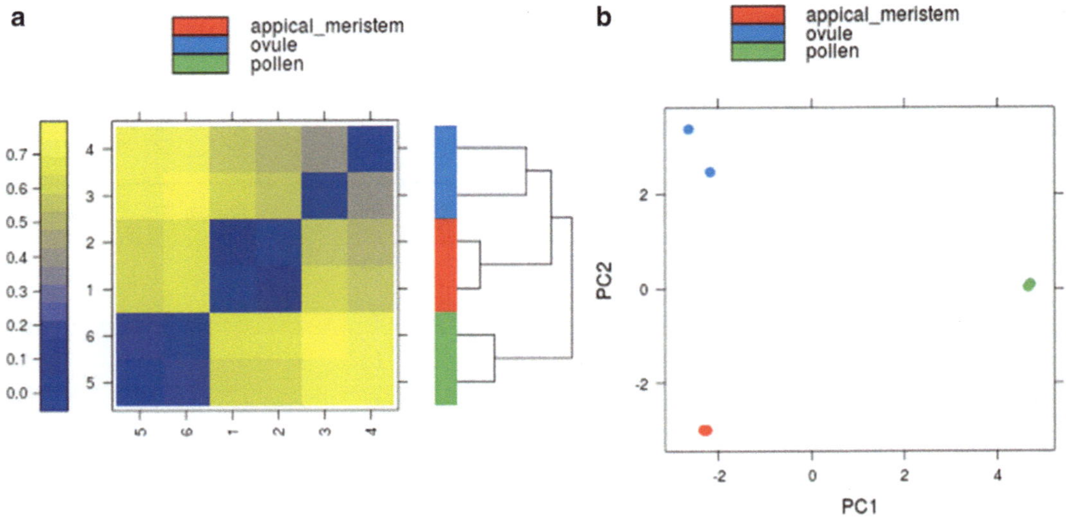

Fig. 9 The relationship between miRNA expression profiles obtained from the apical meristem, pollen, and ovules of Brassica rapa and estimated using hierarchical clustering (**a**) and principal component analysis (**b**). The graphs were built using the arrayQualityMetrics Bioconductor package

genes and samples; the expression of individual features can be displayed as a heatmap. A good way to do this in R is to use the heatmap.2 function from the gplots package (Fig. 10):

```
>heatmap.2(norm,dendrogram="both",scale="row",trace="none",
    col=redgreen(75), cexRow=0.5, cexCol=1)
```

Next, we use the DESeq wrapper function to perform an estimation of size factors, count dispersion, and GLM fitting; the technicalities are described in detail in Anders and Huber [31] and in the DESeq2 manual:

```
> dds <- DESeq(dds)
```

The results of statistical comparisons are extracted using the results function which allows us to specify the contrast (comparison) of interest. Let us take a closer look at the results of comparisons between miRNA transcriptional profiles in the apical meristem and pollen:

```
> app_vs_pol <- results(dds, contrast=c("tissue", "apical_meri-
    stem", "pollen"))
```

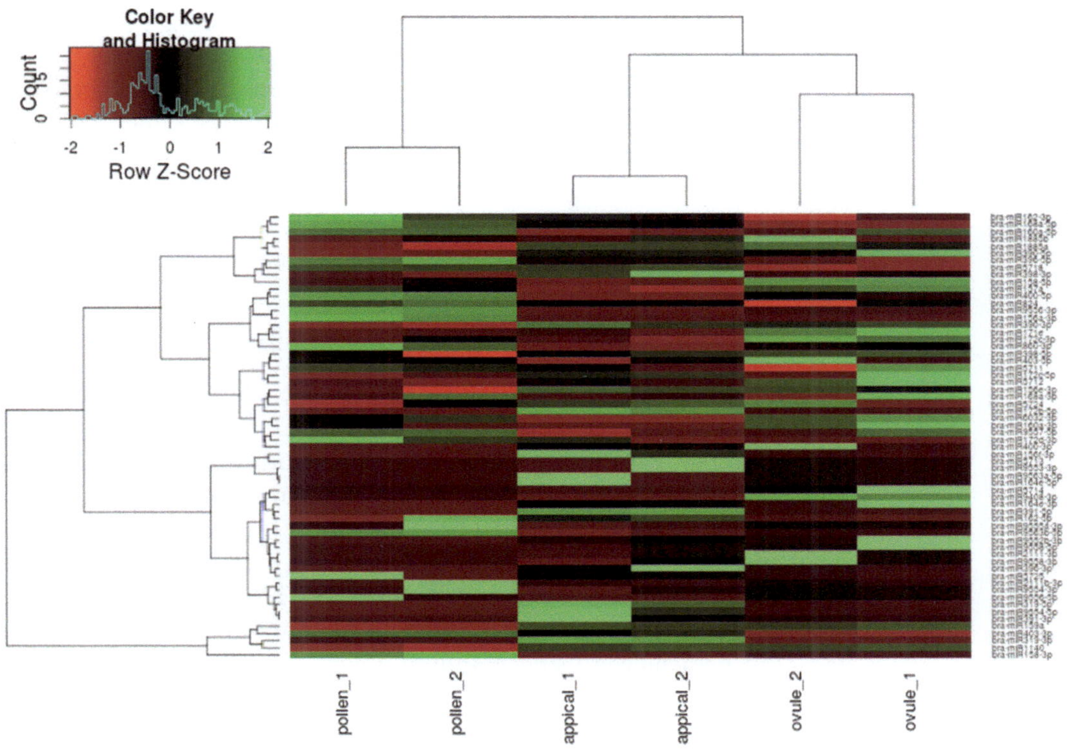

Fig. 10 Heatmap of the scaled and log-transformed expression levels of the known *Brassica rapa* miRNAs from three different tissues. Sample and miRNA clustering was performed using the ward. D method with the Euclidean distance used as a measure of dissimilarity between sets of values. Heatmap was drawn using the gplots R packages

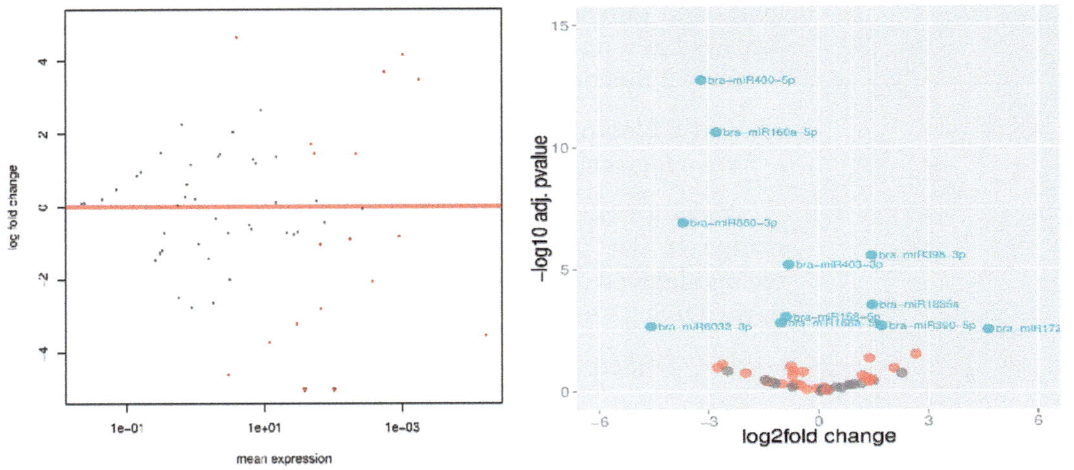

Fig. 11 Displaying the differentially expressed miRNAs between the apical meristem and pollen of *Brassica rapa* on MA (*left*) and the volcano plot (*right*). MiRNAs with the Benjamini-Hochberg-adjusted *p*-values below 0.05 were considered differentially expressed

In the simplest case illustrated here, the contrast is specified as the character vector with three elements: the name of the variable (tissue), the numerator, and the denominator for log2 fold change calculation (the apical meristem and pollen, respectively). The results of other comparisons, the apical meristem versus the ovule or the ovule versus pollen, can be extracted in the same fashion.

The results can be displayed as an X–Y plot where the *X*-axis is the mean expression level of the feature across experimental groups (A), and the *Y*-axis corresponds to log2 fold change in the expressions—MA plot (Fig. 11). DESeq2 offers a convenient function to quickly build MA plots:

```
> plotMA(app_vs_pol, ylim = c(-5, 5), alpha=0.05)
```

In this case, all the miRNAs with the FDR-adjusted *p*-values below 0.05 are depicted as red dots (Fig. 11). Another popular way to display the results of gene expression study is a volcano plot—an X–Y scatter plot where log2 fold changes are plotted on the *X*-axis opposite to the log-transformed *p*-values on the *Y*-axis. Features considered DE based on the user-defined *p*-value and fold-change threshold are shown in a different color (Fig. 11). Unfortunately, DESeq2 offers no built-in function to draw volcano plots, but this can be achieved using graphical libraries written for R. Here, we present an example of the code used to draw a volcano plot with the function implemented in an excellent ggplot2 package [35]:

```
> app_vs_pol <- as.data.frame(app_vs_pol) # convert comparison
    results to data frame
> threshold <- as.vector(app_vs_pol$padj < 0.05) # specify signifi-
    cance threshold
```

```
> app_vs_pol$threshold <- threshold # add defined thresholds as a column
> labels <- rownames(app_vs_pol)[which(threshold == TRUE)] # create a vector of names for DE miRNAs
> idx <- rownames(app_vs_pol) %in% labels
> app_vs_pol$names <- rownames(app_vs_pol)
> app_vs_pol$names[!idx] <- NA # label non DE miRNAs as NAs
# draw a volcano plot using ggplot2
> ggplot(data=app_vs_pol,
aes(x=log2FoldChange, y=-log10(pvalue),
colour=threshold)) +
geom_point(alpha=0.7, size=5) +
theme(legend.position = "none", axis.text.x=element_text(size=15),
axis.text.y=element_text(size=15), axis.title.x=element_text(size=20), axis.title.y=element_text(size=20)) +
xlim(c(-6, 6)) + ylim(c(0, 15)) +
xlab("log2fold change") + ylab("-log10 adj. pvalue") +
geom_text(aes(x=log2FoldChange+0.2, label=names), size=4, hjust=0)
```

In order to save the results, they have to be converted to data frame and saved as a comma-separated file that can be viewed and analyzed in Excel:

```
> app_vs_pol <- as.data.frame(app_vs_pol)
> write.csv(app_vs_pol, file="apical_meristem_vs_pollen.csv", quote=F)
```

The table with the results has six columns:

1. Base mean—mean of normalized counts for all samples.
2. A log2 fold change of "treated" versus "control".
3. The standard error: condition "treated" versus "untreated".
4. The Wald statistic: condition "treated" versus "untreated".
5. A Wald test *p*-value: condition "treated" versus "untreated".
6. Benjamini-Hochberg-adjusted *p*-values [36].

The results may contain NA values; this is a normal behavior of the software which assigns NA to features that do not pass independent filtering performed to optimize the number of adjusted *p*-values below the selected critical alpha level [37].

An approach implemented in DESeq2 and other similar methods that utilize raw count data is extremely flexible and can be applied to various scenarios of small RNA expression analysis (*see* **Note 8**). The method described above is most suitable for singular small RNAs produced from well-defined short precursors such as miRNAs; moreover, this approach requires the pre-existing annotation for sRNAs of interest. Yet one may aim to detect the differentially expressed sRNAs with a very different biogenesis. One of the possible scenarios is small RNAs processed from a long precursor, for example, a transposon, a coding gene, or a long noncoding RNA. In this case, the process of preparing raw count data will change: instead of mapping sequencing reads to the focused set of features, we will align them to the whole genome and subsequently count alignments falling within specific genomic intervals. Another possible approach is based on the identification of clusters of sRNA expression without taking into account any annotation information. One example of a computation tool for the identification of sRNA loci—SILOCO—is implemented in the UAE small RNA Workbench at http://srna-workbench.cmp.uea.ac.uk/tools/analysis-tools/siloco/. Reads falling into sRNA expression loci are counted and compared between experimental groups using DESeq2, edgeR, or Bayseq. This approach may provide a distinctive advantage when the researcher is aiming at a global comparison of sRNA expression between groups which will include all classes of sRNA captured during the library construction process regardless of their biogenesis, novel classes of small RNAs, and loci located within the intergenic regions with no clear functional significance.

8 Notes

1. Illumina uses the following naming scheme: **<sample name>_<barcode sequence>_L<lane (0-padded to 3 digits)>_R<read number>_<set number (0-padded to 3 digits>.fastq.gz** (http://support.illumina.com/).
2. A valid Illumina fastq file name for multiplexed run will look like this: 1_ATCACG_L005_R1_001.fastq.gz. gzipped files can be conveniently concatenated using the *zcat* shell command in the Unix environment to simplify further processing. Some short read aligners, such as Burrows Wheeler Aligner [38] or Tophat [39], can accept the gzipped files directly, which helps to decrease the data storage requirements.
3. A quality of 30 corresponds to 1/1000 probability of a wrong base call.
4. The genomic fasta files prepackaged with aligner indexes, annotations, and contaminating sequences can be downloaded from Illumina iGENOME site at http://support.illumina.com/sequencing/sequencing_software/igenome.html; currently, plant species include Arabidopsis, corn, rice, and sorghum.

5. Sequence Alignment/Map (SAM) format is a generic format used to store and describe large nucleotide alignments (Li et al. [40]).
6. The Bowtie authors use the notion of "stratum" to describe a set of all valid alignments with an equal number of mismatches in the read or in the seed region if -n mode is enabled. When the --best option is specified, Bowtie guarantees reporting only the best alignments in terms of quality and number of mismatches; moreover, the use of --best increases sensitivity because it widens a scope of search for valid alignments and eliminates strand bias. We recommend using --best in every case because an increase in run-time caused by it is practically negligible. For other details on Bowtie functionality and behavior, please consult the user's manual http://bowtie-bio.sourceforge.net/manual.shtml.
7. This step is extremely time consuming; on our system it took over 24 h.
8. For more information see an example of DESeq2 workflow put together by the authors of DESeq2: http://www.bioconductor.org/help/workflows/rnaseqGene/.

References

1. Martin M (2011) Cutadapt removes adapter sequences from high-throughput sequencing reads. EMBnet J 17:1, Gener. Seq. Data Anal
2. Axtell MJ (2013) Classification and Comparison of Small RNAs from Plants. Annu Rev Plant Biol 64:137–159
3. Calarco JP, Borges F, Donoghue MTA, Van Ex F, Jullien PE, Lopes T, Gardner R, Berger F, Feijó JA, Becker JD, Martienssen RA (2012) Reprogramming of DNA Methylation in Pollen Guides Epigenetic Inheritance via Small RNA. Cell 151:194–205
4. Lilljebjorn H, Rissler M, Lassen C, Heldrup J, Behrendtz M, Mitelman F, Johansson B, Fioretos T (2012) Whole-exome sequencing of pediatric acute lymphoblastic leukemia. Leukemia 26:1602–1607
5. Carroll TS, Liang Z, Salama R, Stark R, de Santiago I (2014) Impact of artifact removal on ChIP quality metrics in ChIP-seq and ChIP-exo data. Front Genet 5:75
6. Lindner R, Friedel CC (2012) A Comprehensive Evaluation of Alignment Algorithms in the Context of RNA-Seq. PLoS One 7, e52403
7. Langmead B, Trapnell C, Pop M, Salzberg SL (2009) Ultrafast and memory-efficient alignment of short DNA sequences to the human genome. Genome Biol 10:R25
8. Yang X, Li L (2011) miRDeep-P: a computational tool for analyzing the microRNA transcriptome in plants. Bioinformatics 27:2614–2615
9. Xie F, Xiao P, Chen D, Xu L, Zhang B (2012) miRDeepFinder: a miRNA analysis tool for deep sequencing of plant small RNAs. *Plant Mol.* Biol 80:75–84
10. Moxon S, Schwach F, Dalmay T, MacLean D, Studholme DJ, Moulton V (2008) A toolkit for analysing large-scale plant small RNA datasets. Bioinformatics 24:2252–2253
11. Axtell MJ (2013) ShortStack: Comprehensive annotation and quantification of small RNA genes. RNA 19:740–751
12. An J, Lai J, Lehman ML, Nelson CC (2013) miRDeep*: an integrated application tool for miRNA identification from RNA sequencing data. Nucleic Acids Res 41:727–737
13. Lorenz R, Bernhart SH, HönerzuSiederdissen C, Tafer H, Flamm C, Stadler PF, Hofacker IL (2011) ViennaRNA Package 2.0. Algor Mol Biol 6:26
14. Bologna NG, Schapire AL, Zhai J, Chorostecki U, Boisbouvier J, Meyers BC, Palatnik JF (2013) Multiple RNA recognition patterns during microRNA biogenesis in plants. Genome Res 23:1675–1689

15. Allen E, Xie Z, Gustafson AM, Carrington JC (2005) microRNA-Directed Phasing during Trans-Acting siRNA Biogenesis in Plants. Cell 121:207–221
16. Xie Z, Allen E, Wilken A, Carrington JC (2005) DICER-LIKE 4 functions in trans-acting small interfering RNA biogenesis and vegetative phase change in Arabidopsis thaliana. Proc Natl Acad Sci U S A 102:12984–12989
17. Chen H-M, Li Y-H, Wu S-H (2007) Bioinformatic prediction and experimental validation of a microRNA-directed tandem trans-acting siRNA cascade in Arabidopsis. Proc Natl Acad Sci U S A 104:3318–3323
18. Stocks MB, Moxon S, Mapleson D, Woolfenden HC, Mohorianu I, Folkes L, Schwach F, Dalmay T, Moulton V (2012) The UEA sRNA workbench: a suite of tools for analysing and visualizing next generation sequencing microRNA and small RNA datasets. Bioinformatics 28:2059–2061
19. Quinlan AR, Hall IM (2010) BEDTools: a flexible suite of utilities for comparing genomic features. Bioinformatics 26:841–842
20. Robinson JT, Thorvaldsdottir H, Winckler W, Guttman M, Lander ES, Getz G, Mesirov JP (2011) Integrative genomics viewer. Nat Biotechnol 29:24–26
21. Li, H., Handsaker, B., Wysoker, A., Fennell, T., Ruan, J., Homer, N., Marth, G., Abecasis, G., Durbin, R., 1000 Genome Project Data Processing Subgroup (2009) The Sequence Alignment/Map format and SAMtools. Bioinformatics 25:2078–2079
22. Mochida K, Shinozaki K (2010) Genomics and Bioinformatics Resources for Crop Improvement. Plant Cell Physiol 51:497–523
23. Martinez M (2013) From plant genomes to protein families: computational tools. Comput Struct Biotechnol J 8, e201307001
24. Yu X, Wang H, Lu Y, de Ruiter M, Cariaso M, Prins M, van Tunen A, He Y (2012) Identification of conserved and novel microRNAs that are responsive to heat stress in Brassica rapa. J Exp Bot 63:1025–1038
25. Burge SW, Daub J, Eberhardt R, Tate J, Barquist L, Nawrocki EP, Eddy SR, Gardner PP, Bateman A (2013) Rfam 11.0: 10 years of RNA families. Nucleic Acids Res 41:D226–D232
26. Cheng F, Liu S, Wu J, Fang L, Sun S, Liu B, Li P, Hua W, Wang X (2011) BRAD, the genetics and genomics database for Brassica plants. BMC Plant Biol 11:136
27. Cordero F, Beccuti M, Arigoni M, Donatelli S, Calogero RA (2012) Optimizing a Massive Parallel Sequencing Workflow for Quantitative miRNA Expression Analysis. PLoS One 7, e31630
28. Emde A-K, Grunert M, Weese D, Reinert K, Sperling SR (2010) MicroRazerS: rapid alignment of small RNA reads. Bioinformatics 26:123–124
29. Rumble SM, Lacroute P, Dalca AV, Fiume M, Sidow A, Brudno M (2009) SHRiMP: Accurate Mapping of Short Color-space Reads. PLoS Comput Biol 5, e1000386
30. Hardcastle TJ, Kelly KA (2010) baySeq: Empirical Bayesian methods for identifying differential expression in sequence count data. BMC Bioinformatics 11:422
31. Anders S, Huber W (2010) Differential expression analysis for sequence count data. Genome Biol 11:R106
32. Kozomara A, Griffiths-Jones S (2014) miRBase: annotating high confidence microRNAs using deep sequencing data. Nucleic Acids Res 42:D68–D73
33. Robinson MD, McCarthy DJ, Smyth GK (2010) edgeR: a Bioconductor package for differential expression analysis of digital gene expression data. Bioinformatics 26:139–140
34. Kauffmann A, Gentleman R, Huber W (2009) arrayQualityMetrics—a bioconductor package for quality assessment of microarray data. Bioinformatics 25:415–416
35. Wickham H (2009) ggplot2: elegant graphics for data analysis., Springer New York
36. Benjamini Y, Hochberg Y (1995) Controlling the False Discovery Rate: A Practical and Powerful Approach to Multiple Testing. J R Stat Soc Ser B Methodol 57:289–300
37. Bourgon R, Gentleman R, Huber W (2010) Independent filtering increases detection power for high-throughput experiments. Proc Natl Acad Sci U S A 107:9546–9551
38. Li H, Durbin R (2009) Fast and accurate short read alignment with Burrows–Wheeler transform. Bioinformatics 25:1754–1760
39. Trapnell C, Pachter L, Salzberg SL (2009) TopHat: discovering splice junctions with RNA-Seq. Bioinforma Oxf Engl 25:1105–1111
40. Li H, Handsaker B, Wysoker A, Fennell T, Ruan J, Homer N, Marth G, Abecasis G, Durbin R, 1000 Genome Project Data Processing Subgroup (2009) The Sequence Alignment/Map format and SAMtools. Bioinformatics 25:2078–2079

Chapter 17

Increasing a Stable Transformation Efficiency of *Arabidopsis* by Manipulating the Endogenous Gene Expression Using Virus-Induced Gene Silencing

Andriy Bilichak and Igor Kovalchuk

Abstract

Virus-induced gene silencing (VIGS) is a powerful epigenetic tool that allows in a relatively short period of time to down-regulate the expression of an endogenous gene in infected plants for either monitoring the resulting phenotype or enhancing/modifying a particular trait associated with the gene. Here, we describe the utilization of *Tobacco rattle virus* (TRV) as a vector for the VIGS technique in *Arabidopsis* plants. The unique ability of TRV to infect both somatic tissues and gametes allows deciphering the role of genes in these tissues simultaneously. As an example, we demonstrate the utilization of TRV to down-regulate the expression of *AGO2* and *NRPD1a* genes in ovules of *Arabidopsis* plants in order to boost the stable transformation efficiency by floral dip.

Key words VIGS, TRV, Transient down-regulation, *Arabidopsis*, *Agrobacterium*, Posttranscriptional gene silencing

1 Introduction

The advancement of plant transgenesis techniques for the delivery and expression of foreign genes in plant cells resulted in the engineering of transgenic plants with enhanced traits [1, 2]. Among transformation techniques available at the present time, the *Agrobacterium tumefaciens* (*Agrobacterium*)-mediated transformation is considered to be the most efficient and reproducible method of dicotyledonous plant (dicot) transformation [3]. The ability of *Agrobacterium* to deliver a portion of its DNA to the plant genome has been widely exploited for the transient and stable plant transformation by using oncogene-free or disarmed T-DNAs. Unfortunately, the transgenes within the T-DNA are often either poorly expressed or not expressed at all in the plant cells, predominantly due to the RNA silencing machinery [4].

Previously, we and others have reported that certain *Arabidopsis* mutants demonstrate a higher susceptibility to the

Igor Kovalchuck (ed.), *Plant Epigenetics: Methods and Protocols*, Methods in Molecular Biology, vol. 1456, DOI 10.1007/978-1-4899-7708-3_17,

stable *Agrobacterium*-mediated transformation than wild-type plants [5, 6]. Nevertheless, the practical utilization of such knowledge for elevating transformation efficiency in model species and economically important crops requires the generation of stable mutants of the required genes that is laborious and time consuming. The transient down-regulation using virus-induced gene silencing (VIGS) overcomes the limitations associated with the generation of stable mutants. VIGS is an epigenetic method for the transient down-regulation of gene expression that relies on the natural ability of viruses to trigger the plant's immune response directed at posttranscriptional gene silencing of viral genes. Cloning of a gene fragment into the replicating viral vector allows for the transient homology-dependent down-regulation of the targeted endogenous gene. Among plant viruses such as *apple latent spherical virus*, *cabbage leaf curl virus*, *potato virus X*, and *turnip yellow mosaic virus* [7] that can trigger VIGS in somatic *Arabidopsis* tissues, only *tobacco rattle virus* has been shown to penetrate into an embryo sac of *Nicotiana benthamiana* plants [8]. TRV is a bipartite, positive-sense, single-stranded RNA virus that is one of the most efficient and widely used gene silencing tools in *Arabidopsis* plants at present time [9, 10].

The ability of TRV to penetrate into the embryo sac of infected plants has been proven to be very efficient for transient down-regulation of endogenous genes, the so-called suppressors of stable plant transformation. This, in turn, renders *Arabidopsis* wild-type plants infected with the recombinant virus to demonstrate a plant phenotype susceptible to *Agrobacterium* [5].

The genome of TRV is conveniently cloned into two binary vectors—pTRV1 and pTRV2. The pTRV1 plasmid carries genes encoding the 194 and 134 kDa replicase proteins, the 29 kDa movement protein, and the 16 kDa cysteine-rich protein, the function of which is not fully known. The pTRV2 plasmid has genes encoding the coat protein and nonstructural proteins [11]. To be used as a VIGS vector, the two nonstructural protein-encoding genes in pTRV2 were replaced with multiple cloning sites for inserting fragments of the target gene to be silenced.

Here, we describe a simple and efficient protocol for transient down-regulation of the expression of an endogenous gene in somatic and reproductive tissues of *Arabidopsis* plants using TRV. Consequently, transient down-regulation of target genes can aid in increasing the efficiency of either transient or stable plant transformation and can help characterize the overall phenotype of the infected plant that is partially deficient in the expression of a particular gene (Fig. 1).

2 Materials

2.1 Vectors of pTRV Series

1. pTRV1 vector plasmid DNA (Arabidopsis Biological Resource Center (ABRC) stock no. CD3-1039, vector name YL192),

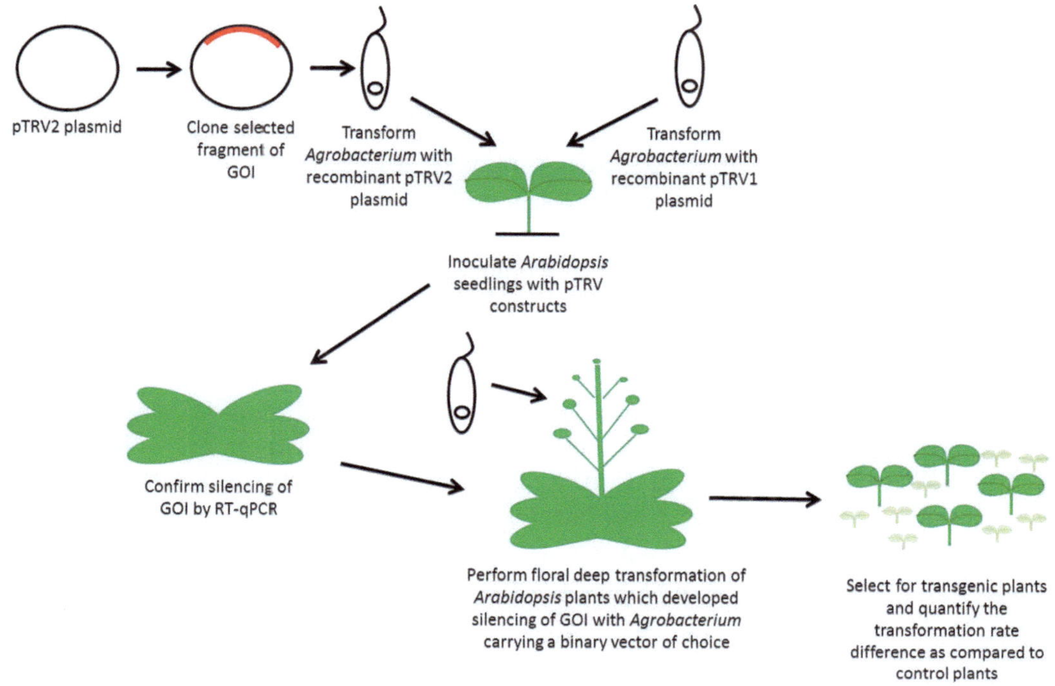

Fig. 1 A general overview of the VIGS protocol for the *Agrobacterium*-mediated stable transformation. A fragment of a gene of interest (GOI) is selected using the "RNAi Scan" program [12], amplified from cDNA, and cloned into the pTRV2 binary vector to generate the pTRV2::GOI vector. Then, pTRV2::GOI and pTRV1 plasmids are transformed separately into competent *Agrobacterium* cells. *Arabidopsis* seedlings are inoculated with the mixture of *Agrobacterium* carrying pTRV2::GOI and pTRV1 plasmids. Later, the development of silencing is confirmed by RT-qPCR, and the plants with the lowest mRNA abundance of the target gene are let to flower. The flowering plants are transformed using floral dip with *Agrobacterium* carrying the preferred binary vector. Following the transformation, seeds are harvested individually from all the plants, and transgenic plants are selected using the required herbicide

TRV1 National Center for Biotechnology Information (NCBI) accession no. AF406990.

2. pTRV2 vector plasmid DNA (ABRC stock no. CD3-1040, vector name YL156). TRV2 NCBI accession no. AF406991.
3. pTRV2::AtPDS—control vector used to visually monitor the progress of VIGS in infected plants (ABRC stock no. CD3-1047, vector name YL154).
4. A set of enzymes for PCR and conventional cloning: Phusion High-Fidelity DNA polymerase, FastDigest Pack, Rapid DNA Ligation kit, FastAP Thermosensitive alkaline phosphatase.
5. *Escherichia coli*—chemically competent cells (*see* **Note 1**).
6. Liquid Luria Broth (LB) media, LB agar plates, 1000× stocks of antibiotics—25 mg/ml of rifampicin, 50 mg/ml of kanamycin, and 25 mg/ml of gentamycin.

2.2 Inoculation of Arabidopsis with Agrobacterium and Stable Floral-Dip Transformation

1. *A. tumefaciens* chemically competent cells, strain GV3101 (*see* **Note 2**).
2. *Agrobacterium* cultivation medium for inoculation experiment (YEBi medium): Prepare and autoclave the following solution: 0.5% beef extract, 0.1% yeast extract, and 0.5% bacto-peptone. Before the inoculation of *Agrobacterium*, divide a sterile solution into 100 ml aliquots (the number of aliquots depends on the number of pTRV2 recombinant constructs used) and add the following filter-sterilized solutions: 100 μl of 2 M $MgSO_4$ (pH 7.0), 10 μl of 200 mM acetosyringone (*see* **Note 3**), and 10 ml of 100 mM MES, and antibiotics in the required concentrations.
3. The agro-infiltration inoculation solution (MMAi solution): 0.5% MS salts (Phytotechnology Laboratories, KS, USA), 100 mM MES, 2% sucrose, and 200 μM acetosyringone, pH 5.6 (*see* **Note 4**).
4. The floral-dip inoculation medium—5.0% sucrose and 0.05% (w/w) Silwet L-77.
5. A thermo-shaker.
6. Incubators for 28 and 37 °C.
7. A variable-speed centrifuge.
8. A benchtop pH meter with a probe.
9. A spectrophotometer.
10. A mini vortexer.
11. A needleless syringe, 1 ml.
12. *Arabidopsis thaliana* ecotype Col-0 plants grown as described in Subheading 3. Individually potted plants (2–3 leaf stage) are used in the VIGS inoculation procedure (*see* **Note 5**).
13. *All purpose potting soil* mixed with vermiculite in the proportion 4:1 in 2×2 in. square pots.
14. Miracle-Gro fertilizer.
15. A 4 °C refrigerator to break *Arabidopsis* seed dormancy and growth chambers for seed germination and the initial seedling growth.
16. A bleach solution (like Clorox, available from local stores).

2.3 Verification of Gene Down-Regulation

1. A mortar and pestle.
2. SsoFast EvaGreen Supermix for qPCR (*see* **Note 6**).
3. An RNeasy mini plant RNA extraction kit.
4. An iScript cDNA synthesis kit.
5. A plasmid Plus mini kit.
6. Custom-synthesized gene-specific primer pairs for RT-qPCR analysis; *see* **Note 7** for design considerations.
7. The real-time PCR detection system.

3 Methods

3.1 The pTRV2 Recombinant Plasmids

1. In order to minimize the off-target effects, analyze the coding sequence of a desired gene to be down-regulated for any cross homology with other endogens using the "RNAi Scan" program and *Arabidopsis thaliana* mRNA database [12]. Choose the region of around 500 bp in a gene's CDS that is not predicted to generate any off-target smRNAs.
2. Extract RNA from the *Arabidopsis* Col-0 plant tissue using the RNeasy mini kit according to the manufacturer's instructions and including DNase treatment in the procedure (*see* **Note 8**).
3. Following RNA isolation, use 500 ng of total RNA and perform cDNA synthesis using the cDNA synthesis kit following the manufacturer's instruction.
4. Design the primers containing restriction sites at the 5′ end preceded by the required number of nucleotides for an efficient digestion by selected restriction enzymes in order to clone the generated PCR fragment into a multiple cloning site of pTRV2 plasmid.
5. PCR amplify the fragment of the target gene using the synthesized cDNA as a template.
6. Verify the generated recombinant constructs by sequencing. The resulting pTRV2::Gene of interest (GOI) plasmid will be used for agro-inoculation.

3.2 Plant Cultivation

1. Sow *Arabidopsis* seeds on *All Purpose Potting Soil* mixed with vermiculite in proportion 4:1 in 2 × 2 in. square pots.
2. Pre-soak the soil mixture in the pots once with Miracle-Gro fertilizer and maintain it continuously moist with tap water.
3. Incubate the seeds at 4 °C in darkness for 2 days to break dormancy, and then move to a growth chamber.
4. Two to three days post-germination, thin the seedlings and continue to cultivate them in the growth chamber. Grow the plants in high-light conditions (32.8 $\mu E/m^2/s$) at 22/18 °C under a 16/8-h light/dark regime, respectively, and under a constant humidity of 65 %.

3.3 Transformation of Agrobacterium with pTRV Series Plasmids

1. Prepare plasmid DNA for pTRV2::GOI (a silencing construct), pTRV2::MCS (an empty viral vector control), pTRV2::PDS (a vector control to visually monitor the development of silencing), and pTRV1 constructs by using the Plasmid Plus mini kit according to the manufacturer's instructions.
2. Separately transform 1 μl of plasmid DNA (concentration 50 ng/μl) of each construct into chemically competent cells of the *A. tumefaciens* strain GV3101 (80 μl of competent cells per plasmid).

3. Separately set up 1 ml of culture in LB liquid medium without antibiotics for each transformation event. Grow the cultures at 28 °C in a shaker at 250 g for 3 h.
4. Separately spread 100 µl of each culture on LB agar plates containing rifampicin (25 µg/ml), gentamycin (25 µg/ml), and kanamycin (50 µg/ml). Incubate the plates at 28 °C for 2 days. Individual bacterial colonies carrying each construct can be streaked on a fresh LB agar plate containing antibiotics and grown as described in this step; after bacterial growth, these plates can be stored for 1 week at 4 °C. Liquid cultures of bacteria can be stored as 10% glycerol stocks at −80 °C for several years.

3.4 Inoculation of Arabidopsis Seedlings with Agrobacterium Carrying pTRV Plasmids

1. Inoculate separate colonies of *Agrobacterium* carrying the pTRV1 and recombinant pTRV2 constructs into 3 ml of LB medium containing the required antibiotics and grow at 28 °C and shaking at 250 × *g* overnight. Transfer 1 ml of every culture into 100 ml of YEBi medium and grow until OD_{600} reading reaches between 0.8 and 1.2 (*see* **Note 9**).
2. Collect *Agrobacterium* cells by centrifugation for 10 min at 3000 × *g* (*see* **Note 10**).
3. Resuspend the bacterial pellet in the MMAi solution into suspensions of OD_{600} = 4.0 followed by shaking at a speed of 50 × *g* for 2 h.
4. Mix the suspensions of *Agrobacterium* carrying the pTRV1 and pTRV2 recombinant constructs at a 1:1 ratio to lead a final OD_{600} = 2.0 of each mixture (*see* **Note 11**).
5. The delivery of *Agrobacterium* suspension is performed with a needleless 1 ml syringe into two leaves of two- to three-leaf-stage plants, infiltrating the entire leaf from the abaxial side of the leaf (*see* **Note 12**). About 0.2 ml of culture is delivered into each leaf. Repeat in the rest of constructs in separate plants. Use ten plants per construct for infiltration, and set aside ten plants to use as an uninfected control (*see* **Note 13**).
6. Following inoculation, leave the plants covered with a plastic wrap overnight in the chamber. The symptoms of viral infection will become apparent approximately 10 days after virus introduction. The appearance of new bleached leaves in the pTRV::PDS-infected plants will indicate the successful infection and silencing of the *PDS* gene (*see* **Notes 14** and **15**, Fig. 2).

3.5 Monitoring the Down-Regulation of GOI Using RT-qPCR

1. After successful establishment of silencing, verify the down-regulation of GOI using RT-qPCR. Collect 2–3 leaves from the infected and uninfected plants at 3 weeks post-germination stage (*see* **Note 16**).
2. Extract RNA from plant tissue using the RNeasy mini kit according to the manufacturer's instructions and including DNase treatment in the procedure (*see* **Note 8**).

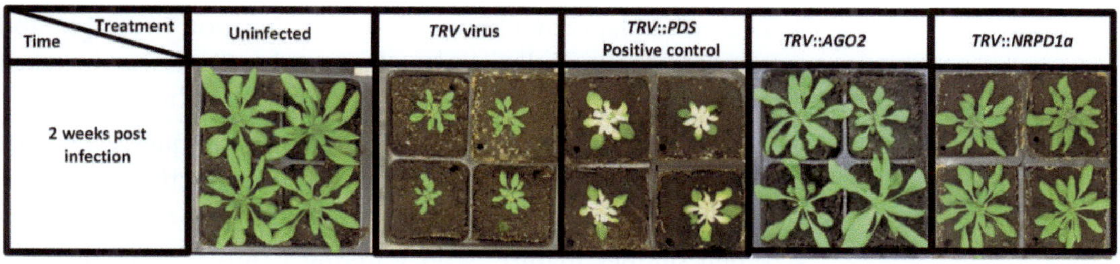

Fig. 2 The optimization of TRV-mediated VIGS in *Arabidopsis* plants. Seedlings of *Arabidopsis* Col-0 plants at the two- to three-leaf stage were infected with either of the following constructs: the empty TRV virus, TRV virus carrying a fragment of either *PDS* (positive control), *AGO2*, or *NRPD1a* genes. Symptoms of the virus-mediated suppression of gene activity became apparent at 2 weeks post-infection (TRV-PDS, 2 wpi), albeit plants infected with the empty TRV virus demonstrated a severe delay in growth as compared to uninfected controls. Later, most of the empty TRV-infected plants recovered from infection (a picture not shown)

3. Following RNA isolation, use 500 ng of total RNA and perform cDNA synthesis using the cDNA synthesis kit. Quantify gene expression of GOIs using qPCR with the previously synthesized primers. Perform normalization against two or more housekeeping genes (*see* **Note 17**).
4. Calculate the fold difference of the target gene's transcript levels as compared to the uninfected plants. The relative expression of the target gene can be calculated on the basis of the cycle threshold (Ct) values of the target gene and reference gene transcripts. The relative expression values are used to calculate the fold difference. Plants with the lowest abundance of GOI transcripts can be selected for further experiments.
5. Continue watering plants and allow them to flower to perform stable floral dip transformation.

3.6 Floral Dip Transformation of Infected Plants

1. *Agrobacterium*-mediated floral-dip transformation of *Arabidopsis* is essentially done as described in [13]. Inoculate *Agrobacterium* carrying any plant transformation binary vector established in the lab into 200 ml of sterilized LB medium with antibiotics (rifampicin—25 μg/ml, gentamycin—25 μg/ml, and kanamycin 50 μg/ml) and grow to a stationary phase (typically overnight) at 28 °C, 250 × *g* (*see* **Note 18**).
2. Harvest cells by centrifugation for 20 min at room temperature at 5500 × *g* (*see* **Note 10**).
3. Resuspend the bacterial pellet in the floral-dip infiltration medium to a final OD_{600} of approximately 0.80.
4. Add the inoculum to a 200 ml beaker, and invert the plants into this suspension such that all inflorescence shoots carrying flowers are submerged. Remove the plants after 3–5 s of a gentle agitation (*see* **Notes 14** and **19**).

5. Leave the plants in a low-light or dark location overnight and return to the growth chamber the next day. Care must be taken to keep the domed plants out of a direct sunlight.
6. Grow the plants for further 3–5 weeks until siliques are brown and dry, keeping the bolts from each pot together and separated from the neighboring pots using tape and/or wax paper.
7. When siliques mature, stop watering plants and let seeds to dry. Harvest seeds from individually transformed plants, and if necessary dry overnight at 37 °C.
8. Perform the selection of transgenic plants by planting an equal amount of seeds (based on weight) using the required antibiotics/herbicides for the respective T-DNA integration (*see* **Note 20**, Fig. 3).
9. Count the number of transgenic plants obtained from around 1 g of seeds (~40,000 plants) and calculate changes in transformation efficiency of the infected plants in relation to uninfected control plants.

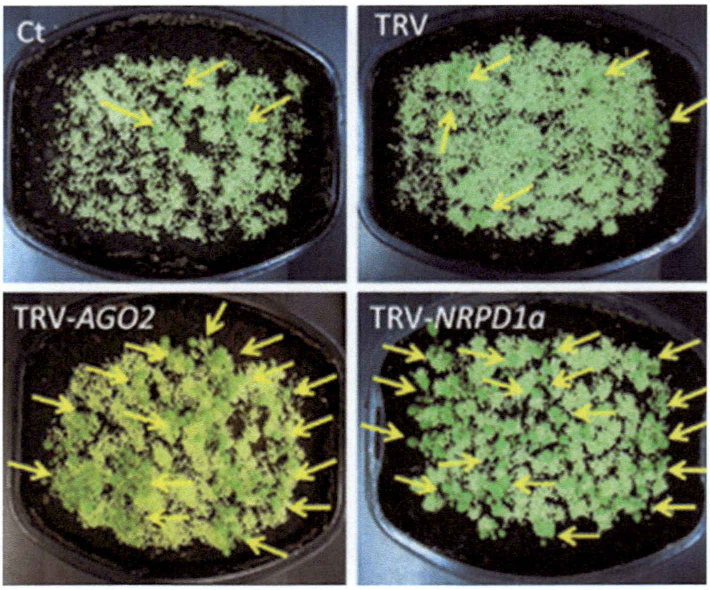

Fig. 3 *In solium* selection of transgenic plants obtained by using a method for the stable transformation mediated by virus-induced gene silencing. Selection plates with transgenic plants obtained after the transformation of untreated plants (Ct), plants infected with the empty *Tobacco rattle virus* (TRV) virus (TRV), plants infected with TRV virus carrying an *AGO2* fragment (TRV-AGO2), and those ones infected with TRV virus carrying an *NRPD1a* fragment (TRV-NRPD1a), respectively. The *arrows* point at the representative transgenic plants that were resistant to glufosinate ammonium and scored in every plate

4 Notes

1. Any commercially available or in-house-prepared chemically competent cells can be used.
2. Other disarmed *Agrobacterium* strains, such as GV2260, EHA105, and LBA4404, can also be used.
3. Acetosyringone stock solution preparation—dissolve 39.24 mg of acetosyringone powder in 1 ml of DMSO and vortex the mixture thoroughly to prepare a 200 mM solution. Store the stock at −20 °C. DMSO should be handled in a fume cupboard, and protective clothing should be worn when handling it. Use pipette tips that are safe to handle solvents or use glass pipettes. Acetosyringone stock should be used within 3 months of the date of preparation.
4. Prepare MMAi solution fresh every time before the inoculation procedure. 50 ml of MMAi solution is enough for agro-infiltration of 50 seedlings.
5. We have optimized the VIGS technique for *Arabidopsis* ecotype Col-0. Nevertheless, it is safe to assume that any other *Arabidopsis* ecotype established in a lab can be used for VIGS experiments.
6. Any commercially available qPCR mix suitable for a particular qPCR instrument can be used.
7. Forward and reverse primers for the quantification of a target gene fragment by RT-qPCR can be designed with PrimerQuest software by IDT technologies (http://www.idtdna.com/Primerquest). Any other similar primer design software can also be used. Some preferred primer characteristics are as follows: Tm ~60 °C, 18–25 nt length, and 40–60% GC content. Product length can be in the range of 60–150 nt. At least one primer should be designed outside the region of the gene that is used in the pTRV2 vector. Primers should not form dimers and should be specific for the target gene without cross-amplification of non-target genes. This is especially important for the amplification of a specific gene from a gene family [11].
8. Any other plant RNA extraction protocol can also be used, but it is critical to include DNase treatment in the extraction protocol in order to avoid the amplification from traces of gDNA. Confirm RNA quality by measuring the OD at 260 and 280 nm with a spectrophotometer before proceeding to cDNA synthesis.
9. Every *Agrobacterium* culture carrying pTRV2 recombinant constructs will be mixed with the equal amount of culture harboring pTRV1 plasmid before inoculation; hence the volume of culture for *Agrobacterium* carrying pTRV1 plasmid should correspond to the total volume of all cultures with the pTRV2 construct. For

instance, if three pTRV2 recombinant constructs are tested, 300 ml of pTRV1 culture should be set up. *Agrobacterium* carrying pTRV1 sometimes grow slower than pTRV2 cultures. In such case, use freshly transformed *Agrobacterium* for inoculation.

10. The supernatant should be treated with an equal volume of bleach (20% (vol/vol)) for 30 min before discarding it in the drain.
11. Up to 5 ml of mixture is enough to perform agro-infiltration of eight seedlings.
12. In our experience, the plant's stage used for agro-infiltration is critical for a successful infection and development of VIGS phenotype in *Arabidopsis* Col-0 plants. As a general rule, the older the plants, the less efficient the silencing of target genes. Additionally, the efficiency of VIGS is reduced at higher (24–26 °C) ambient temperatures.
13. Cultures can accidentally start spraying during infiltration. Use a face shield or other appropriate face protection, and wear a lab coat. Do not allow cultures to contact the neighboring plants (i.e., those that are assigned for a different construct or experiment) because this will lead to cross-contamination [11].
14. All cultures, plants, plant growth media, and apparatuses used for handling *Agrobacterium* cultures must be disposed according to the common biosafety standards and regulations. In general, autoclave all materials that come in contact with *Agrobacterium*-containing TRV before disposal. TRV can infect more than 400 plant species; hence, *Agrobacterium* carrying this virus must be handled with caution, and the environmental release of this virus must be prevented [11].
15. Typically, changes in a phenotype due to target gene silencing are expected to occur 2–3 weeks after inoculation. A visible phenotypic change due to target gene silencing is not expected for every gene-silenced plant. Hence, the confirmation of gene silencing should also be based on the down-regulation of a target gene and its expected molecular or biochemical changes [11].
16. Use an appropriate plant tissue produced after TRV inoculation for RNA extraction. For example, a newly emerged leaf after the inoculation should be used. Do not use the TRV-inoculated leaves [11].
17. *TUBULINE* (At5g62690) and *RCE1* (At4g36800) reference genes can be used for the normalization of qPCR data [14, 15]. If the construct used for VIGS has been predicted to cause any off-target gene silencing on the basis of a bioinformatics analysis, the transcript levels of the predicted off-target genes should be assessed by RT-qPCR to determine whether (and to what extent) the unintended down-regulation of those genes has occurred. This will facilitate the interpretation of results of a subsequent plant analysis.

18. We routinely use the pCAMBIA3301 binary vector with the *BAR* herbicide-resistance gene under a strong 35S promoter with a double enhancer that confers the resistance to glufosinate ammonium.
19. In order to compare the stable transformation efficiency, the transformation of all infected plants and control uninfected plants should be done on the same day using the same *Agrobacterium* solution.
20. When the pCAMBIA3301 plasmid or any other vector carrying the *BAR* gene is used for a stable transformation, the transgenic selection can be done as follows: plant seeds obtained from plants transformed by floral dip at a high density in 2-in. dip pots (100 mg of seeds per pot) filled with soil mix and presoaked in a fertilizer. Two or 3 days post-germination, spray the seedlings with the 1000× dilution of Liberty 150CN Herbicide and Crop Desiccant stock solution (150 g/L concentration of glufosinate ammonium in the stock solution; Aventis CropScience Canada Co, Ontario, Canada) once a day for 7 days in a row. Transplant the healthy-looking transgenic plants into separate pots and grow for 2 weeks followed by tissue collection for transgene analysis.

Acknowledgment

We acknowledge the grants from the Natural Sciences and Engineering Research Council of Canada and Agricultural Research Consortium of Alberta. We thank Nina Kepeshchuk for technical assistance and Valentina Titova for proofreading the manuscript.

References

1. Tsaftaris AS, Polidoros AN, Kapazoglou A, Tani E, Kovačević NM (2008) Epigenetics and plant breeding. In: Plant breeding reviews. Wiley, New York, pp 49–177
2. Herrera-Estrella L, Simpson J, Martinez-Trujillo M (2005) Transgenic plants: an historical perspective. Methods Mol Biol 286:3–32
3. Windels P, Buck S, Depicker A (2008) Agrobacterium tumefaciens-mediated transformation: patterns of T-DNA integration into the host genome. In: Tzfira T, Citovsky V (eds) Agrobacterium: from biology to biotechnology. Springer, New York, pp 441–481
4. Dunoyer P, Himber C, Voinnet O (2006) Induction, suppression and requirement of RNA silencing pathways in virulent Agrobacterium tumefaciens infections. Nat Genet 38(2): 258–263
5. Bilichak A, Yao Y, Kovalchuk I (2014) Transient down-regulation of the RNA silencing machinery increases efficiency of Agrobacterium-mediated transformation of Arabidopsis. Plant Biotechnol J 12(5):590–600
6. Endo M, Ishikawa Y, Osakabe K, Nakayama S, Kaya H, Araki T et al (2006) Increased frequency of homologous recombination and T-DNA integration in Arabidopsis CAF-1 mutants. EMBO J 25(23):5579–5590
7. Purkayastha A, Dasgupta I (2009) Virus-induced gene silencing: a versatile tool for discovery of gene functions in plants. Plant Physiol Biochem 47(11-12):967–976
8. Marton I, Zuker A, Shklarman E, Zeevi V, Tovkach A, Roffe S, Ovadis M, Tzfira T, Vainstein A (2010) Nontransgenic genome modification in plant cells. Plant Physiol 154(3):1079–1087

9. Pflieger S, Blanchet S, Camborde L, Drugeon G, Rousseau A, Noizet M, Planchais S, Jupin I (2008) Efficient virus-induced gene silencing in Arabidopsis using a 'one-step' TYMV-derived vector. Plant J 56(4):678–690
10. Deng X, Kelloniemi J, Haikonen T, Vuorinen AL, Elomaa P, Teeri TH, Valkonen JP (2013) Modification of Tobacco rattle virus RNA1 to serve as a VIGS vector reveals that the 29K movement protein is an RNA silencing suppressor of the virus. Mol Plant Microbe Interact 26(5):503–514
11. Senthil-Kumar M, Mysore KS (2014) Tobacco rattle virus-based virus-induced gene silencing in Nicotiana benthamiana. Nat Protoc 9(7): 1549–1562
12. Xu P, Zhang Y, Kang L, Roossinck MJ, Mysore KS (2006) Computational estimation and experimental verification of off-target silencing during posttranscriptional gene silencing in plants. Plant Physiol 142(2): 429–440
13. Clough SJ, Bent AF (1998) Floral dip: a simplified method for Agrobacterium-mediated transformation of Arabidopsis thaliana. Plant J 16(6):735–743
14. Fridman E, Zamir D (2003) Functional divergence of a syntenic invertase gene family in tomato, potato, and Arabidopsis. Plant Physiol 131(2):603–609
15. Seki M, Narusaka M, Ishida J, Nanjo T, Fujita M, Oono Y et al (2002) Monitoring the expression profiles of 7000 Arabidopsis genes under drought, cold and high-salinity stresses using a full-length cDNA microarray. Plant J 31(3):279–292

Chapter 18

The Random Oligonucleotide-Primed Synthesis Assay for the Quantification of DNA Strand Breaks

Andriy Bilichak and Igor Kovalchuk

Abstract

DNA strand breaks arise from normal cellular processes such as replication, transcription, and DNA repair as well as spontaneous DNA damage caused by cell metabolic activities. In addition, strand breaks occur due to direct or indirect DNA damage produced by various abiotic and biotic stresses. Strand breaks are among the most problematic DNA lesions because unrepaired strand breaks may lead to cell cycle arrest, gross chromosome rearrangements, or even cell death. Thus, the measurement of the relative number of strand breaks can provide an informative picture of genome stability of a given cell, tissue, or organism. Here, we describe the use of random oligonucleotide-primed synthesis (ROPS) assay for the detection and quantification of the level of strand breaks in tissue samples. The applications of the assay for a quantitative detection of 3′OH, 3′P, or DNA strand breaks at a cleavage site of the deoxyribose residue are discussed.

Key words Random oligonucleotide-primed synthesis, ROPS, Single-strand breaks, SSBs, Double-strand breaks, DSBs, Genome stability

1 Introduction

DNA strand breaks occur upon the disruption of one or two strands of the DNA double helix that leads to the creation of a single-strand break (SSB) or a double-strand break (DSB). Strand breaks are an essential component for many cellular processes, such as homologous recombination during gametogenesis [1] and the generation of antigen-receptor and immunoglobulin diversity during the development of T and B lymphocytes in vertebrates [2]. DNA strand breaks arise through many cellular activities, including replication, transcription, and DNA repair; they also occur as a result of cell metabolic activity and oxidative damage to DNA. In addition, external stresses such as ionizing radiation or exposure to chemicals may disrupt DNA strands directly or generate an excessive amount of free radicals resulting in the damage to nucleotides or the sugar-phosphate backbone [3, 4]. Recent reports also suggest that many stressors can lead to the formation of SSBs and

Igor Kovalchuck (ed.), *Plant Epigenetics: Methods and Protocols*, Methods in Molecular Biology, vol. 1456,
DOI 10.1007/978-1-4899-7708-3_18, © Springer Science+Business Media New York 2017

DSBs through an indirect mechanism, including changes in DNA methylation and production of small noncoding RNAs [5, 6].

Strand breaks are taken care of by nonhomologous end-joining and homologous recombination—two repair mechanisms that have many sub-pathways [7]. Unrepaired strand breaks are deleterious to the cell, and their accumulation may lead to apoptosis and cell death [8]. Strand break repair often results in mutations (point mutations, deletions, and insertions of various sizes) and gross chromosomal rearrangements which negatively affect genome stability [9–11].

Exposure to many environmental factors can cause the damage to DNA and the destabilization of the genome. Thus, the measurement of levels of DNA strand breaks may be a reflection of genotoxicity of various environmental factors and chemicals. Measuring DNA damage would require an effective and sensitive assay that permits a quantitative detection of DNA strand breaks. In this chapter, we describe a rapid, cheap, and sensitive assay for the quantification of DNA breaks [12].

The method developed by Basnakian and James [12] represents a modification of a previously reported random oligonucleotide-primed synthesis (ROPS) assay [13]. Using this method, Ausubel et al. [13] produced the uniformly labeled DNA fragments with the Klenow fragment polymerase. In contrast, Basnakian and James [12] used the inability of the Klenow fragment enzyme to discriminate between complementary primers and primers with mismatches in their terminal regions. This allowed the development of a quantitative assay for the detection of DNA damage. Specifically, using the incorporation of radioactively labeled nucleotides, it is possible to detect and quantify 3′OH single-stranded gaps and 3′OH single- and double-stranded breaks in DNA [12]. Further, the non-radioactively labeled non-3′OH breaks can be assayed by treatment with phosphatase or exonuclease III.

In brief, the assay is based on the separation of double-stranded DNA containing nicks and breaks into single-stranded sequences by heat denaturation (Fig. 1). Cooling the samples allows a random reassociation of DNA fragments. During reassociation, the relatively short DNA fragments play the role of primers and associate with an excess of larger DNA fragments that serve as a template. During the next step of DNA synthesis performed by the Klenow fragment polymerase, the radioactively labeled nucleotides are incorporated. This incorporation is proportional to the number of breaks and reflects the relative level of strand break accumulation [5]. The assay is highly sensitive and capable of the detection of the low frequency of strand breaks [14]. The assay is work efficient and permits the analysis of several hundred samples per day.

Fig. 1 (continued) the high-molecular-weight DNA play the role of primers for DNA synthesis performed by the Klenow fragment polymerase. During DNA synthesis, the Klenow fragment polymerase incorporates the radioactively labeled nucleotides at a rate that is proportional to the number of original breaks

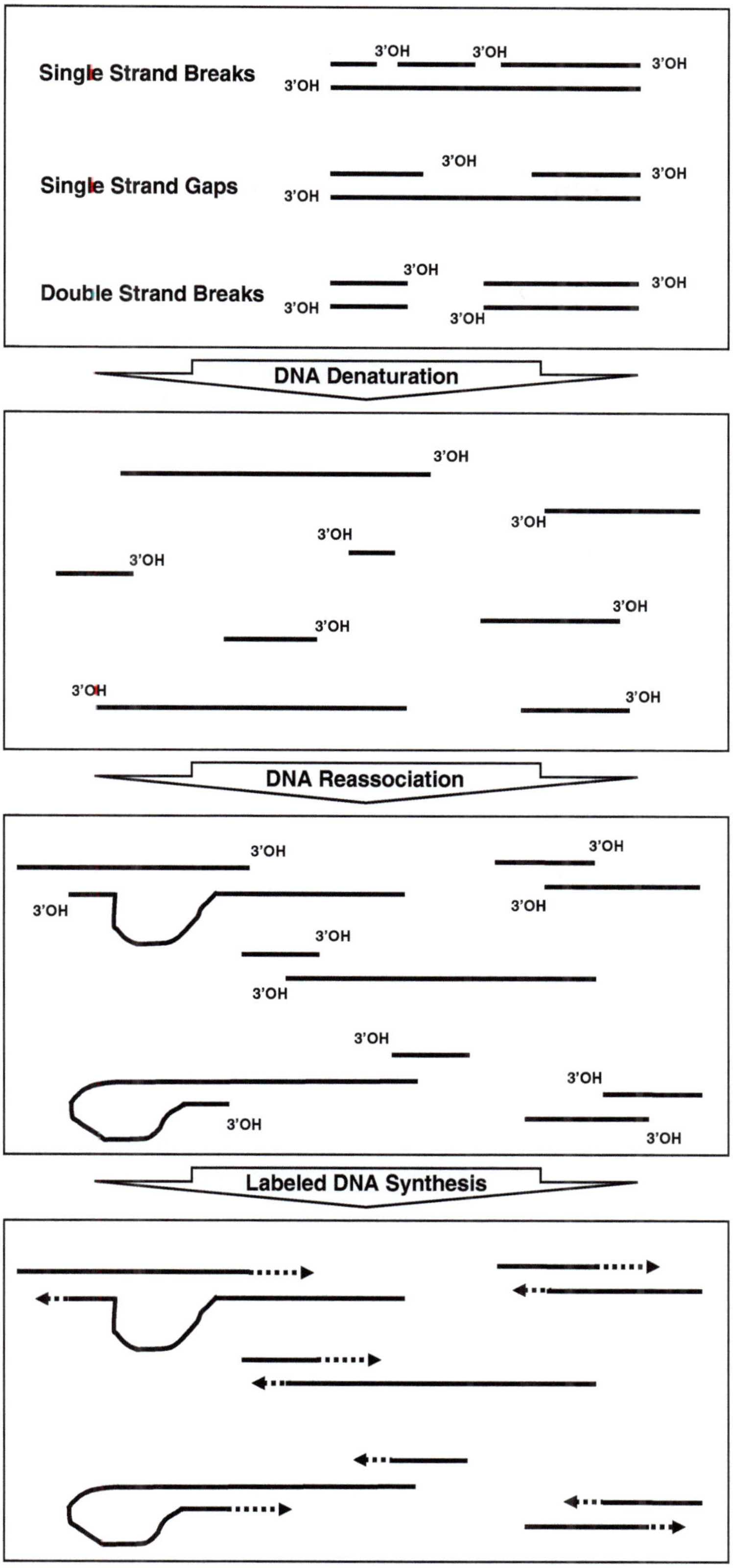

Fig. 1 The mechanism of the detection of DNA strand breaks using a modified ROPS assay. The assay permits the detection of various types of DNA damage, including nicks, single-stranded gaps, and single- and double-stranded DNA breaks that have a 3′OH group at their ends. Following heat denaturation, the single-stranded DNA fragments randomly reassociate with each other. Next, the relatively short DNA fragments that are associated with

2 Materials

1. Nuclease-free water.
2. Agarose, electrophoresis grade.
3. 1× TBE: 90 mM Tris–HCl, pH 8.0, 90 mM boric acid, 2 mM EDTA.
4. 6× DNA gel loading buffer.
5. 0.5 mM 3dNTP mix (dATP, dGTP, dTTP).
6. 33 μM dCTP.
7. 10× Klenow fragment buffer.
8. The Klenow fragment polymerase.
9. [^{3}H]dCTP. **CAUTION: Radiation protection measures must be taken for handling 3H and all derived materials.** Store in a shielded container in a dedicated freezer at −20 °C.
10. 12.5 mM EDTA, pH 8.0.
11. Whatman DE-81 ion-exchange filters.
12. 500 mM Na-phosphate buffer, pH 7.0.

3 Methods

1. Using nuclease-free water, prepare 0.25 μg genomic DNA aliquots in a final volume of 10 μl. Keep the samples on ice (*see* **Notes 1–3**).
2. Denature the DNA at 100 °C for 5 min. Chill the samples on ice immediately.
3. Add a 15 μl of reaction mixture to each sample, while the samples are on ice. Mix well. The reaction mixture for ten samples (150 μl) should contain 25 μl of 0.5 mM 3dNTP mix (dATP, dGTP, dTTP), 25 μl of 10× Klenow fragment buffer, 4.5 μl of 33 μM dCTP, 5 U of the Klenow fragment polymerase, and [^{3}H]dCTP (42.9 Ci/mmol) (*see* **Note 4**).
4. Incubate the samples at room temperature for 1 h (*see* **Note 5**).
5. Place the samples on ice and stop the reaction by adding 25 μl of 12.5 mM EDTA, pH 8.0.
6. Apply each reaction on Whatman DE-81 ion-exchange filters. Air-dry the filters (*see* **Note 6**).
7. Wash the filters in 500 mM Na-phosphate buffer, pH 7.0 at room temperature for 10 min.
8. Repeat wash twice.
9. Air-dry the filters and process them in a scintillation counter.
10. Express the results as relative [^{3}H]dCTP incorporation/0.25 μg of DNA.

4 Notes

1. The assay design allows a quantitative detection of 3′OH DNA strand breaks only. However, if the detection of other types of DNA strand breaks (3′P or breaks at a cleavage site of the deoxyribose residues) is required, then additional steps may be included to expose the 3′OH ends. The 3′P ends can be removed by phosphatase treatment. Similarly, the 3′-glycosyl ends can be eliminated by *E. coli* exonuclease III treatment. Both treatments will result in the exposed 3′OH ends that can be quantified using the ROPS assay.
2. The quality of genomic DNA is essential for the assay. The samples can be quantified using a spectrophotometer. However, we recommend to check equal sample loading using gel electrophoresis and if necessary adjust it accordingly. Moreover, the analysis of highly fragmented DNA may require an addition of a non-degraded high-molecular-weight DNA template to the samples, as recommended by Basnakian and James [14]. An excess of the high-molecular-weight DNA template is a major requirement for the assay because high amounts of DNA fragmentation will result in a quick reaction of saturation affecting the frequency of [^{32}P] or [^{3}H] incorporation. DNA fragmentation can be checked using a 1 % agarose gel: at least 50 % of DNA should be located in the initial band [14].
3. It is important to insure that DNA preparation is not contaminated with SDS, EDTA, proteinase K, or phenol because these chemicals can significantly inhibit the activity of the Klenow fragment polymerase.
4. The method was originally developed for the application of [^{32}P]dCTP. Considering a short half-life of [^{32}P], the assay was modified for using [^{3}H]dCTP instead. Regardless of an isotope used, both [^{32}P]dCTP and [^{3}H]dCTP should be supplied to the reaction in a mixture with unlabeled dCTPs. In our work, we primarily used [^{3}H]dCTP.
5. Incubation time may be decreased to 30 min, if the frequency of [^{3}H]dCTP incorporation is too high. Similarly, incubation temperature can be decreased to 16 °C to reduce variability between samples. Assay sensitivity can be increased either by increasing the amount of radioactively labeled dCTP or by decreasing the amount of unlabeled dCTP.
6. Using Whatman DE-81 ion-exchange filters is essential because it drastically reduces DNA contamination with unincorporated nucleotides.

References

1. Richardson C, Horikoshi N, Pandita TK (2004) The role of the DNA double-strand break response network in meiosis. DNA Repair 3(8-9):1149–1164
2. Rooney S, Chaudhuri J, Alt FW (2004) The role of the non-homologous end-joining pathway in lymphocyte development. Immunol Rev 200:115–131
3. Al-Emam A, Arbon D, Kysela B (2014) Deacetylation of Ku70 regulates ionizing-radiation induced DNA damage responses in human cells. BMC Genomics 15(Suppl 2):P24
4. Cooke MS, Evans MD, Dizdaroglu M, Lunec J (2003) Oxidative DNA damage: mechanisms, mutation, and disease. FASEB J 17(10): 1195–1214
5. Cuozzo C, Porcellini A, Angrisano T, Morano A, Lee B, Di Pardo A et al (2007) DNA damage, homology-directed repair, and DNA methylation. PLoS Genet 3(7):e110
6. Yamanaka S, Siomi H (2014) diRNA-Ago2-RAD51 complexes at double-strand break sites. Cell Res 24(5):511–512
7. Mao Z, Bozzella M, Seluanov A, Gorbunova V (2008) Comparison of nonhomologous end joining and homologous recombination in human cells. DNA Repair 7(10):1765–1771
8. Shrivastav M, De Haro LP, Nickoloff JA (2008) Regulation of DNA double-strand break repair pathway choice. Cell Res 18(1):134–147
9. Orel N, Kyryk A, Puchta H (2003) Different pathways of homologous recombination are used for the repair of double-strand breaks within tandemly arranged sequences in the plant genome. Plant J 35(5):604–612
10. Dudas A, Chovanec M (2004) DNA double-strand break repair by homologous recombination. Mutat Res 566(2):131–167
11. Puchta H (2005) The repair of double-strand breaks in plants: mechanisms and consequences for genome evolution. J Exp Bot 56(409):1–14
12. Basnakian AG, James SJ (1994) A rapid and sensitive assay for the detection of DNA fragmentation during early phases of apoptosis. Nucleic Acids Res 22(13):2714–2715
13. Ausubel et al (1989) In: Ausubel M, Brent R, Kingston RE, Moore DD, Seidman JG, Smith JA, Struhl K (eds) Current protocols in molecular biology, vols 1 and 2. Wiley, Media, PA. Mol Reprod Dev 1(2):146
14. Basnakian AG, James SJ (1996) Quantification of 3'OH DNA breaks by random oligonucleotide-primed synthesis (ROPS) assay. DNA Cell Biol 15(3):255–262

Chapter 19

Profiling Transposable Elements and Their Epigenetic Effects in Non-model Species

Christian Parisod

Abstract

Taking transposable elements into consideration in surveys of genetic and epigenetic variation remains challenging in species lacking a high-quality reference genome. Here, molecular techniques reducing genome complexity and specifically targeting restructuring and methylation changes in TE genome fractions are described. In particular, methyl-sensitive transposon display (MSTD) uses isoschizomers and PCR amplifications to assess the methylation environment of TE insertions. MSTD offers reliable insights into genome-wide epigenetic changes associated with TEs, especially when used together with similar techniques tracking random sequences.

Key words Transposons, Transposable elements, Epigenetic effects, Plants, Methyl-sensitive transposon display

1 Introduction

Transposable elements (TEs, also called jumping genes) represent a major and labile fraction of eukaryotic genomes [1]. In addition to being highly mutagenic by inserting across the genome, TEs are epigenetically silenced by overlapping mechanisms including DNA methylation that may spread and control the expression of neighboring sequences [2]. Fueling genome structural and epigenetic reorganization, TEs likely impact on phenotypes and represent key evolutionary players [3]. Surveys of variation within and among taxa should thus take TEs into consideration for further integration of the genotypic, phenotypic, and fitness landscapes.

Current technological advances in high-throughput sequencing offer tremendous opportunities to shed light on the genetic and epigenetic variation associated with TEs [4]. Currently producing reads that are typically much shorter than TEs, such approaches however rely on the mapping of TE variation on accurately sequenced and assembled genome references (*see*, e.g., [5] and references therein). Accordingly, for studies requiring large

Igor Kovalchuck (ed.), *Plant Epigenetics: Methods and Protocols*, Methods in Molecular Biology, vol. 1456, DOI 10.1007/978-1-4899-7708-3_19, © Springer Science+Business Media New York 2017

sample sizes or for species lacking high-quality reference genomes, molecular techniques reducing genome complexity and specifically targeting TE genome fractions remain most appropriate [6].

Among existing transposon display strategies (i.e., high-resolution TE-anchored PCR strategy allowing the simultaneous detection of multiple insertions), sequence-specific amplified polymorphism (SSAP) is one of the most easily applicable and reliable (*see* ref. 7 and references therein). Briefly, the SSAP procedure is derived from the amplified fragment length polymorphism (AFLP) strategy, but specifically targets TE insertions. It relies on the amplification of digested genomic DNA with primers designed at the border of TEs, generating a pool of labeled fragments containing the termini of inserted copies of a given TE and its flanking genomic region (Fig. 1). Provided that the appropriate TE-specific primers are used (*see* **Note 3**), SSAP usually generates highly polymorphic markers that reliably assess and partition the variation in TE genome fractions [8, 9]. As SSAP polymorphism reflects molecular changes at insertion sites, it is worth noting that the modification of restriction sites or the size of the amplification product is also highlighted [10]. Thus, the variation in SSAP banding patterns among related lineages provides useful insights into the dynamics of TE genome fractions but benefits from systematic comparison with banding patterns of random sequences, such as produced by AFLP (*see* **Note 6**).

Here, updated information on the methyl-sensitive transposon display (MSTD) approach is described (Fig. 1a). This protocol represents a minor modification of the SSAP protocol, involving the widely used isoschizomers *Msp*I and *Hpa*II at the digestion step, to provide useful knowledge on the methylation environment of TE insertions [11, 12]. This couple of enzymes has been used for methyl-sensitive displays tracking random sequences (e.g., methyl-sensitive amplified polymorphism, MSAP [13]). They indeed recognize the same tetranucleotide sequence (5′-CCGG-3′), but *Hpa*II is sensitive to methylation of any cytosine at both strands (5′-CCGG-3′), while *Msp*I cuts the methylated internal cytosine (5′-C^{5m}CGG-3′; [14]). These properties allow assessing the methylation status of internal cytosine at restriction sites (CpG methylation; Fig. 1b). As *Hpa*II cleaves when the external cytosine is methylated on one strand, while *Msp*I does not, hemi-methylated CpCpG sites can also be detected with this MSTD. Note, however, that *Msp*I is sensitive to methylation of the external cytosine (5′-^{5m}CCGG-3′), and methylation of the external cytosine on both strands (CpCpG methylation: 5′-^{5m}CCGG-3′ and 5′-^{5m}C^{5m}CGG-3′) may not produce bands with this MSTD.

In addition to the limits inherent to SSAP [7–9], MSTD profiles have to be interpreted carefully because CpCpG methylation on both strands prevents the enzymes from cutting (e.g., in heavily methylated portions of the genome). MSTD data thus offer insights

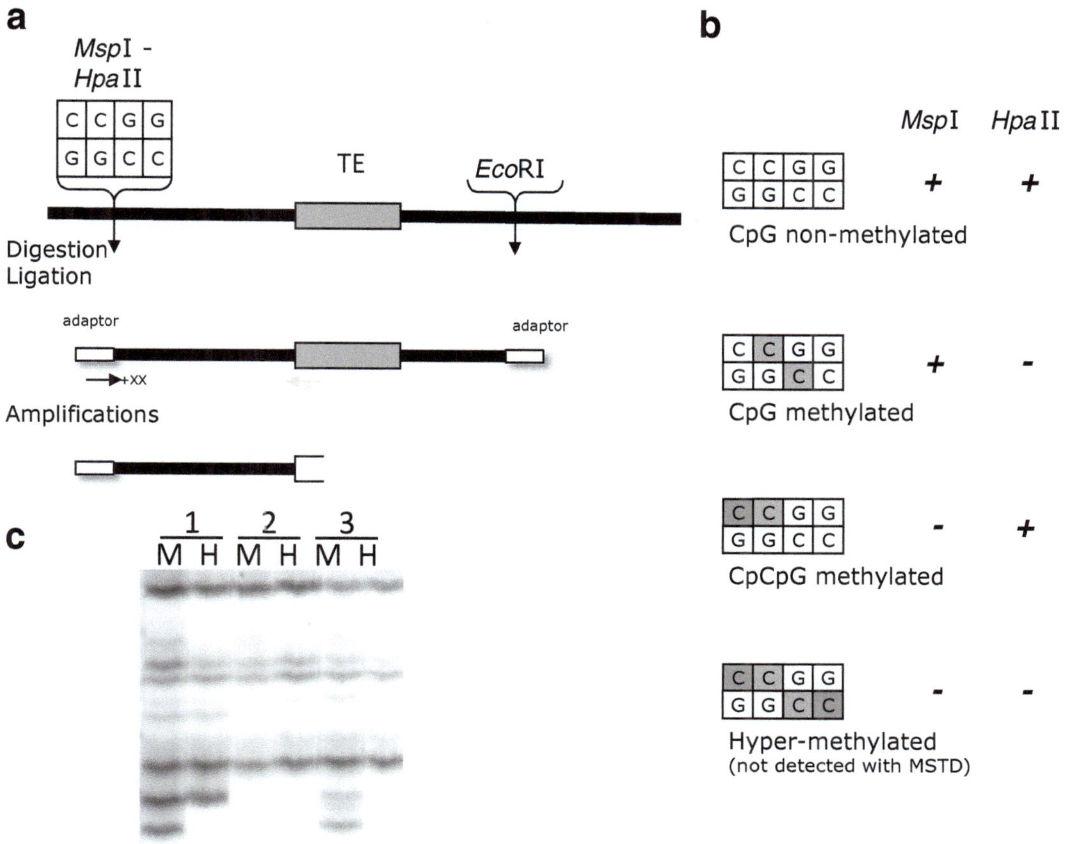

Fig. 1 Principle of the methyl-sensitive transposon display (MSTD). (**a**) Schematic representation of the high-resolution TE-anchored PCR strategy allowing the simultaneous detection of multiple insertions. After digestion of genomic DNA with rare cutter (e.g., *MspI/HpaII*) and frequent cutter (e.g., *EcoRI*) restriction enzymes, adaptors are ligated to DNA fragments. PCR amplifications are carried out using a primer complementary to the rare cutter adaptor and a labeled (*) primer specific to the targeted transposable element (TE). (**b**) Methylation sensitivity of isoschizomer enzymes (*MspI* and *HpaII*) and the interpretation of the resulting banding patterns according to the presence (+)/absence (–) of a given MSTD band. (**c**) An example of MSTD banding pattern for three samples (*1–3*). A comparison of band presence/absence in *MspI* (M) and *HpaII* (H) profiles reveals the methylation state of restriction sites flanking the corresponding TE insertion

about the methylation status of sequences flanking a particular TE insertion (i.e., the CCGG site next to a TE insertion) and may be biased toward the non-heavily methylated regions. Furthermore, the restructuring of a given TE insertion or increased methylation in the vicinity of this TE insertion might similarly result in the absence of bands, whereas transposition events or demethylation in the vicinity of a TE insertion might result in a band specific to selected samples. Accordingly, MSTD data benefit from the comparison with random sequences such as provided by MSAP (*see* **Note 6**; [12, 15]).

2 Materials

2.1 Digestion

Tango buffer (10×): 33 mM Tris-acetate (pH 7.9), 10 mM Mg-acetate, 66 mM K-acetate; 0.1 mg/ml BSA.

*Eco*RI (10 U/μl): Rare cutter enzyme (5′-GAATTC-3′) (*see* **Note 1**).

*Msp*I/*Hpa*II (10 U/μl): Frequent cutter enzymes recognizing the same tetranucleotide sequence (5′-CCGG-3′) but displaying differential sensitivity to DNA methylation.

2.2 Ligation

*Eco*RI-adaptors (100 μM): 5′-CTCGTAGACTGCGTACC-3′ and 5′-AATTGGTACGCAGTCTAC-3′. Preparation: Mix equal volumes of the two adaptors (a final concentration: 50 μM), warm up to 95 °C for 5 min, and then let cool down at room temperature. Then dilute at 1/10 for a final concentration of 5 μM.

*Msp*I/*Hpa*II-adaptors (100 μM): 5′-GACGATGAGTCTAGAA-3′ and 5′-CGTTCTAGACTCATC-3′. Preparation: Mix equal volumes of the two adaptors (final concentration: 50 μM), warm up to 95 °C for 5 min, and then let cool down at room temperature.

ATP (20 mM).

T4 DNA ligase (5 U/μl).

2.3 PCR Preselective Amplification

*Eco*RI + A primer (10 μM): 5′-GACTGCGTACCAATTC*A*-3′.
*Msp*I/*Hpa*II + C primer (10 μM): 5′-GATGAGTCTAGAACGG*C*-3′.
Rxn Buffer (10×): 200 mM Tris pH 8.4 + 500 mM KCl.
Equimolar dNTPs (10 mM).
$MgCl_2$ (25 mM).
Taq polymerase (5 U/μl).

2.4 PCR Selective Amplification

Labeled TE-specific primers (*see* **Notes 2** and **3**).

MspI/HpaII selective primers were similar to preselective primers, with the addition of two variable nucleotides (=*Msp*I/*Hpa*II + CXX primer).

Otherwise similar to Subheading 2.3

3 Methods

3.1 Digestion (See Note 4)

1. Add 5 μl of Tango buffer to 12.7 μl of sterile water.
2. Add 0.1 μl (1 U) of *Eco*RI.
3. Add 0.2 μl (2 U) of *Msp*I (alternatively, *Hpa*II) and gently mix.
4. Add 250 ng of DNA in 7 μl to this mix (a final volume 25 μl) and gently mix.
5. Incubate at 37 °C for 3 h.
6. Inactivate restriction enzymes at 70 °C for 15 min.

3.2 Ligation (See Note 4)

1. Add 3 μl of Tango buffer to 8.5 μl of sterile water.
2. Add 1 μl of ATP.
3. Add 1 μl of *Eco*RI-adaptors.
4. Add 1 μl of *Msp*I/*Hpa*II-adaptors and vortex.
5. Add 0.5 μl (5 U) of T4 DNA ligase and gently mix.
6. Add this ligation mix (15 μl) to 25 μl of digestion mix (a final volume 40 μl) and gently mix.
7. Incubate at room temperature (22 °C) overnight.
8. (Optional) 5 μl of the product can be electrophoresed on 1% agarose gels and stained with ethidium bromide in order to verify the success of digestion.
9. Dilute the digestion-ligation mix four times (e.g., 30 μl of product in 90 μl of sterile water) (*see* **Note 5**).

3.3 PCR Preselective Amplification

1. Add 2 μl of Rxn buffer to 12.7 μl of sterile water.
2. Add 0.5 μl of dNTP.
3. Add 1.6 μl of $MgCl_2$.
4. Add 0.5 μl of *Eco*RI + A primer.
5. Add 0.5 μl of *Msp*I/*Hpa*II + C primer and vortex.
6. Add 0.2 μl of Taq polymerase (1 U).
7. Add 18 μl of this preselective mix to 2 μl of diluted digestion-ligation mix (a final volume 20 μl).
8. Place in a thermocycler to perform this PCR amplification: 94 °C for 180 s, followed by 28 cycles at 94 °C for 30 s, 60 °C for 60 s, and 72 °C for 60 s, and a final extension at 72 °C for 180 s.
9. Dilute the preselective amplification products 1:10 with sterile water (e.g., 10 μl of the product in 190 μl of water).

3.4 PCR Selective Amplification (See Note 6)

1. Add 2 μl of Rxn buffer to 11.1 μl of sterile water.
2. Add 0.5 μl of dNTP.
3. Add 1.6 μl of $MgCl_2$.
4. Add 0.8 μl of TE-specific primer.
5. Add 0.8 μl of *Msp*I/*Hpa*II + CXX primer and vortex.
6. Add 0.2 μl of Taq polymerase (1 U).
7. Add 17 μl of this selective mix to 3 μl of the diluted preselective amplification product (a final volume of 20 μl).
8. Place in a thermocycler to perform this touch-down PCR amplification: 94 °C for 120 s, followed by 13 cycles at 94 °C for 30 s, 65–56 °C (decreasing by 0.7 °C per cycle) for 30 s, and 72 °C for 60 s, followed by 25 cycles at 94 °C for 30 s, 56 °C for 30 s, and 72 °C for 60 s, and a final extension at 72 °C for 300 s.
9. Prepare the amplification products according to your electrophoresis protocol (*see* **Note 2**).

4 Notes

1. *Selection of isoschizomers*: EcoRI is very widely used, but it can be variable in its sensitivity to CpG methylation [14]. SSAP protocols using Csp6 instead of EcoRI have been developed [7] and might be profitably used for MSTD. Unfortunately, Csp6 is a four-base restriction enzyme (i.e., a frequent cutter) and might provide too many SSAP bands for proper scoring in complex genomes. Furthermore, with such frequent cutter, it might happen that the TE itself presents restriction sites, thus inducing the amplification of bands internal to the TE and confusing results.
2. *Labeling and detection of TE loci*: It is vital to label the TE primer in order to highlight bands containing the termini of an inserted TE and its flanking genomic region. The TE primer can be radioactively labeled with P33 or with fluorochromes. Amplification products labeled with P33 can be visualized after electrophoresis on a 6% Long Ranger denaturing gel for 5 h (75 V, limited to 2000 W) by autoradiography. Amplification products labeled with fluorochromes can be visualized with automatic sequencers after electrophoresis. One of the advantages of the latter strategy is that it particularly enables multiplexing of different TEs labeled with different fluorochromes and reliably distinguishes fragments with shorter length differences. Accordingly, it quickly generates large amounts of high-resolution data.
3. *Design of appropriate SSAP primers*: A reliable amplification of genomic regions flanking insertions of a given TE lineage necessitates TE-specific primers designed at the vicinity of the TE border (ideally within 100 bp). The primers are preferably designed based on the knowledge of sequence variations among copies of the targeted TE lineage. The primers matching the highly conserved TE regions (e.g., reported in divergent species) will comprehensively track the corresponding insertions and likely amplify old insertions to a large extent [16]. The recent TE dynamics is appropriately surveyed with primers designed in more variable TE regions, enabling the amplification of selected subsets of insertions [9]. Note that superficial genome sequencing offers excellent opportunities to target TE insertions at the appropriate resolution in non-model species. The reader willing to know more about such a promising approach would profitably read the work of Senerchia et al. on the complex wild wheat genomes [17]. There, genome skimming (i.e., sequencing of as low as 2% of the genome) first inferred the dynamics of several TE families [17] and then used sequence variation to design the appropriate SSAP primers tracking the corresponding insertions in different species [18] and their hybrids [15].

4. *Digestion/ligation as a sensible step*: MspI and HpaII have to be used on the same samples in parallel in order to provide a methyl-sensitive transposon display (MSTD). Accordingly, preparing two mixes in parallel at each step (one for MspI reactions and one for HpaII reactions) can improve reliability and comparability of MSTD profiles. Although SSAP and MSTD approaches generate reliable consistent patterns, it is strongly advisable to perform the protocol several times on the selected samples in order to estimate the error rate (*see* ref. 19 for further details).
5. *Assay storage*: The diluted digestion-ligation product can be stored at −20 °C.
6. *TE-specific vs. background genomic variation*: In order to highlight the TE-specific variation, selective amplifications for SSAP/MSTD and AFLP/MSAP can be performed on the same pre-amplification products by appropriately using TE-specific and adaptor-specific primers. Scoring of large numbers of SSAP/MSTD loci together with a systematic comparison to AFLP/MSAP variation minimizes inference biases due to (1) segregating polymorphisms, (2) molecular changes at the insertion site modifying the size of the amplified product rather than indicating transposition events, (3) the sensitivity of EcoRI restriction enzyme to rare cytosine methylation states, and (4) preferential amplifications inherent to competitive PCRs. As these techniques share very similar features, statistical comparisons of SSAP/MSTD (i.e., tracking TE insertions) and AFLP/MSAP (i.e., tracking random sequences) offer a reliable account of variation that is specific to a given TE lineage. Molecular events that are not TE specific will indeed yield a similar variation in both SSAP/MSTD and AFLP/MSAP profiles, whereas TE-specific events will appear to be significant in the corresponding SSAP/MSTD profiles only (*see* refs. 15, 18). Note that this approach is blind to the mechanisms underlying polymorphism at specific loci whose elucidation requires cloning and locus-specific PCR-based assays.

References

1. Hua-Van A, Le Rouzic A, Boutin TS, Filee J, Capy P (2011) The struggle for life of the genome's selfish architects. Biol Direct 6:19
2. Hollister JD, Smith LM, Guo YL, Ott F, Weigel D, Gaut BS (2011) Transposable elements and small RNAs contribute to gene expression divergence between Arabidopsis thaliana and Arabidopsis lyrata. Proc Natl Acad Sci U S A 108:2322–2327
3. Bonchev G, Parisod C (2013) Transposable elements and microevolutionary changes in natural populations. Mol Ecol Resour 13:765–775
4. Treangen TJ, Salzberg SL (2012) Repetitive DNA and next-generation sequencing: computational challenges and solutions. Nat Rev Genet 13:36–46
5. Jiang C, Chen C, Huang Z, Liu R, Verdier J (2015) ITIS, a bioinformatics tool for accurate identification of transposon insertion sites using next-generation sequencing data. BMC Bioinformatics 16:72

6. Kalendar R, Flavell AJ, Ellis THN, Sjakste T, Moisy C, Schulman AH (2011) Analysis of plant diversity with retrotransposon-based molecular markers. Heredity 106:520–530
7. Syed NH, Flavell AJ (2006) Sequence-specific amplification polymorphisms (SSAPs): a multi-locus approach for analyzing transposon insertions. Nat Protoc 1:2746–2752
8. Melayah D et al (2004) Distribution of the Tnt1 retrotransposon family in the amphidiploid tobacco (Nicotiana tabacum) and its wild Nicotiana relatives. Biol J Linn Soc Lond 82:639–649
9. Petit M et al (2007) Differential impact of retrotransposon populations on the genome of allotetraploid tobacco (Nicotiana tabacum). Mol Genet Genomics 278:1–15
10. Petit M et al (2010) Mobilization of retrotransposons in synthetic allotetraploid tobacco. New Phytol 186:135–147
11. Kashkush K, Khasdan V (2007) Large-scale survey of cytosine methylation of retrotransposons and the impact of readout transcription from long terminal repeats on expression of adjacent rice genes. Genetics 177:1975–1985
12. Parisod C, Salmon A, Zerjal T, Tenaillon M, Grandbastien MA, Ainouche M (2009) Rapid structural and epigenetic reorganization near transposable elements in hybrid and allopolyploid genomes in Spartina. New Phytol 184:1003–1015
13. Cervera MT, Ruiz-Garcia L, Martinez-Zapater JM (2002) Analysis of DNA methylation in Arabidopsis thaliana based on methylation-sensitive AFLP markers. Mol Genet Genomics 268:543–552
14. Roberts RJ, Vincze T, Posfai J, Macelis D (2010) REBASE - a database for DNA restriction and modification: enzymes, genes and genomes. Nucleic Acids Res 38:D234–D236
15. Senerchia N, Felber F, Parisod C (2015) Genome reorganization in F1 hybrids uncovers the role of retrotransposons in reproductive isolation. Proc R Soc Lond B Biol Sci 282
16. Gustafsson ALS et al (2014) Genetics of cryptic speciation within an arctic mustard, Draba nivalis. PLoS One 9:e93834
17. Senerchia N, Wicker T, Felber F, Parisod C (2013) Evolutionary dynamics of LTR retrotransposons in wild wheats assessed with high throughput sequencing. Genome Biol Evol 5:1010–1020
18. Senerchia N, Felber F, Parisod C (2014) Contrasting evolutionary trajectories of multiple retrotransposons following independent allopolyploidy in wild wheats. New Phytol 202:975–985
19. Bonin A, Bellemain E, Bronken Edeisen P, Pompanon F, Brochmann C, Taberlet P (2004) How to track and assess genotyping errors in population genetics studies. Mol Ecol 13:3261–3273

Index

Igor Kovalchuck (ed.), *Plant Epigenetics: Methods and Protocols*, Methods in Molecular Biology, vol. 1456,
DOI 10.1007/978-1-4899-7708-3,

MIX
Papier aus verantwortungsvollen Quellen
Paper from responsible sources
FSC® C105338

If you have any concerns about our products,
you can contact us on
ProductSafety@springernature.com

In case Publisher is established outside the EU,
the *EU authorized representative is:*
Springer Nature Customer Service Center GmbH
Europaplatz 3, 69115 Heidelberg, Germany

Printed by Libri Plureos GmbH
in Hamburg, Germany